Food Packaging: Advanced Materials, Technologies and Innovations

Food Packaging: Advanced Materials, Technologies and Innovations

Edited by Skyler Watts

SYRAWOOD
PUBLISHING HOUSE
New York

Published by Syrawood Publishing House,
750 Third Avenue, 9th Floor,
New York, NY 10017, USA
www.syrawoodpublishinghouse.com

Food Packaging: Advanced Materials, Technologies and Innovations
Edited by Skyler Watts

International Standard Book Number: 978-1-64740-370-6 (Hardback)

Cataloging-in-publication Data

Food packaging : advanced materials, technologies and innovations / edited by Skyler Watts.
p. cm.
Includes bibliographical references and index.
ISBN 978-1-64740-370-6
1. Food--Packaging. 2. Active food packaging. 3. Coated foods. I. Watts, Skyler.
TP374 .F66 2023
664.09--dc23

TABLE OF CONTENTS

PREFACE

Over the recent decade, advancements and applications have progressed exponentially. This has led to the increased interest in this field and projects are being conducted to enhance knowledge. The main objective of this book is to present some of the critical challenges and provide insights into possible solutions. This book will answer the varied questions that arise in the field and also provide an increased scope for furthering studies.

Food packaging is the process that involves enclosing food in such a way that it does not get damaged, contaminated, spoiled, attacked by pests, or tampered during transport, storage and retail sale. The package contains nutritional facts label and other information such as the amount of the contents, ingredients and shelf life. There are various materials used for food packaging such as aseptic packages, plastic trays, bags, bottles, boxes, cartons, pallets, and wrappers. Apart from these, atmosphere packaging, bags-in-boxes, and wine box packaging are three specialized techniques of packaging. Vacuum packaging or inert atmosphere packaging is a common method in which foods are packaged at reduced pressure or by the use of an inert gas such as nitrogen. Bags-in-boxes is another type of packaging technique that is used for packing soft drink syrups, other liquid products, and meat products. To ensure sustainability and avoid the use of plastic, the food packaging industry is shifting its focus to manufacture of antimicrobial active, biodegradable, edible, and biopolymer film packaging. This book discusses the advanced materials, technologies and innovations related to food packaging. It is appropriate for students seeking detailed information in this area as well as for experts.

I hope that this book, with its visionary approach, will be a valuable addition and will promote interest among readers. Each of the authors has provided their extraordinary competence in their specific fields by providing different perspectives as they come from diverse nations and regions. I thank them for their contributions.

Editor

1

Thin-Film Coated Plastic Wrap for Food Packaging

Hsin-Yu Wu [1], Ting-Xuan Liu [2], Chia-Hsun Hsu [3], Yun-Shao Cho [3], Zhi-Jia Xu [3], Shu-Chuan Liao [4,5], Bo-Han Zeng [4], Yeu-Long Jiang [1] and Shui-Yang Lien [3,*]

1 Graduate Institute of Optoelectronic Engineering and Department of Electrical Engineering, National Chung Hsing University, Taichung 40227, Taiwan; xinmarchentic@gmail.com (H.-Y.W.); yljiang@nchu.edu.tw (Y.-L.J.)
2 Department of Electrical Engineering, Da-Yeh University, Chunghua 51591, Taiwan; f0003209@gmail.com
3 Department of Materials Science and Engineering, Da-Yeh University, Chunghua 51591, Taiwan; cstcaptive@gmail.com (C.-H.H.); ysc@mail.dyu.edu.tw (Y.-S.C.); s313326229@gmail.com (Z.-J.X.)
4 Bachelor Program for Design and Materials for Medical Equipment and Devices, Da-Yeh University, Changhua 51591, Taiwan; liaozizi@mail.dyu.edu.tw (S.-C.L.); z0978588769@gmail.com (B.-H.Z.)
5 Institute of Biomedical Engineering, National Taiwan University, Taipei 106, Taiwan
* Correspondence: syl@mail.dyu.edu.tw.

Abstract: In this study, the antimicrobial property and food package capability of polymethylpentene (PMP) substrate with silicon oxdie (SiO_x) and organic silicon (SiC_xH_y) stacked layers deposited by an inductively coupled plasma chemical vapor deposition system were investigated. The experimental results show that the stacked pair number of SiO_x/SiC_xH_y on PMP is limited to three pairs, beyond which the films will crack and cause package failure. The three-pair SiO_x/SiC_xH_y on PMP shows a low water vapor transmission rate of 0.57 g/m^2/day and a high water contact angle of 102°. Three-pair thin-film coated PMP demonstrates no microbe adhesion and exhibits antibacterial properties within 24 h. Food shelf life testing performed at 28 °C and 80% humidity reports that the three-pair thin-film coated PMP can enhance the food shelf-life to 120 h. The results indicate that the silicon-based thin film may be a promising material for antibacterial food packaging applications to extend the shelf-life of food products.

Keywords: polymethylpentene; food package; inductively coupled plasma chemical vapor deposition; water vapor transmission rate; shelf-life

---

1. Introduction

There has been growing interest in recent years to develop materials with antibacterial properties which improve food safety and shelf-life. Antibacterial packaging is one of the most promising active packaging systems [1–5]. In addition to the basic properties of packaging materials, antimicrobial food package should prevent the gain of moisture, hinder microbial contamination, and act as a barrier against the permeation of water vapor, oxygen, carbon dioxide, and other volatile compounds.

Basic packaging materials, such as paper and paperboard, plastic, glass, and metal, are used to fulfill the capabilities and requirements of packaged foods, depending on their type. However, there has been an ever-increasing endeavor to develop different kinds of food packaging materials in order to extend the shelf-life with improved water vapor resistivity and antibacterial property. Silicon oxide (SiO_x) barrier films with high density and mechanical resistance, as well as a low water permeation rate can be deposited using plasma-enhanced chemical vapor deposition at low temperatures [6,7]. It has been proven that SiO_x is useful in improving the gas barrier properties of some food packaging materials such as polyethylene terephthalate [8,9] and polyvinyl chloride [10]. Polymethylpentene (PMP) is a thermoplastic polyolefin with excellent mechanical and thermal stabilities [11]. However,

PMP plastic food wraps have a high moisture permeation rate and there are few research studies related to the improvement of the water vapor resistivity and antibacterial property of PMP.

In this paper, inorganic SiO_x and organic silicon (SiC_xH_y) stacked layers were prepared on a PMP substrate by using inductively coupled plasma chemical vapor deposition (ICPCVD). Silicon oxide-based thin film is widely known as a great barrier against water vapor [12], but SiO_x may likely have cracks induced by the accumulation of internal residual compressive stress when the film thickness increases. SiC_xH_y films with good tensile stress were thus stacked with SiO_x to reduce the total internal stress. The stacked layers were continually deposited in the same chamber without breaking the vacuum conditions. The pair of the SiO_x/SiC_xH_y stacked layers was varied in order to reduce the water vapor transmission rate (WVTR). Finally, the food package capability of PMP with SiO_x/SiC_xH_y stacked layers was evaluated, and the PMP with the stacked layers demonstrated greatly improved antibacterial properties and extended the food shelf-life.

2. Materials and Methods

2.1. Materials

Silicon-based thin films, such as SiO_x and SiC_xH_y layers, were prepared on PMP substrates by a 13.56-MHz radio frequency ICPCVD system. A gas mixture of oxygen (O_2) and tetramethylsilane (TMS) was used to deposit SiO_x, while argon (Ar) and TMS were used to deposit SiC_xH_y. For SiO_x layers, the gas flow ratio of O_2 to TMS was 4. For SiC_xH_y layers, the gas flow ratio of Ar to TMS was 0.03. The radio frequency power for SiO_x and SiC_xH_y were 600 and 400 W, respectively. The deposition pressure was 5 mTorr and the temperature was kept at 90 °C. In our previous study, a minimal internal stress was obtained when the thicknesses of SiO_x and SiC_xH_y were 300 and 30 nm, respectively. The pair number of the silicon-based stacked layers was varied from one to three.

2.2. Antibacterial Efficacy Test

The bacterial strains used in this study were *Escherichia coli* (*E. coli*) (ATCC, strain 25922) from Creative Life Science Co., Ltd. (Taipei, Taiwan). The antimicrobial effect was tested by the Kirby-Bauer diffusimetrical method. The bacteria was incubated for 17 h with vigorous shaking (250 rad/min) at 37 °C, and were obtained from a bacterial inoculum, which was standardized according to the McFarland scale, yielding to 10^7 CFU/mL. Luria-Bertani (LB) medium was inoculated with that inoculum and afterwards, different pairs of the SiO_x/SiC_xH_y films with a diameter of 1 cm were applied on the surface of the medium. The antimicrobial property against the strains of *E. coli* was assessed by measuring the ratio of microbes to sample area after 0, 2, 4, 8, 12, 24, 48, and 72 h of incubation at 37 °C.

2.3. Bread Packaging Test

Four fresh toast slices (size: 10 cm × 5 cm) without preservatives were packed in the PMP without and with different pairs of SiO_x/SiC_xH_y films. The toast slices were stored at a relatively humidity of 80% and a temperature of 28 °C. Photographs of the toast slices were taken after 120 h, and ImageJ computing software (National Institute of Mental Health, Maryland, MD, USA) [13,14] was used to evaluate the area of fungal contamination.

2.4. Characterization

The topographical morphologies of the SiO_x/SiC_xH_y films were observed by optical microscopy (OM, MICROTECH MX810-RF, M&T Optics, Co., Ltd., Taipei, Taiwan). The WVTR of the films was measured by a WVTR permeation instrument (PERMATRAN-WR Model 3/61, Mocon Inc., Minneapolis, MN, USA) under the conditions of 40 °C and 100% relative humidity (RH) [15,16]. Water contact angle measurements were performed with a pocket goniometer (Model PGX, Testing Machines Inc., Veenendaal, The Netherlands) using 2 μL droplets, with at least five repetitions for each sample.

The morphologies of *E. coli* grown on PMP without and with different pairs of SiO_x/SiC_xH_y films were observed using high resolution thermal field emission scanning electron microscopy (HRFEG-SEM, JSM-7610F, JEOL, Tokyo, Japan) at 10 kV.

3. Results and Discussion

3.1. Properties of Silicon-Based Thin Films

The OM images of PMP with different pairs of SiO_x/SiC_xH_y films are shown in Figure 1. The PMP without, with one-, two-, and three-pair SiO_x/SiC_xH_y stacked layers show smooth surface and good structural integrity without significant pores or cracks. The four-pair SiO_x/SiC_xH_y stacked layer exhibits crack patterns on the film surface. One possible reason for the cracks might be attributed to the heat accumulation on the substrate during the deposition process. PMP substrate is very likely to curl under long plasma deposition as the ion bombardment will heat the PMP substrate. Another reason might be due to the increased internal residual stress as the pair number increases. Overall, the maximal pair number of SiO_x/SiC_xH_y on PMP was limited to three in order to avoid film cracking and packaging failure.

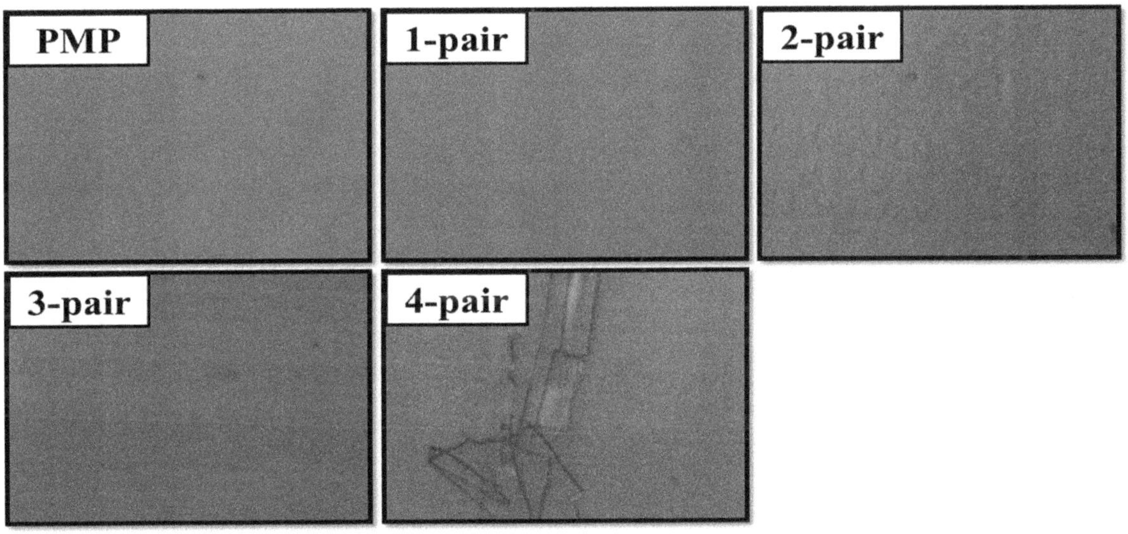

Figure 1. Optical microscope images of the polymethylpentene (PMP) without, with one-, two-, three- and four-pair SiO_x/SiC_xH_y films.

Figure 2 shows the WVTR at 40 °C/100% RH of the SiO_x/SiC_xH_y stacked layers with one, two, and three pairs. It can be seen that the WVTR of the PMP substrate is as high as 775 $g/m^2/day$. The one-pair SiO_x/SiC_xH_y significantly reduced WVTR values to 8.82 $g/m^2/day$, which further decreases to 1.04 $g/m^2/day$ for the two-pair SiO_x/SiC_xH_y. The three-pair SiO_x/SiC_xH_y demonstrates a WVTR value of 0.57 $g/m^2/day$. The reduction in WVTR can be related to the increase of the total thickness of the gas barrier, which causes the water vapor to spend more time traveling through the barrier layer to the other side. A WVTR value ranging from 0.01 to 1 $g/m^2/day$ is suggested for food packaging [17]. The PMP substrate can satisfy this requirement if the three-pair SiO_x/SiC_xH_y film is coated.

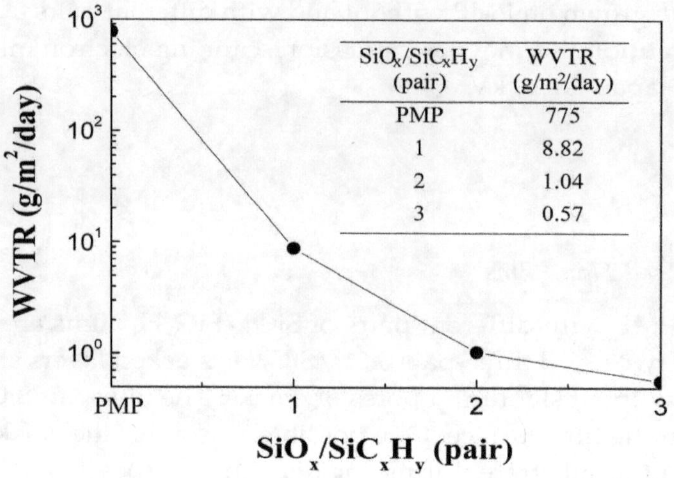

Figure 2. Water vapor transmission rate for the PMP without and with one-, two-, and three-pair SiO_x/SiC_xH_y layers at 40 °C/100% relative humidity (RH).

The hydrophobicity of a packaging material is a major concern, as it limits water vapor attachments. However, the SiO_x/SiC_xH_y stacked films may exhibit a hydrophilic surface with the water contact angle <90°, due to the hydrophilicity of the silicon oxide nature. In this study, a 30 nm SiC_xH_y layer was intentionally added to the top of the stacked structure to produce a hydrophobic surface. The water contact angle measurement results for the PMP without, and with SiC_xH_y/three-pair SiO_x/SiC_xH_y stack are shown in Figure 3. The water contact angles for the PMP without and with the stacked layer are approximately 65° and 102°, respectively. Hydrophobic properties are known to be observed at contact angles above 90° [18]. The PMP with the stacked layer is confirmed to attain a desirable level of hydrophobicity.

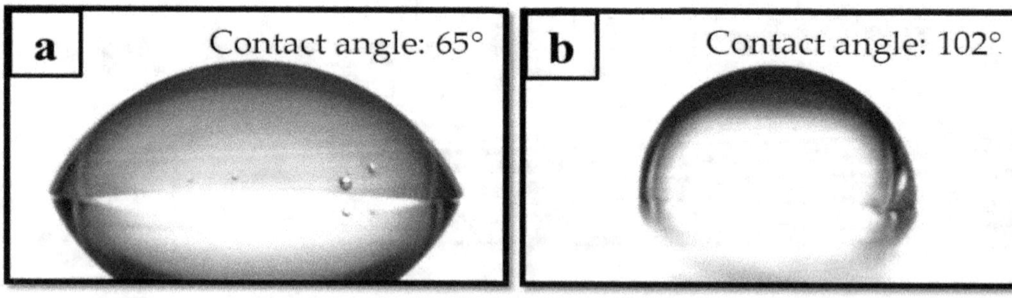

Figure 3. Water contact angle images for the PMP (**a**) without and (**b**) with SiC_xH_y/three-pair SiO_x/SiC_xH_y stack.

3.2. Evaluation of the Antibacterial Effect

The adhesion of the microbes covered by PMP without and with stacked layers is observed by SEM, and the ratio of microbes to sample area as a function of time is shown in Figure 4. For the PMP substrate without the stacked layers, the microbe ratio is negligible for the first 8 h. Afterwards, the ratio rapidly increases to 80% at 72 h. With the stacked layers, the microbe ratio significantly decreases as compared to that of the PMP without stacked layers. The one- and two-pair stacked layers can lead to a microbe ratio of 0% within 12 h, while the PMP with the three-pair stacked layer demonstrates no microbe adhesion within 24 h. This result indicates the great antibacterial function of the PMP with the SiO_x/SiC_xH_y stacks. The SEM images at a certain position on each sample at 72 h are shown in Figure 5. The microbe ratios are 79.3%, 25.9%, 18.5%, and 3.9% for the PMP without, and with one-, two-, and three-pair stacked layers, respectively.

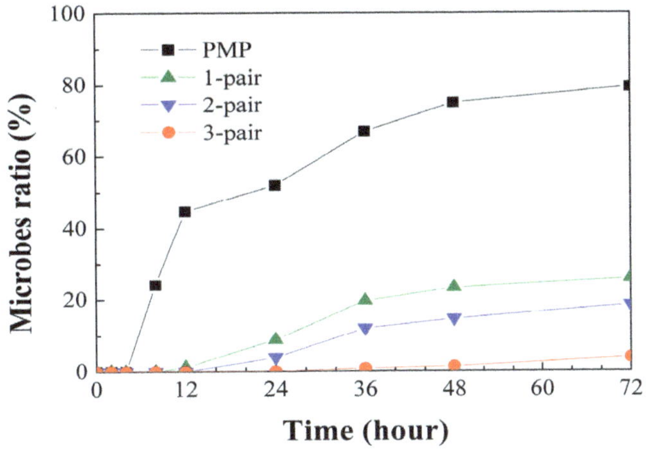

Figure 4. Ratio of microbe adhesion area to sample area for PMP without and with different pairs of SiO_x/SiC_xH_y stacked films as a function of time.

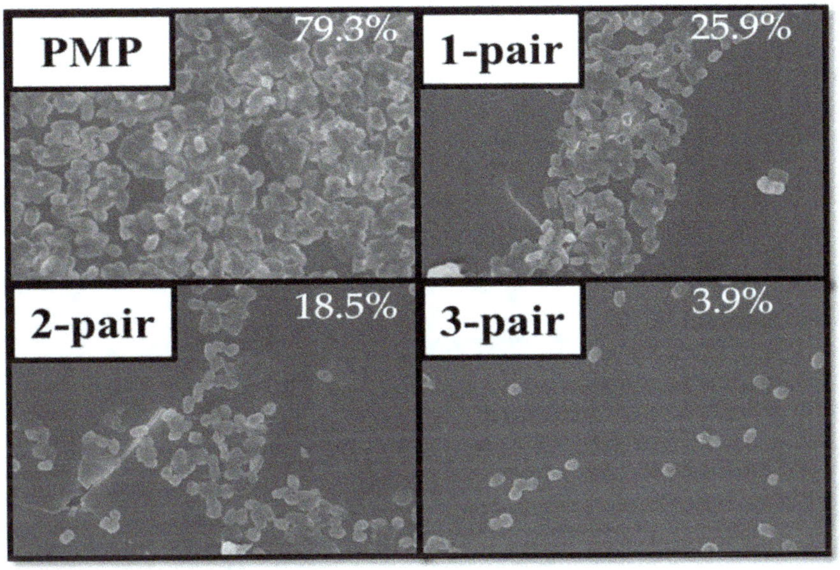

Figure 5. SEM images for *Escherichia coli* adhesion on PMP without and with different pairs of SiO_x/SiC_xH_y stacked films at 72 h. The percentage represents the ratio of microbes to total area.

The PMP substrates without and with different pairs of SiO_x/SiC_xH_y stacked films are applied to food packaging, and the appearances of the packed toasts stored at 80% RH and 28 °C after 120 h are shown in Figure 6. In this study, we choose the toast without any added preservatives. It can be observed that fungi grow on the toasts packed by the PMP without, as well as with one-pair and two-pair SiO_x/SiC_xH_y stacked films. While the toast packed with the three-pair stacked film shows the absence of fungi growth. The extent of the fungi growth is evaluated as listed in Table 1. The area ratio of the fungi to sample surface is greater than 2% for the PMP substrate without the stacked films. The ratio is significantly reduced to less than 1% for the PMP with one pair of stacked film. The two-pair stacked film further decreases the ratio to 0.7%. Finally, no observable fungi are presented for the three-pair stacked film. The trend of this result can be directly linked to the WVTR values. Lower WVTR value means decreased water and vapor pass through the packaging layer, and this reduces fungi growth.

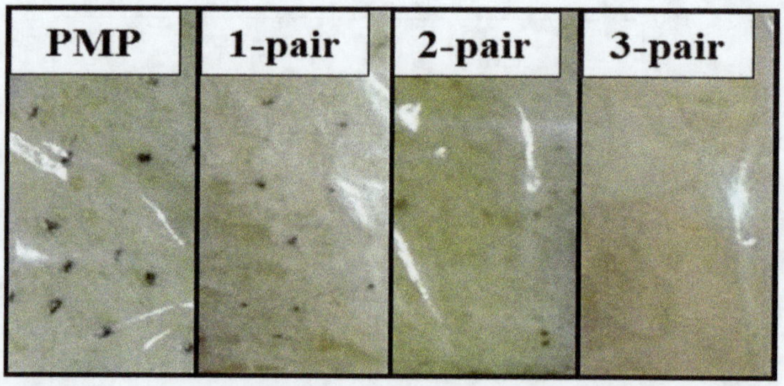

Figure 6. Images of food packed by PMP without and with different pairs of SiO_x/SiC_xH_y stacked film.

Table 1. Extent of fungi growth area ratio for food packed by PMP without and with different pairs of SiO_x/SiC_xH_y stacked film.

Type of Packaging	PMP	1-Pair	2-Pair	3-Pair
Fungi growth [1]	+++	++	+	–

[1] absence of fungal growth, + fungal growth on <0.7% of the surface, ++ fungal growth on <1% of the surface, +++ fungal growth on >2% of the surface.

The package technique in this work can be used in large area (25 cm × 25 cm) or roll-to-roll deposition, as shown in Figure 7. The PMP with the three-pair stacked film can easily be stretched and rolled without obvious film cracking, and the film retains its flexibility. The SiO_x/SiC_xH_y stacked film demonstrates great potential to be used in food package applications.

Figure 7. Photograph of stretched and rolled PMP with three-pair SiO_x/SiC_xH_y stack.

4. Conclusions

Silicon-based SiO_x and SiC_xH_y layers are prepared on PMP using the ICPCVD system. The PMP with the three-pair SiO_x/SiC_xH_y stack shows a low WVTR value of 0.57 $g/m^2/day$ and a high contact angle value of 102°. The three-pair stack exhibits a remarkable antibacterial ability against *E. coli* within 24 h. The food package testing shows that the PMP with the stacked films can extend the food shelf-life to 120 h at 28 °C and 80% RH. These results suggest that the SiO_x/SiC_xH_y stack has great moisture barrier property as well as antimicrobial characteristics, making it useful in food packaging applications.

Acknowledgments: This work is sponsored by the Ministry of Science and Technology of the Republic of China under contract Nos. 105-2632-E-212-001- and 104-2221-E-212-002-MY3.

Author Contributions: Hsin-Yu Wu and Shui-Yang Lien designed the experiments; Hsin-Yu Wu, Ting-Xuan Liu, Zhi-Jia Xu and Bo-Han Zeng performed the experiments and measurements; Hsin-Yu Wu, Ting-Xuan Liu, Chia-Hsun Hsu, Yun-Shao Cho, Zhi-Jia Xu, Shu-Chuan Liao, Bo-Han Zeng, Yeu-Long Jiang and Shui-Yang Lien analyzed the results; Hsin-Yu Wu, Chia-Hsun Hsu and Shu-Chuan Liao wrote the paper.

References

1. Tripathi, S.; Mehrotra, G.K.; Dutta, P.K. Physicochemical and bioactivity of cross-linked chitosan–PVA film for food packaging applications. *Int. J. Biol. Macromol.* **2009**, *45*, 372–376. [CrossRef] [PubMed]
2. Silvestre, C.; Duraccio, D.; Marra, A.; Strongone, V.; Cimmino, S. Development of antibacterial composite films based on isotactic polypropylene and coated ZnO particles for active food packaging. *Coatings* **2016**, *6*, 4. [CrossRef]
3. Krepker, M.; Shemesh, R.; Poleg, Y.D.; Kashi, Y.; Vaxman, A.; Segal, E. Active food packaging films with synergistic antimicrobial activity. *Food Control* **2017**, *76*, 117–126. [CrossRef]
4. Vasile, C.; Sivertsvik, M.; Mitelut, A.C.; Brebu, M.A.; Stoleru, E.; Rosnes, J.T.; Tanase, E.E.; Khan, W.; Pamfil, D.; Cornea, C.P.; et al. Comparative Analysis of the Composition and active property evaluation of certain essential oils to assess their potential applications in active food packaging. *Materials* **2017**, *10*, 45. [CrossRef]
5. Soumya, B.N.; Jyothi, A.N.; Sajeev, M.S. Chitosan-konjac glucomannan-cassava starch-nanosilver composite films with moisture resistant and antimicrobial properties for food-packaging applications. *Starch-Starke* **2017**, *69*, 1600210.
6. Martin, S. Atmospheric pressure PE-CVD of silicon based coatings using a glow dielectric barrier discharge. *Surf. Coat. Technol.* **2004**, *177*, 693–698. [CrossRef]
7. Scopece, P.; Viaro, A.; Sulcis, R.; Kulyk, I.; Patelli, A.; Guglielmi, M. SiO_x-based gas barrier coatings for polymer substrates by atmospheric pressure plasma jet deposition. *Plasma Process. Polym.* **2009**, *6*, S705–S710. [CrossRef]
8. Plog, S.; Schneider, J.; Walker, M.; Schulz, A.; Stroth, U. Investigations of plasma polymerized SiO_x barrier films for polymer food packaging. *Surf. Coat. Technol.* **2011**, *205*, S165–S170. [CrossRef]
9. Deilmann, M.; Grabowski, M.; Theiß, S.; Bibinov, N.; Awakowicz, P. Permeation mechanisms of pulsed microwave plasma deposited silicon oxide films for food packaging applications. *J. Phys. D Appl. Phys.* **2008**, *41*, 135207. [CrossRef]
10. Fei, F.; Wang, Z.; Chen, Q.; Liu, Z.; Sang, L. Study of functional barrier layer on PVC by PECVD for migrations resistant. *Surf. Coat. Technol.* **2013**, *228*, S61–S66. [CrossRef]
11. Michaljanicova, I.; Slepicka, P.; Kasalkova, N.S.; Sajdl, P.; Svorcik, V. Plasma and laser treatment of PMP for biocompatibility improvement. *Vacuum* **2014**, *107*, 184–190. [CrossRef]
12. Kima, H.; Sohn, S. Permeabilities of inorganic thin composite $(SiO_2)_{1-x}(ZnO)_x$ Films prepared by RF-magnetron sputtering method. *ECS Trans.* **2009**, *19*, 795–802.
13. Schneider, C.A.; Rasband, W.S.; Eliceiri, K.W. NIH image to imageJ: 25 years of image analysis. *Natl. Methods* **2012**, *9*, 671–675. [CrossRef]
14. *ImageJ*, version 1.50i; Software for Image Processing and Analysis; National Institutes of Mental Health: Bethesda, MD, USA, 2016.
15. Koo, W.H.; Jeong, S.M.; Choi, S.H.; Baik, H.K.; Lee, S.M.; Lee, S.J. Water Vapor Barrier Properties of Transparent SnO_2-SiO_x Composite Films on Polymer Substrate. *J. Phys. Chem. B* **2004**, *108*, 18884–18889. [CrossRef]
16. Dimitrov, V.; Komatsu, T. Classification of Simple Oxides: A Polarizability Approach. *J. Sol. State Chem.* **2002**, *163*, 100–112. [CrossRef]
17. Spee, D.A.; Rath, J.K.; Schropp, R.E.I. Using hot wire and initiated chemical vapor deposition for gas barrierthin film encapsulation. *Thin Solid Films* **2015**, *575*, 67–71. [CrossRef]
18. Block, S.; Hupfield, P.; Itami, Y.; Kitaura, E.; Kleyer, D.; Masutani, T.; Nakai, Y. New Anti-Fingerprint Coatings. *PCI Mag.* **2008**, *24*, 88–93.

2

Atomic Layer Deposited TiO_2 and Al_2O_3 Thin Films as Coatings for Aluminum Food Packaging Application

Vanessa Dias [1,2], Homero Maciel [1,3], Mariana Fraga [4], Anderson O. Lobo [5], Rodrigo Pessoa [1,3,*] and Fernanda R. Marciano [2,3,*]

1 Centro de Ciência e Tecnologia de Plasmas e Materiais (PlasMat), Instituto Tecnológico de Aeronáutica (ITA), São José dos Campos, SP 12228-900, Brazil; van_ametista@yahoo.com.br (V.D.); homero@ita.br (H.M.)

2 Laboratório de Nanotecnologia Biomédica, Universidade do Vale do Paraíba (Univap), São José dos Campos, SP 12244-000, Brazil

3 Instituto Científico e Tecnológico, Universidade Brasil, São Paulo, SP 08230-030, Brazil

4 Instituto de Ciência e Tecnologia, Universidade Federal de São Paulo (Unifesp), São José dos Campos, SP 12231-280, Brazil; mafraga@ieee.org

5 LIMAV-Laboratório Interdisciplinar de Materiais Avançados, Universidade Federal do Piauí (UFPI), Teresina, PI 64049-550, Brazil; aolobo@pq.cnpq.br

* Correspondence: rspessoa@ita.br (R.P.); femarciano@gmail.com (F.R.M.)

Abstract: Titanium dioxide (TiO_2) and aluminum oxide (Al_2O_3) coatings have been investigated in a wide range of bio-applications due to their biodegradation and biocompatibility properties, that are key parameters for their use in the food packaging and biomedical devices fields. The present study evaluates and compares the electrochemical behavior of the non-coated, commercial resin-coated, TiO_2-coated and Al_2O_3-coated aluminum in commercial beer electrolyte. For this, TiO_2 and Al_2O_3 thin films were deposited on aluminum (Al) substrates using atomic layer deposition (ALD). The evaluation of the corrosion barrier layer properties was performed by linear sweep voltammetry (LSV) during 10 min and electrochemical impedance spectroscopy (EIS). In addition, profilometry, grazing incidence X-ray diffractometry (GIXRD), scanning electron microscopy (SEM) and Fourier-transform infrared spectroscopy (FT-IR) analyses were performed to investigate the physical and chemical properties of the pristine and / or corroded samples. TiO_2 and Al_2O_3 films presented an amorphous structure, a morphology that follows Al substrate surface, and a thickness of around 100 nm. Analysis of LSV data showed that ALD coatings promoted a considerable increase in corrosion barrier efficiency being 86.3% for TiO_2-coated Al and 80% for Al_2O_3-coated Al in comparison with 7.1% of commercial resin-coated Al. This is mainly due to the lower electrochemical porosity, 11.4% for TiO_2-coated Al and 20.4% for Al_2O_3-coated Al in comparison with 96% of the resin-coated Al, i.e. an increase of up to twofold in the protection of Al when coated with TiO_2 compared to Al_2O_3. The EIS results allow us to complement the discussions about the reduced corrosion barrier efficiency of the Al_2O_3 film for beer electrolyte once SEM and FT-IR analyzes did not show drastic changes in both investigated ALD films after the corrosion assays. The above results indicate that ALD TiO_2 and Al_2O_3 films may be a viable alternative to replace the synthetic resin coatings frequently used in aluminum cans of use in the food industry.

Keywords: corrosion barrier; titanium dioxide; aluminum oxide; atomic layer deposition; linear sweep voltammetry; electrochemical impedance spectroscopy

1. Introduction

The food industry is always looking for new technologies for improving packaging techniques in order to maintain the food quality and assuring food safety. At the same time, many studies have been devoted to for the use of nanotechnology in food science because of the growing need by healthier and low-cost products.

Aluminum (Al) is widely used in the food packaging industry because it is lightweight, recyclable, protects the stored liquid from various external factors such as moisture, microbiological contamination, among others [1]. However, some internal factors, such as the chemical processes promoted by the stored liquid, might result in degradation of Al surface resulting in corrosion and migration of metals [2–4].

The CO_2-containing beverages such as beer and carbonated drinks have acidity and chlorides in the composition and, when packed in Al cans, can promote the corrosion process by coming into contact with the Al material. To mitigate this problem, commercial Al cans are internally spray coated by synthetic resins, generally natural or synthetic epoxy, that aims to protect the Al surface from direct contact with the beverage [4–6]. However, when the resin is not applied correctly or when the Al can undergoes mechanical impact or elevated temperatures, the occurrence of localized corrosion is verified due to the presence of pinholes, cracks or grains boundaries in the coatings [7]. Moreover, some resins contain bisphenol A (BPA), which is introduced into the coating during production. There is a concern that BPA may leach into the liquid contained inside beverage cans [8]. These problems may cause loss of product quality and integrity. In addition, if the contaminated product reaches the final consumer it can be harmful to human health [8,9].

Currently, need for rapid development of new internal coatings on Al cans has arisen in the food and beverage industry, mainly as a response to proposed volatile organic compound regulations and a desire to contain new and more aggressive food chemistries [10,11]. Acidity of commercial soft drinks varies greatly; the pH of beverages is typically between 2.3 to 4.0. For instance, Coke™ is pH 2.38 and a beer is pH around 4.0 [12]. Thus, a coating should behave as a barrier throughout the range of pH solutions, avoiding reaching the underlying surface of the aluminum.

Ceramic thin films have attracted much attention as protective barrier coatings due to their resistance to heat, corrosion, and wear [13–15]. Al_2O_3 and TiO_2 thin films have been the focus of extensive research in recent years [16,17]. Al_2O_3 is the most readily studied thin film for corrosion protection, because it presents low porosity that prevents the solution from accessing the metal [18]. TiO_2 is other important ceramic material used as corrosion protection layer of metals [19,20]. However, TiO_2 has issues during film nucleation leading to high porosity [21]. The grain boundaries present in crystalline TiO_2 are susceptible to species diffusion, which promote surface corrosion [22]. A possible solution is to deposit amorphous TiO_2 thin films.

There are many well-studied methods for applying corrosion protective coatings including reactive sputtering, spray pyrolysis, electrochemical deposition, and chemical vapor deposition (CVD). Nevertheless, such preparation methods usually leave cracks or pinholes in the coatings [23]. Recently, several groups have explored atomic layer deposition (ALD) technique as a mean to form various metal oxide corrosion protection layers such as Al_2O_3, TiO_2, ZnO, HfO_2, etc., on various types of surfaces. ALD enables to obtain pinhole-free thin films with precise control over composition and thickness [20]. Furthermore, the ALD stands out in relation to CVD and PVD technologies, because it allows to deposit more conformal layers even on complex 3D shaped substrates [24].

The use of ALD thin films for the corrosion protection of metallic substrates was speculated for the first time in the 1990s by Matero et al., which demonstrated that ALD TiO_2/Al_2O_3 films deposited on stainless steel have good corrosion resistance [22]. Since then, much research has been done to investigate the limits of application of this technique. Shan et al. investigated the improvement of corrosion resistance of CrN coated stainless steel with ALD TiO_2 film [20,23]. Du et al. reported on the chemical corrosion protection of aluminum mirrors by ALD SiO_2 coatings [25]. Marin et al. evaluated the long-term corrosion resistance performances of three different ALD single layer strategies (TiO_2,

Al_2O_3 and AlN) applied on AISI 316 substrates [24]. Recently, Daubert et al. investigated the corrosion protection of the copper using ALD Al_2O_3, TiO_2, ZnO, HfO_2, and ZrO_2 thin films [18].

In food packaging, the application of ALD technique has already been discussed for the protection and increase of shelf life of products such as paperboard with Al_2O_3 and TiO_2 coatings [26,27]. However, the synthesis of protective thin films using the ALD process in the field of aluminum cans for carbonated beverages has not been reported.

In this article, due to the lack of studies on the use of ALD Al_2O_3 and TiO_2 films as corrosion protection of Al in low pH environment, we investigated the feasibility of these coatings for such application showing their preparation and initial corrosion resistance (10 min) when deposited on aluminum substrates. Profilometry, grazing incidence X-ray diffractometry (GIXRD), scanning electron microscopy (SEM) and Fourier-transform infrared spectroscopy (FT-IR) analyses were performed to investigate the physical and chemical properties of the pristine and / or corroded samples. The evaluation of the corrosion resistance of the protected samples was performed by linear sweep voltammetry (LSV). Moreover, electrochemistry impedance spectroscopy (EIS) measurements were performed to investigate the electrical characteristics of the electrode–solution interface.

2. Materials and Methods

2.1. Substrate Preparation

Commercial bare and resin-coated aluminum (AA3104 alloy, Latapack-Ball, Jacareí, Brazil) cans were cut into pieces of 1×1 cm^2 and separated in four groups: (i) control or bare Al, (ii) resin-coated, (iii) TiO_2-coated Al, and (iv) Al_2O_3-coated Al. For ALD coating, all samples were cleaned with distilled water and isopropyl alcohol, and then dried with N_2 gas before being placed in the ALD reactor. The TiO_2 and Al_2O_3 films were prepared by a Beneq TFS-200 equipment (Beneq Oy, Espoo, Finland), operating in thermal mode, at 100 °C and in the condition of 1000 reaction cycles. The choice of the process temperature is related to the good anticorrosive results of ALD metal oxides films reported by Sammelselg et al. [28]. Titanium tetrachloride ($TiCl_4$) and trimethylaluminum (TMA) were used as metallic precursors and deionized water (H_2O) as oxidant. Nitrogen (N_2) of 99.999 % purity was used as purge gas. The corresponding ALD cycle time parameters for TiO_2 films were 0.25, 2, 0.25 and 2 s for $TiCl_4$ pulse, purge, H_2O pulse and purge. While for Al_2O_3 films the cycle times were 0.15, 0.75, 0.15 and 0.75 s for the TMA pulse, purge, H_2O pulse and another purge, respectively. These are optimal cycle time conditions investigated in early studies [29–34]. The vapors of $TiCl_4$, TMA and H_2O were led into the reaction chamber from external reservoirs kept with liquid $TiCl_4$ (99.95 %, Sigma-Aldrich, São Paulo, Brazil), liquid TMA (97%, Sigma-Aldrich) and deionized water at temperature of 21 °C. A capillary tube, adapted to the reactor, was used to control the precursor flow injected into the ALD chamber, by action of the precursor vapor pressure only, i.e., no bubbling system was used. The base pressure of the reactor was lower than 10^{-2} mbar and during the deposition the gas pressure was maintained around 1.0 mbar through the insertion of 300 sccm of N_2.

2.2. Film Characterization

The thickness of the as-deposited films was measured using a KLA Tencor P-7 profilometer (KLA Corporation, Milpitas, CA, USA). To characterize the structure of the as-deposited films, grazing incidence X-ray diffraction (GIXRD) method was used. GIXRD patterns were obtained at room temperature in a Shimadzu XRD 6000 goniometer (Shimadzu Corporation, Kyoto, Japan) using a copper target (CuKα radiation 1.5418 Å), 2θ from 20°–80°, at a scanning speed of 0.02° s^{-1}, a voltage of 40 kV, and a current of 30 mA. To investigate the chemical bonds of the pristine and corroded samples, infrared measurements were performed on an ATR-FTIR PerkinElmer 400 IR spectrometer (PerkinElmer Brasil, São Paulo, Brazil) at a resolution of 2 cm^{-1}. Each ATR spectrum was recorded with the blank ATR cell (PerkinElmer Brasil, São Paulo, Brazil) as the background. In addition, some analyzes of the surface morphology of the pristine and corroded samples were made with a field emission scanning electron microscope (FE-SEM) Tescan Mira 3 FEG (TESCAN Brno, s.r.o., Kohoutovice, Czech Republic) operated at 5 kV.

2.3. Electrochemical Measurements

All electrochemical measurements were conducted on Autolab 302N potentiostat/galvonostat (Metrohn Autolab B.V., Utrecht, the Netherlands) controlled by Nova 2.0 software (Metrohn Autolab B.V., Utrecht, the Netherlands). A standard three-electrode electrochemical cell (Metrohn Autolab B.V., Utrecht, the Netherlands) was used in electrochemical experiments. In the setup, saturated Ag/AgCl (3M KCl) (Metrohn Autolab B.V., Utrecht, the Netherlands) was used as a reference electrode, pure platinum coiled wire (Metrohn Autolab B.V., Utrecht, the Netherlands) as a counter electrode and the samples (i-iv) as working electrode. The exposed surface area of the tested samples was 0.78 cm^2 and their back side have been sanded to improve the ohmic contact. For each experiment, the electrochemical cell was filled with 250 mL commercial bottled lager beer (Brasil Kirin, Itu, Brazil) (pH = 4.12 $\pm$ 0.02). All electrochemical assays were performed after 1 h immersion of pristine and coated Al samples in beer at the open circuit potential (OCP). The impedance spectra were obtained over the 100 mHz to 100 kHz frequency range with sine wave potential of 30 mV. The experiments were conducted at room temperature and a Kramers–Kronig routine was performed to ensure the linearity, causality and stability of the system through the measurements [35]. The EIS data was then modeled using equivalent circuit models, and curve fitting was performed using the Nova 2.0 software package. The fitted parameters have an uncertainty of $\pm$15%.

The LSV tests were performed in voltage range of -0.65 to -0.35 V at scan rate of 1 mV s^{-1}. The corrosion current (i_{corr}) and corrosion potential (V_{corr}) were obtained from LSV and Tafel plots; i_{corr} was the current obtained from intercept of tangent lines on cathodic and anodic branches of Tafel plots at V_{corr} [36]. The potential corresponding to zero current (current transition from cathodic to anodic) was assigned as V_{corr}. These three parameters were easily obtained through signal analysis option of Nova software.

From these LSV parameters is possible to calculate other electrochemical parameters namely protection efficiency (P_{eff}), polarization resistance (R_p), and electrochemical porosity (EP). The protection efficiency (P_{eff}) was calculated using the Equation (1).

$$P_{eff} = 100\left(1 - \frac{j_{corr}}{j_{corr}{}^{0}}\right) \tag{1}$$

where j_{corr} and $j_{corr}{}^{0}$ indicate the corrosion current densities in the presence and absence of the coating, respectively [37,38]. The polarization resistance (R_p) can be calculated using the Stern-Geary equation [18].

$$R_p = \frac{b_a b_c}{2.3\, i_{corr}\, (b_a + b_c)} \tag{2}$$

where b_a and b_c are the slopes of the anodic and cathodic branches of the Tafel plot.

Finally, the electrochemical porosity (EP) of the samples (ii-iv) was calculated using Equation (3) [18].

$$EP = \left(\frac{R_{pcontrol}}{R_{pcoated}}\right) \times 10^{-\left(\frac{\Delta V_{corr}}{b_a}\right)} \times 100 \tag{3}$$

where $R_{pcontrol}$ and $R_{pcoated}$ are the polarization resistances of the control and coated samples, respectively, and ΔV_{corr} is the difference in the corrosion potentials inferred from the polarization curve.

3. Results and Discussion

3.1. Material Characterization and Potentiodynamic Polarization

The measured film thicknesses were about 97 ± 4 nm for TiO_2 and 105 ± 3 nm for Al_2O_3 on Al substrate. The thickness of the TiO_2 film is in agreement with previous work of Chiappim et al. that deposited by thermal ALD TiO_2 thin films on Si(100) surface at 100 °C using $TiCl_4$ + H_2O precursors [29]. For Al_2O_3 thin film on Al substrate, the measured thickness of 105 ± 3 nm was slightly higher than reported in literature, i.e. around 100 nm [39]. It is important to highlight that, although not focused here, our earlier studies on thickness dependence with number of cycles demonstrate that the process at 100 °C behaves as a real ALD [33].

The structural properties of TiO_2 and Al_2O_3-coated aluminum were characterized by GIXRD (spectra not show here). The peaks observed at angle 2θ = 44.73°, 65.13° and 78.23° can be assigned to the diffraction from aluminum (JCPDS#04-0787), while no peaks coming from the TiO_2 and Al_2O_3 films were detected, which suggests that both materials are amorphous. In fact, as observed by Chiappim et al. for TiO_2 films [29] and by Miikkulainen et al. for Al_2O_3 films [40], ALD performed at temperature about 100 °C is not capable to initiate the crystallization process during the film growth.

After the first considerations about the TiO_2 and Al_2O_3 thin films, the next analysis concerns the corrosion assays in beer electrolyte. LSV tests were used to investigate the corrosion resistance of the coatings. The electrochemical parameters obtained from the polarization curves (Figure 1) using Tafel plot and equations (1–3) are summarized in Table 1. The smooth shape of the polarization curves of the samples, with only a single peak at the corrosion potential, indicates that the film is electrochemically inert [18]. For bare Al, the corrosion potential (V_{corr}) is about −0.40 V, while the value is −0.46 V for the commercial resin-coated Al, −0.50 V for TiO_2-coated Al, and −0.55 V for Al_2O_3-coated Al sample. The shift of the V_{corr} to more negative potential indicates an improved corrosion resistance. The ALD coatings led a considerable increase in protection efficiency (P_{eff}), i.e. 86.3% for TiO_2-coated Al and 80% for Al_2O_3-coated Al. Note here that the P_{eff} of the resin-coated Al is only 7.1%. Concerning the electrochemical porosity (*EP*), a direct correlation with P_{eff} can be observed from Table 1, where TiO_2-coated Al had the lowest value (11.4%). In comparison with the work of Daubert et al. [18], where TiO_2 with *EP* of 0.9% and Al_2O_3 with *EP* of 0.1% were obtained by ALD at 150 °C on copper substrate, our results indicate the deposition of films on aluminum with higher *EP*. One possible explanation is due to lower ALD process temperature of our work (100 °C)—it is known that at low process temperature the ALD films tend to have lower densities [41]; however, more studies are needed to better explain this difference between *EP* values.

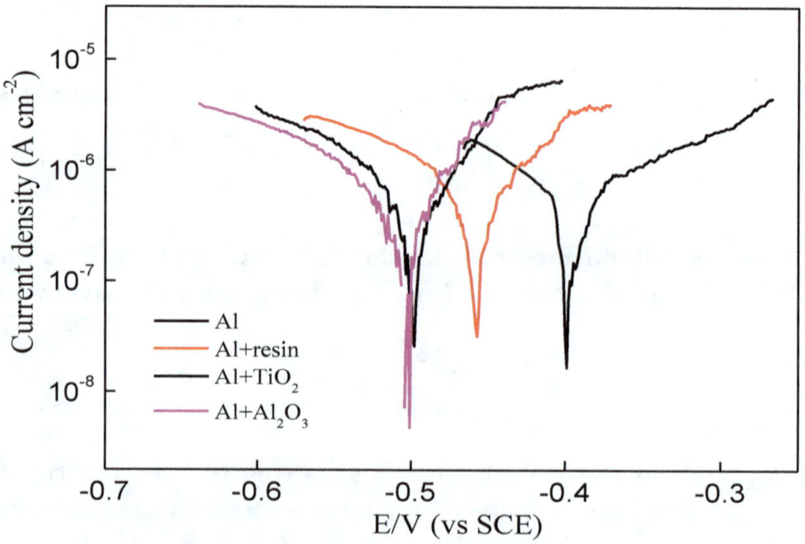

Figure 1. Polarization curves of the studied samples.

Table 1. Corrosion properties of the studied samples obtained from polarization curves.

Samples	V_{corr} (V vs. Ag/AgCl)	j_{corr} (10^{-6} A.cm^2)	R_p (10^4 Ω/cm^2)	EP (%)	P_{eff} (%)
Al	−0.40	9.06	2.77	100.0	-
Al + resin	−0.46	7.56	2.20	96.0	7.1
Al + TiO_2	−0.50	1.24	2.19	11.4	86.3
Al + Al_2O_3	−0.55	1.79	2.27	20.4	80.0

Figure 2 shows the SEM micrographs of the pristine and corroded samples. The micrograph analyzes corroborate with the results obtained from the polarization curves. For the bare Al sample, a rough surface is observed where the grooves resulting from the rolling process of aluminum cans (Figure 2a). By other hand, for resin-coated Al, it can be observed a smooth surface, but with presence of pores (Figure 2c). After the corrosion period of 10 min, both bare and resin-coated Al maintained the initial morphology (Figure 2b,d). The formation of the TiO_2 layer promoted a slight modification of the Al surface morphology with formation of some grains along the surface (Figure 2e). Indeed, thin amorphous TiO_2 films tend to follow the surface roughness of substrate being coated [29]. Although the as-deposited surface appears rough, the 97 nm TiO_2 coating promoted a reduction in *EP* by almost 90% compared with bare Al and a resulting P_{eff} of 86.3%. This decreasing of *EP* can be result of corrosion process, changing the morphology of TiO_2 film surface (Figure 2f). It is important to highlight that the TiO_2 film thickness oscillated within the standard deviation after corrosion process. As with TiO_2-coated Al, the alumina-coated sample reflects the substrate morphology with grain formation along the surface (Figure 2g). The 105 nm Al_2O_3 coating promoted a decreasing of *EP* of around 80% and a resulting P_{eff} of 80%. The LSV results also corroborates with SEM micrographs for Al_2O_3-coated Al that shows a smoother surface with no considerable modification in the initial film thickness after corrosion process (Figure 2h). Although inferior to the TiO_2 result, the corrosive protection of the alumina-coated sample was superior to that of the resin-coated sample, which was only 7.1%.

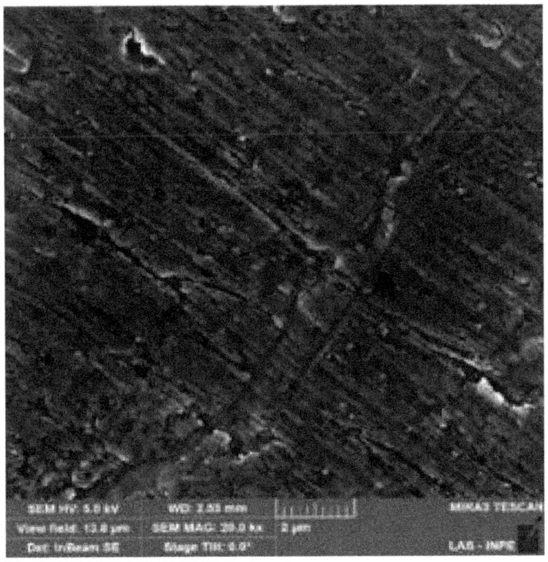

(a)

(b)

Figure 2. *Cont.*

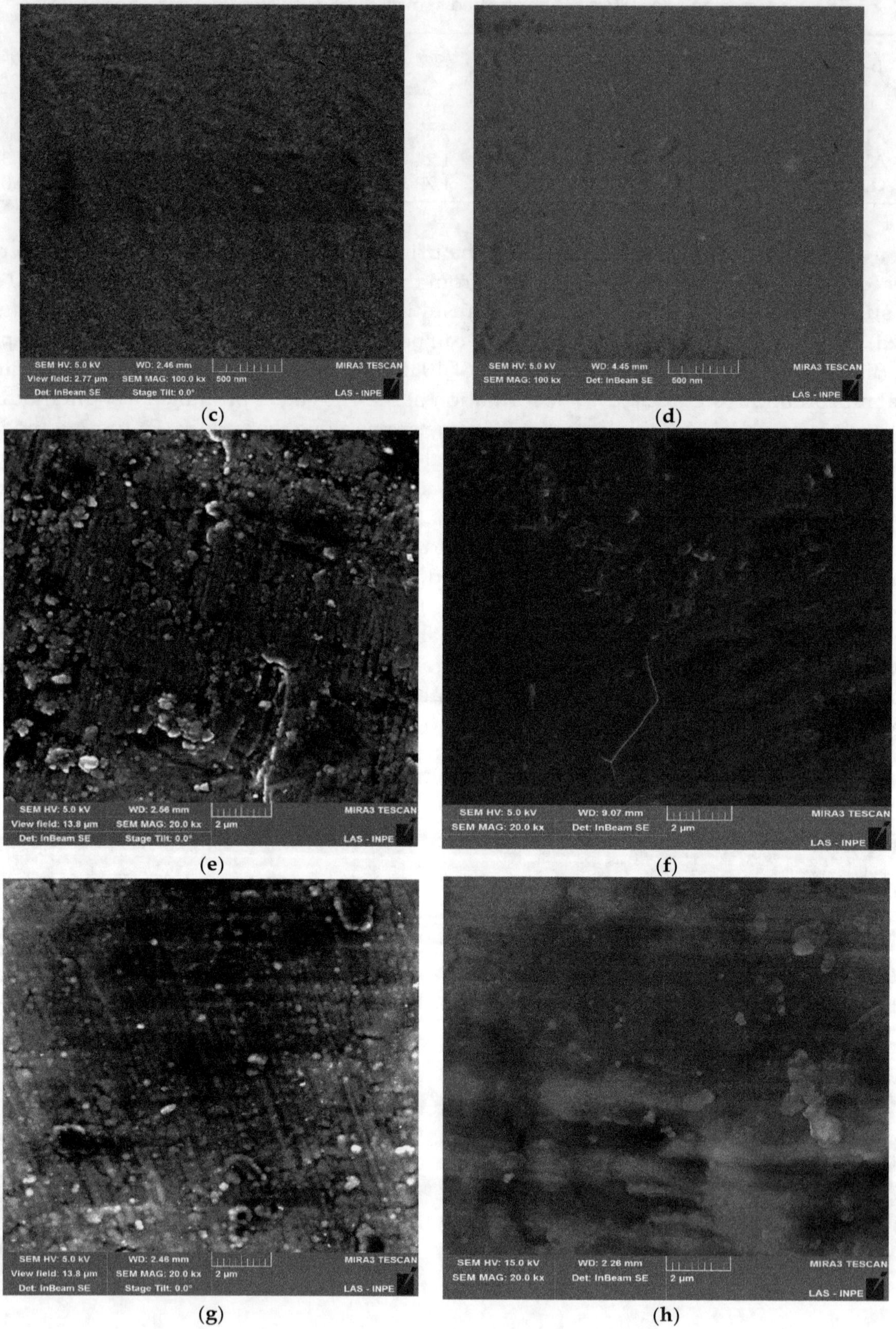

Figure 2. Scanning electron microscopy (SEM) micrographs of the studied samples: (**a**) bare Al (**b**) commercial resin-coated Al, (**c**) TiO_2-coated Al, (**d**) Al_2O_3-coated Al, in the condition of as-deposited; (**e**) bare Al, (**f**) commercial resin-coated Al, (**g**) TiO_2-coated Al, (**h**) Al_2O_3-coated Al in the condition after corrosion process. Here, the corrosion time of the samples was of 10 min.

To complement the LSV and SEM results, FT-IR measurements were performed in pristine and corroded samples. The resulting spectra are shown in Figure 3. Initially, the analysis of the pure aluminum substrate was performed before and after the corrosion process in beer electrolyte (Figure 3a). The spectra show a wide O–H stretch band in the range of 3000 to 3700 cm^{-1} and in the band centered at 1614 cm^{-1}. The strong and wide absorption band, centered at 581 cm^{-1}, probably results from the Al–O stretching vibrations and the Al–OH vibrational mode [42]. Based on the spectra of Figure 3a, it is possible to observe that the aluminum after undergoing the process of corrosion in the beer media, increased the adsorption of water species, as well as formed a superficial oxide. However, aluminum is not protected by an adherent oxide layer at low pH because the superficial aluminum oxide is not stable. Hence, if the coating is not a perfect barrier, and low pH solution is able to reach the underlying aluminum, then it should fail rapidly from corrosion [8].

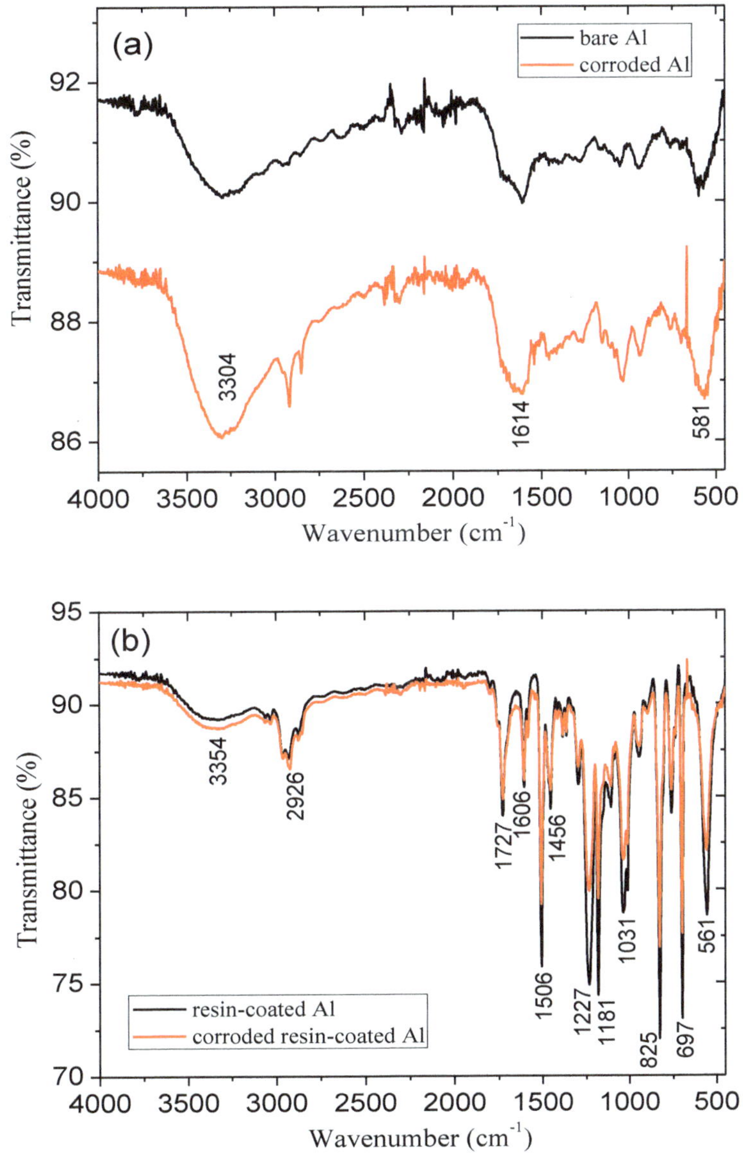

Figure 3. *Cont.*

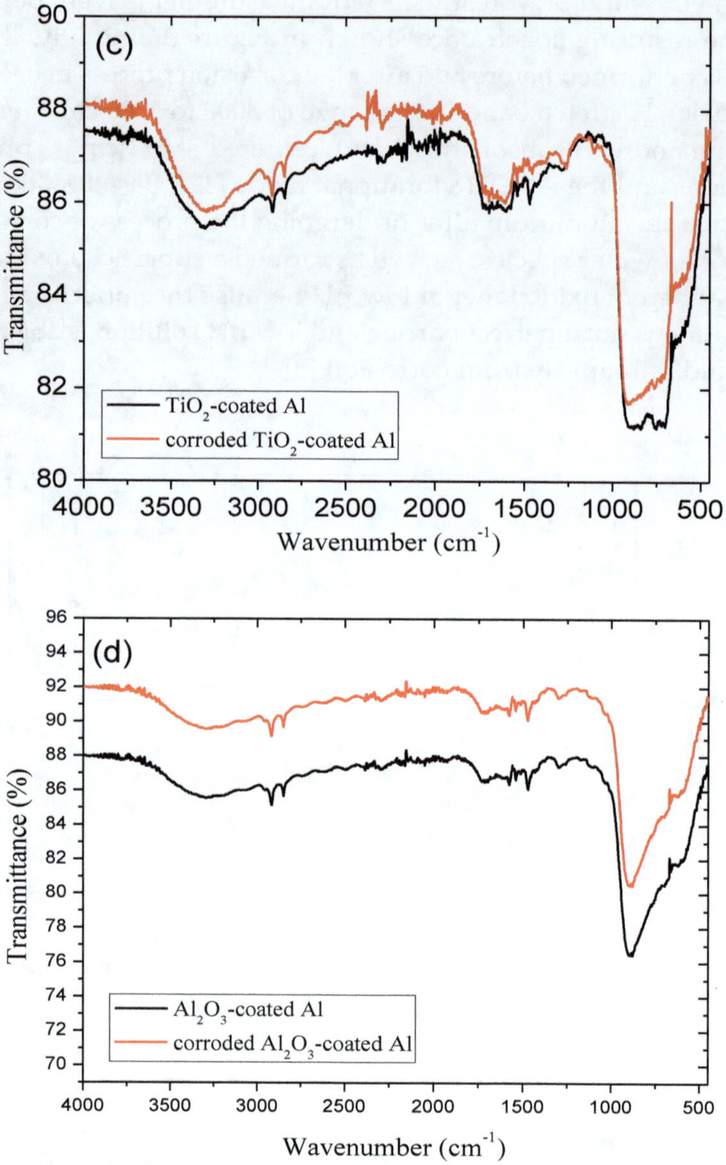

Figure 3. FT-IR spectra of the studied samples in the condition of as-deposited and after corrosion process. (**a**) control or bare Al, (**b**) commercial resin-coated Al, (**c**) TiO_2-coated Al, (**d**) Al_2O_3-coated Al.

The spectra shown in Figure 3b are relative to the aluminum substrate coated with epoxy resin. The specific elongation vibration band for the C–O of the epoxy ring is observed at 825 cm^{-1}. Bands formed between 1587 and 1668 cm^{-1} are assigned to the asymmetric C–O and C=O bonds, respectively. The C–O–C stretch is observed at 1031 cm^{-1}, relative to the reaction of hydroxyl groups with epoxy ring. A small characteristic band at O–H appeared at 3300–3500 cm^{-1}, relative to the reaction of the epoxy with the amine to form hydroxyl groups [43]. Note that the intensity of the peaks relative to the epoxy decreased after the corrosion process in beer electrolyte, indicating a possible reduction in the amount of material. However, as the pH of the medium changed from pH = 4.12 $\pm$ 0.02 to pH = 4.17 $\pm$ 0.03 after the immersion time of 1h10min (OCP + corrosion time), this amount of material may be very lower, even for bare Al substrate.

With the coating of the aluminum substrate with TiO_2 film (Figure 3c), a significant change in the bands between 500–1000 cm^{-1} compared to the bare-Al sample could be observed (Figure 3a). These bands are probably related to Ti–O bonds. Note that there was a small variation in the intensity of the O–H bonds after the corrosion process, indicating good chemical stability of the TiO_2 film. This result is confirmed by the LSV data. In relation to aluminum sample coated with alumina (Figure 3d), a behavior very similar to TiO_2 was observed.

3.2. *Electrochemical Impedance Spectroscopy*

The EIS allows a detailed understanding of the electrical characteristics of the electrode–solution interface, by means of the application of a small alternating voltage disturbance in the electrochemical system and recording from which the impedance is calculated [44]. Figures 4 and 5 present the complex plane (Nyquist) plot and the Bode plots of the samples, respectively. Examination of Figure 4 shows that all films are very effective in terms of increasing impedance of the Al substrate. The samples show capacitive arc characteristic that indicates corrosion protection, although some samples also present resistive characteristic in the region of high frequencies, which is attributed to non-ideality of the coatings in complete blocking the corroding species (such as water, oxygen, and ions) [36]. Concern the capacitive arc diameter, the resin-coated sample increased by approximately 11 times the diameter in comparison with bare Al, indicating the ability of the resin to protect the aluminum in contact with the beer electrolyte. When compared with uncoated aluminum, the TiO_2 film provided an increase by a factor of 14 the arc diameter. Instead, the Al_2O_3 film provided an increase of the arc diameter by 12.

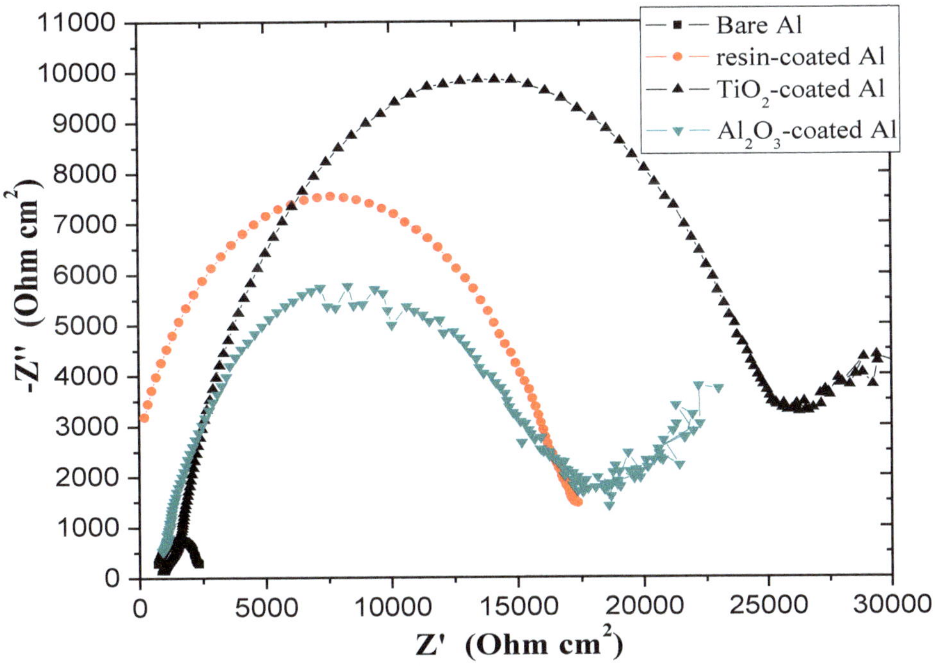

Figure 4. Nyquist plots for the studied samples.

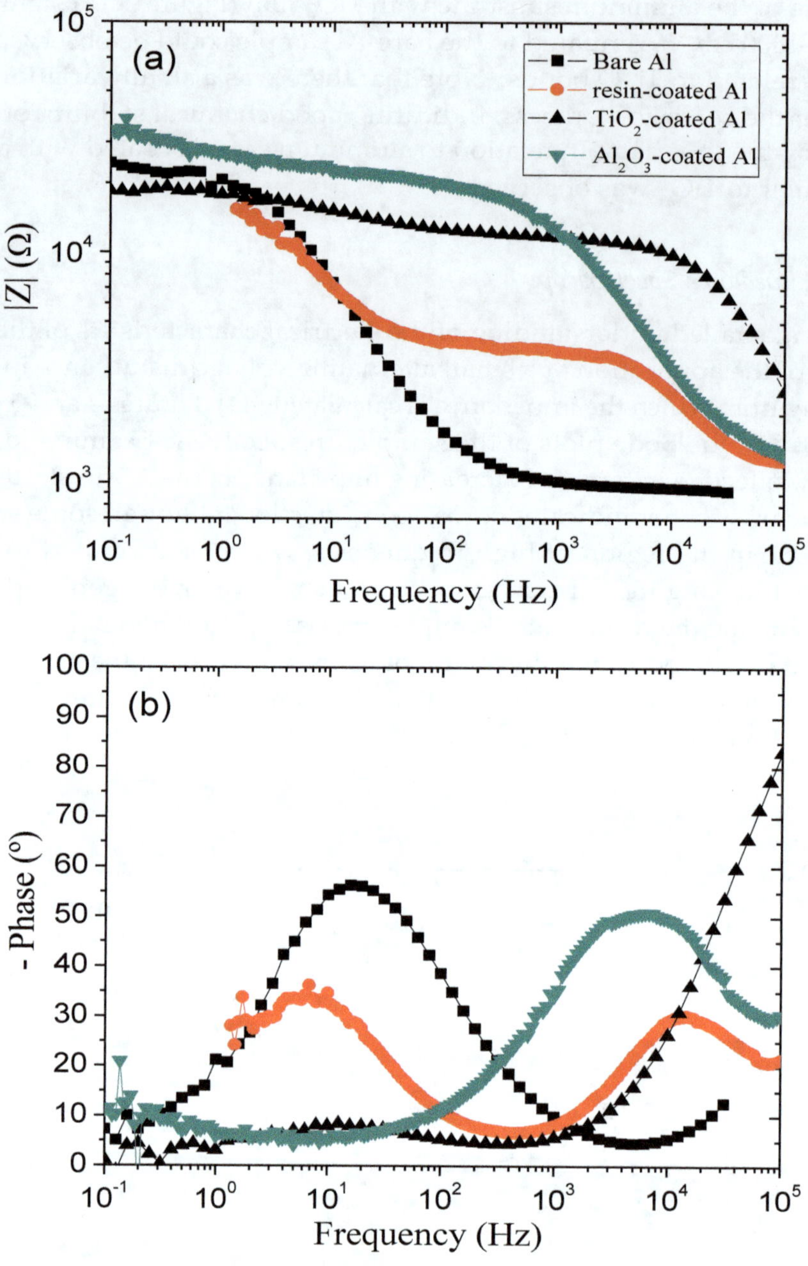

Figure 5. Bode magnitude (**a**) and Bode phase (**b**) plots of the electrochemical impedance spectroscopy for the studied samples.

For a better analysis of the EIS data, corresponding Bode magnitude and Bode phase plots are presented in Figure 5a,b, respectively. Bode magnitude plot represents the ratio of the amplitude of alternating voltage to alternating current versus frequency, irrespective of their phase shift. On the other hand, the Bode phase plot represents the phase difference between alternating voltage and alternating current versus frequency, irrespective of their amplitudes [36]. Both plots contain information which is not directly extractable from Nyquist plots. Figure 5a shows an increase in impedance magnitude for Al_2O_3-covered sample. However, the plots for all samples approach to each other at the lowest frequency, reaching values between 20–35 kΩ. s. Bode phase plots in Figure 5b are additional feature of EIS data. A characteristic of Bode phase plot is its ability to identify the predominant electrical behavior of the system in a given frequency range. For example, capacitive, resistive and mixed capacitive-resistive behavior appear as −90, 0, and 0 to −90 degrees lines, respectively [36]. Bode phase plots in Figure 5b shows completely different behavior for samples coated and uncoated. The first knee observed for covered samples, and more evident for commercial resin-coated Al, corresponds to the capacitive-resistive transition of the Bode phase plots [36]. In addition, the resin-coated Al plot resembles to two-time constants with same order of magnitude. These two processes are ascribable to coating and electrolyte–Al interface. The low resistance and high capacitance arise from high porosity of the coating [36], corroborating with LSV results. The hase angle of −50 degrees in high frequency for Al_2O_3-covered Al is characteristic of capacitor. Instead, phase angle of −85 degrees in high frequency for TiO_2-covered Al is characteristic of a pure capacitor acting as efficient blocking coating. The lower efficiency of the Al_2O_3 film in comparison with TiO_2 is intriguing and deserves further investigations.

For extraction of more information from EIS data, the data were fitted to an electrical equivalent circuit. A two-time constant circuit was adopted for this purpose [36,45]. However, due to system heterogeneities, the coincidence of experimental and fitted values was poor and therefore, the capacitor elements were replaced by constant phase elements (CPE). In the equivalent circuit, R_s is the solution/electrolyte resistance and R_p is the polarization resistance. The values of CPE, R_s and R_p are presented in Table 2. As can be seen, R_p increased considerably when aluminum was coated, especially with the ALD films investigated here. As R_p is directly related to the coating resistance, again the higher efficiency of the TiO_2 film compared to the Al_2O_3 films is evident. The CPE and R_s parameters are also optimized for TiO_2.

Table 2. Values of the elements of the equivalent electric circuit used to simulate the electrochemical impedance spectroscopy measurements of the samples in the beer electrolyte. CPE—constant phase element. R_s—solution/electrolyte resistance. R_p—polarization resistance. n—circuit fitting parameter.

Sample	CPE (F)	R_s (Ω)	R_p (Ω)	n (%)
Al	7.37×10^{-9}	1056.3	2.841.8	99.8
Al + V	4.66×10^{-9}	979.7	17.905.0	99.8
Al + TiO_2	3.52×10^{-7}	1682.8	32.619.0	99.7
Al + Al_2O_3	7.52×10^{-9}	1220.2	27.442.0	99.6

4. Conclusions

In conclusion, TiO_2 and Al_2O_3 thin films have been coated onto Al substrates using ALD method in order to improve the corrosion resistance for food packaging application. Here, we evaluated for the first time the initial corrosion barrier resistance of these ALD films in commercial beer electrolyte and performed a comparison with commercial resin-coated Al. GIXRD, SEM and profilometry analyzes showed that the films have an amorphous structure, a morphology that follows Al substrate surface, and a thickness of around 100 nm. Additionally, it was possible to see from SEM micrographs that the films entirely shield the substrates. The electrochemical measurements show that the equilibrium corrosion potential shifts from −0.40 V for bare Al to −0.46 V for the commercial resin-coated Al, −0.50 V for TiO_2-coated Al, and −0.55 V for Al_2O_3-coated Al. By means of a comparative evaluation of the samples, it was possible to visualize the behavior pattern of the capacitive arc and the corrosion process in beer electrolyte. From the analysis of LSV data was verified that ALD coatings promoted a considerable increase in corrosion barrier efficiency being 86.3% for TiO_2-coated Al and 80% for Al_2O_3-coated Al in comparison with 7.1% of commercial resin-coated Al. This is mainly due to the lower electrochemical porosity—11.4% for TiO_2-coated Al and 20.4% for Al_2O_3-coated Al, in comparison with 96% of the resin-coated Al, i.e. an increase of up to twofold in the protection of Al when coated with TiO_2, when compared to aluminum coated. The impedance results confirm that TiO_2-coated Al have good corrosion resistance in beer electrolyte. It was observed from Bode plots in high frequency that the phase angle of −50 degrees for Al_2O_3-covered Al is characteristic of capacitor, while the phase angle of −85 degrees for TiO_2-covered Al is characteristic of a pure capacitor. This result shows the TiO_2-covered Al acting as efficient blocking coating. Also, it was showed that the reduced corrosion barrier efficiency of the Al_2O_3 film in comparison with TiO_2 for beer electrolyte once SEM and FT-IR analyses did not show drastic changes in the films after the corrosion assays. A further investigation is being carried out with standard NaCl electrolytes in order to clarify this low efficiency of Al_2O_3 film. The results of our initial analysis of the protective performance of ALD TiO_2 and Al_2O_3 coatings demonstrated that they could be an alternative for the corrosion protection of aluminum food packaging. In future studies, we can also further explore their protective properties for this application by combining Al_2O_3 and TiO_2 to produce Al_2O_3-TiO_2 nanolaminates. Finally, to evaluate the shelf life time of beer or other carbonated beverage, whether in stock in the industry or in supermarket shelves, it is necessary to evaluate the effect of the immersion time over long periods (several hours or days).

Author Contributions: V.D., R.P., M.F. and F.R.M. provide the investigation and methodology, V.D., H.S., M.F., R.P., A.O.L. and F.R.M. write the original draft.

References

1. Anyadike, N. *Aluminium: The Challenges Ahead*; Elsevier: Amsterdam, The Netherlands, 2002.
2. Lamberti, M.; Escher, F. Aluminium Foil as a Food Packaging Material in Comparison with Other Materials. *Food Rev. Int.* **2007**, *23*, 407–433. [CrossRef]
3. Ojha, A.; Sharma, A.; Sihag, M.; Ojha, S. Food packaging–materials and sustainability-A review. *Agric. Rev.* **2015**, *36*, 241–245. [CrossRef]
4. Simal-Gándara, J. Selection of can coatings for different applications. *Food Rev. Int.* **1999**, *15*, 121–137. [CrossRef]
5. Lazarevic, Z.Z.; Mišković-Stankovic, V.B.; Kacarevic-Popovic, Z.; Drazic, D.M. The study of corrosion stability of organic epoxy protective coatings on aluminium and modified aluminium surfaces. *J. Braz. Chem. Soc.* **2005**, *16*, 98–102. [CrossRef]
6. Salas, B.V.; Wiener, M.S.; Stoytcheva, M.; Zlatev, R.; Beltran, M.C. Corrosion in the food industry and its control. In *Food Industrial Processes-Methods and Equipment*; InTech: London, UK, 2012.
7. Qian, Y.; Li, Y.; Jungwirth, S.; Seely, N.; Fang, Y.; Shi, X. The Application of Anti-Corrosion Coating for Preserving the Value of Equipment Asset in Chloride-Laden Environments: A. *Int. J. Electrochem. Sci.* **2015**, *10*, 10756–10780.
8. Lazur, I. Effect of pH on Coatings Used to Protect Aluminum Beverage Cans. Ph.D. Thesis, Carleton University Ottawa, Ottawa, ON, Canada, 2014.
9. Marsh, K.; Bugusu, B. Food packaging—roles, materials, and environmental issues. *J. food sci.* **2007**, *72*, 39–55. [CrossRef] [PubMed]
10. Grandle, J.; Taylor, S. Electrochemical impedance spectroscopy of coated aluminum beverage containers: Part 1—Determination of an optimal parameter for large sample evaluation. *Corrosion* **1994**, *50*, 792–803. [CrossRef]
11. Ray, C.J. The Impact of VOC and Waste Disposal Regulations on Protective Coating Product Development. Corrosion/92. In Proceedings of the NACE Annual Conference, Houston, TX, USA, 27 April 1992; p. 492.
12. LaKind, J.S. Can coatings for foods and beverages: issues and options, International Journal of Technology. *Policy Manag.* **2013**, *13*, 80–95.
13. Nocuń, M.; Burcon, D.; Siwulski, S. Sodium diffusion barrier coatings prepared by sol-gel method. *Opt. Appl.* **2008**, *38*, 171–179.
14. Jadav, S.; Kumar, A. Ceramic coating [TiO_2-ZrO_2] on aluminium 6061T6 for anti wear properties. *IJRET* **2014**, *3*, 43–49.
15. Lampke, T.; Leopold, A.; Dietrich, D.; Alisch, G.; Wielage, B. Correlation between structure and corrosion behaviour of nickel dispersion coatings containing ceramic particles of different sizes. *Surf. Coat. Technol.* **2006**, *201*, 3510–3517. [CrossRef]
16. Koli, D.K.; Agnihotri, G.; Purohit, R. A review on properties, behaviour and processing methods for Al-nano Al_2O_3 composites. *Procedia Mater. Sci.* **2014**, *6*, 567–589. [CrossRef]
17. Pessoa, R.S.; Fraga, M.A.; Santos, L.V.; Massi, M.; Maciel, H.S. Nanostructured thin films based on TiO_2 and/or SiC for use in photoelectrochemical cells: A review of the material characteristics, synthesis and recent applications. *Mater. Sci. Semicond. Process.* **2015**, *29*, 56–68. [CrossRef]
18. Daubert, J.S.; Hill, G.T.; Gotsch, H.N.; Gremaud, A.P.; Ovental, J.S.; Williams, P.S.; Oldham, C.J.; Parsons, G.N. Corrosion Protection of Copper Using Al_2O_3, TiO_2, ZnO, HfO_2, and ZrO_2 Atomic Layer Deposition. *ACS Appl. Mater. interfaces* **2017**, *9*, 4192–4201. [CrossRef] [PubMed]

19. Shen, G.; Chen, Y.; Lin, C. Corrosion protection of 316 L stainless steel by a TiO_2 nanoparticle coating prepared by sol–gel method. *Thin Solid Films* **2005**, *489*, 130–136. [CrossRef]
20. Shan, C.; Hou, X.; Choy, K.-L. Corrosion resistance of TiO_2 films grown on stainless steel by atomic layer deposition. *Surf. Coat. Technol.* **2008**, *202*, 2399–2402. [CrossRef]
21. Abdulagatov, A.; Yan, Y.; Cooper, J.; Zhang, Y.; Gibbs, Z.; Cavanagh, A.; Yang, R.; Lee, Y.; George, S. Al_2O_3 and TiO_2 atomic layer deposition on copper for water corrosion resistance. *ACS appl. mater. interfaces* **2011**, *3*, 4593–4601. [CrossRef] [PubMed]
22. Matero, R.; Ritala, M.; Leskelä, M.; Salo, T.; Aromaa, J.; Forsén, O. Atomic layer deposited thin films for corrosion protection. *Journal Phys. IV* **1999**, *9*, 493–499. [CrossRef]
23. Shan, C.; Hou, X.; Choy, K.-L.; Choquet, P. Improvement in corrosion resistance of CrN coated stainless steel by conformal TiO_2 deposition. *Surf. Coat. Technol.* **2008**, *202*, 2147–2151. [CrossRef]
24. Marin, E.; Lanzutti, A.; Paussa, L.; Guzman, L.; Fedrizzi, L. Long term performance of atomic layer deposition coatings for corrosion protection of stainless steel. *Mater. Corros.* **2015**, *66*, 907–914. [CrossRef]
25. Du, X.; Zhang, K.; Holland, K.; Tombler, T.; Moskovits, M. Chemical corrosion protection of optical components using atomic layer deposition. *Appl. Opt.* **2009**, *48*, 6470–6474. [CrossRef] [PubMed]
26. Hirvikorpi, T.; Vähä-Nissi, M.; Mustonen, T.; Iiskola, E.; Karppinen, M. Atomic layer deposited aluminum oxide barrier coatings for packaging materials. *Thin Solid Films* **2010**, *518*, 2654–2658. [CrossRef]
27. Hirvikorpi, T.; Vähä-Nissi, M.; Nikkola, J.; Harlin, A.; Karppinen, M. Thin Al_2O_3 barrier coatings onto temperature-sensitive packaging materials by atomic layer deposition. *Surf. Coat. Technol.* **2011**, *205*, 5088–5092. [CrossRef]
28. Sammelselg, V.; Netšipailo, I.; Aidla, A.; Tarre, A.; Aarik, L.; Asari, J.; Ritslaid, P.; Aarik, J. Chemical resistance of thin film materials based on metal oxides grown by atomic layer deposition. *Thin Solid Films* **2013**, *542*, 219–224. [CrossRef]
29. Chiappim, W.; Testoni, G.; de Lima, J.; Medeiros, H.; Pessoa, R.S.; Grigorov, K.; Vieira, L.; Maciel, H. Effect of process temperature and reaction cycle number on atomic layer deposition of TiO_2 thin films using $TiCl_4$ and H_2O precursors: correlation between material properties and process environment. *Braz. J. Phys.* **2016**, *46*, 56–69. [CrossRef]
30. Chiappim, W.; Testoni, G.; Moraes, R.; Pessoa, R.; Sagás, J.; Origo, F.; Vieira, L.; Maciel, H. Structural, morphological, and optical properties of TiO_2 thin films grown by atomic layer deposition on fluorine doped tin oxide conductive glass. *Vacuum* **2016**, *123*, 91–102. [CrossRef]
31. Chiappim, W.; Testoni, G.; Doria, A.; Pessoa, R.; Fraga, M.; Galvão, N.; Grigorov, K.; Vieira, L.; Maciel, H. Relationships among growth mechanism, structure and morphology of PEALD TiO_2 films: the influence of O_2 plasma power, precursor chemistry and plasma exposure mode. *Nanotechnology* **2016**, *27*, 305701. [CrossRef] [PubMed]
32. Testoni, G.; Chiappim, W.; Pessoa, R.; Fraga, M.; Miyakawa, W.; Sakane, K.; Galvão, N.; Vieira, L.; Maciel, H. Influence of the Al_2O_3 partial-monolayer number on the crystallization mechanism of TiO_2 in ALD TiO_2/Al_2O_3 nanolaminates and its impact on the material properties. *J. Phys. D Appl. Phys.* **2016**, *49*, 375301. [CrossRef]
33. Pessoa, R.; dos Santos, V.; Cardoso, S.; Doria, A.; Figueira, F.; Rodrigues, B.; Testoni, G.; Fraga, M.; Marciano, F.; Lobo, A. TiO_2 coatings via atomic layer deposition on polyurethane and polydimethylsiloxane substrates: Properties and effects on C. albicans growth and inactivation process. *Appl. Surf. Sci.* **2017**, *422*, 73–84. [CrossRef]
34. Pessoa, R.; Pereira, F.; Testoni, G.; Chiappim, W.; Maciel, H.; Santos, L. Effect of substrate type on structure of TiO_2 thin film deposited by atomic layer deposition technique. *J. Integr. Circuits Syst.* **2015**, *10*, 38–42.
35. Zanin, H.; May, P.; Fermin, D.; Plana, D.; Vieira, S.; Milne, W.; Corat, E. Porous boron-doped diamond/carbon nanotube electrodes. *ACS Appl. Mater. Interfaces* **2014**, *6*, 990–995. [CrossRef] [PubMed]
36. Torknezhad, Y.; Khosravi, M.; Assefi, M. Corrosion protection performance of nanoparticle incorporated epoxy paint assessed by linear polarization and electrochemical impedance spectroscopy. *Materi. Corros.* **2018**, *69*, 472–480. [CrossRef]
37. Kim, H.-G.; Ahn, S.-H.; Kim, J.-G.; Park, S.J.; Lee, K.-R. Electrochemical behavior of diamond-like carbon films for biomedical applications. *Thin Solid Films* **2005**, *475*, 291–297. [CrossRef]
38. Oliveira, C.; Stein, M.; Saito, E.; Zanin, H.; Vieira, L.; Raniero, L.; Trava-Airoldi, V.; Lobo, A.; Marciano, F.

Effect of gold oxide incorporation on electrochemical corrosion resistance of diamond-like carbon. *Diam. Relat. Mater.* **2015**, *53*, 40–44. [CrossRef]

39. Ylivaara, O.M.E.; Liu, X.; Kilpi, L.; Lyytinen, J.; Schneider, D.; Laitinen, M.; Julin, J.; Ali, S.; Sintonen, S.; Berdova, M.; et al. Aluminum oxide from trimethylaluminum and water by atomic layer deposition: The temperature dependence of residual stress, elastic modulus, hardness and adhesion. *Thin Solid Films* **2014**, *552*, 124–135. [CrossRef]
40. Miikkulainen, V.; Leskelä, M.; Ritala, M.; Puurunen, R.L. Crystallinity of inorganic films grown by atomic layer deposition: Overview and general trends. *J. Appl. Phys.* **2013**, *113*, 2. [CrossRef]
41. Cimalla, V.; Baeumler, M.; Kirste, L.; Prescher, M.; Christian, B.; Passow, T.; Benkhelifa, F.; Bernhardt, F.; Eichapfel, G.; Himmerlich, M. Densification of thin aluminum oxide films by thermal treatments. *Mater. Sci. Appl.* **2014**, *5*, 628. [CrossRef]
42. Contreras, C.; Sugita, S.; Ramos, E. Preparation of sodium aluminate from basic aluminum sulfate. *Adv. Technol. Mater. Mater. Process. J.* **2006**, *8*, 122.
43. Sahoo, S.K.; Mohanty, S.; Nayak, S.K. Toughened bio-based epoxy blend network modified with transesterified epoxidized soybean oil: synthesis and characterization. *RSC Adv.* **2015**, *5*, 13674–13691. [CrossRef]
44. Maddala, J.; Sambath, K.; Kumar, V.; Ramanathan, S. Identification of reaction mechanism for anodic dissolution of metals using Electrochemical Impedance Spectroscopy. *J. Electroanal. Chem. Interfacial. Electrochem.* **2010**, *638*, 183–188. [CrossRef]
45. Shi, X.; Nguyen, T.A.; Suo, Z.; Liu, Y.; Avci, R. Effect of nanoparticles on the anticorrosion and mechanical properties of epoxy coating. *Surf. Coat. Technol.* **2009**, *204*, 237–245. [CrossRef]

3

Recent Trends in the use of Pectin from Agro-Waste Residues as a Natural-Based Biopolymer for Food Packaging Applications

Cristina Mellinas, Marina Ramos, Alfonso Jiménez and María Carmen Garrigós *

Department of Analytical Chemistry, Nutrition & Food Sciences, University of Alicante, ES-03690 Alicante, San Vicente del Raspeig, Spain; cristina.mellinas@ua.es (C.M.); marina.ramos@ua.es (M.R.); alfjimenez@ua.es (A.J.)

* Correspondence: mc.garrigos@ua.es

Abstract: Regardless of the considerable progress in properties and versatility of synthetic polymers, their low biodegradability and lack of environmentally-friendly character remains a critical issue. Pectin is a natural-based polysaccharide contained in the cell walls of many plants allowing their growth and cell extension. This biopolymer can be extracted from plants and isolated as a bioplastic material with different applications, including food packaging. This review aims to present the latest research results regarding pectin, including the structure, different types, natural sources and potential use in several sectors, particularly in food packaging materials. Many researchers are currently working on a multitude of food and beverage industry applications related to pectin as well as combinations with other biopolymers to improve some key properties, such as antioxidant/antimicrobial performance and flexibility to obtain films. All these advances are covered in this review.

Keywords: pectin; food packaging; active compounds; agro-waste residues; circular economy

1. Introduction

Biopolymers are gaining their market share in the plastics industry by their intrinsic biodegradable character combined with interesting properties for specific applications. Biopolymers can be obtained/extracted from natural sources, biosynthesized by living organisms or chemically synthesized from biological materials [1]. In addition, their natural-based origin, i.e., from renewable sources represents a great advantage over plastic commodities since their use decrease dependence from petroleum while preserving and even improving important material properties. There has been an increasing interest for the use of biopolymers in packaging, medicine, agriculture, and other sectors. Different types of carbohydrates, such as starch and cellulose, as well as other polysaccharides, such as alginates and pectin, as well as their combinations with animal-protein-based biopolymers, such as silk, wood, gelatin, collagen, chitosan/chitin, gums, plant-based proteins and lipids offer the possibility of rendering interesting applications for these advanced sectors. All these biopolymers offer interesting advantages in their use, such as their renewable origin, biocompatibility, barrier properties to moisture and/or gases, non-toxicity, non-polluting characteristics, mechanical integrity and relative low cost.

The increase in the use of biopolymers has caused that their global market is expected to reach around 10 billion US Dollars by 2021, growing by almost 17% over the forecast period 2017–2021. Western Europe covers the largest market segment, accounting for 41.5% of the global market while other regions are rapidly increasing their market share [2].

In addition, another important possibility offered by the use of biopolymers is their potential to be synthesized from the non-edible parts of plants or animals, avoiding the risk of depleting food from local communities, most of them in under-developed regions. Figure 1 shows the main types of biopolymers that can be obtained from biomass waste as well as some examples of their sources. They can be divided into three large groups: proteins, lipids and polysaccharides. The protein-based biopolymers can be obtained from both animal and vegetable wastes. For example, slaughterhouse wastes are a good source of proteins from animal origin, like gelatin. These wastes comprise the inedible tissues/parts of the animals slaughtered for the production of meat [3]. Among the proteins from plant origin, soy protein isolate is a good option to develop new materials due to its composition and excellent processing ability by gelling, emulsifying ability and water and oil holding capacity [4].

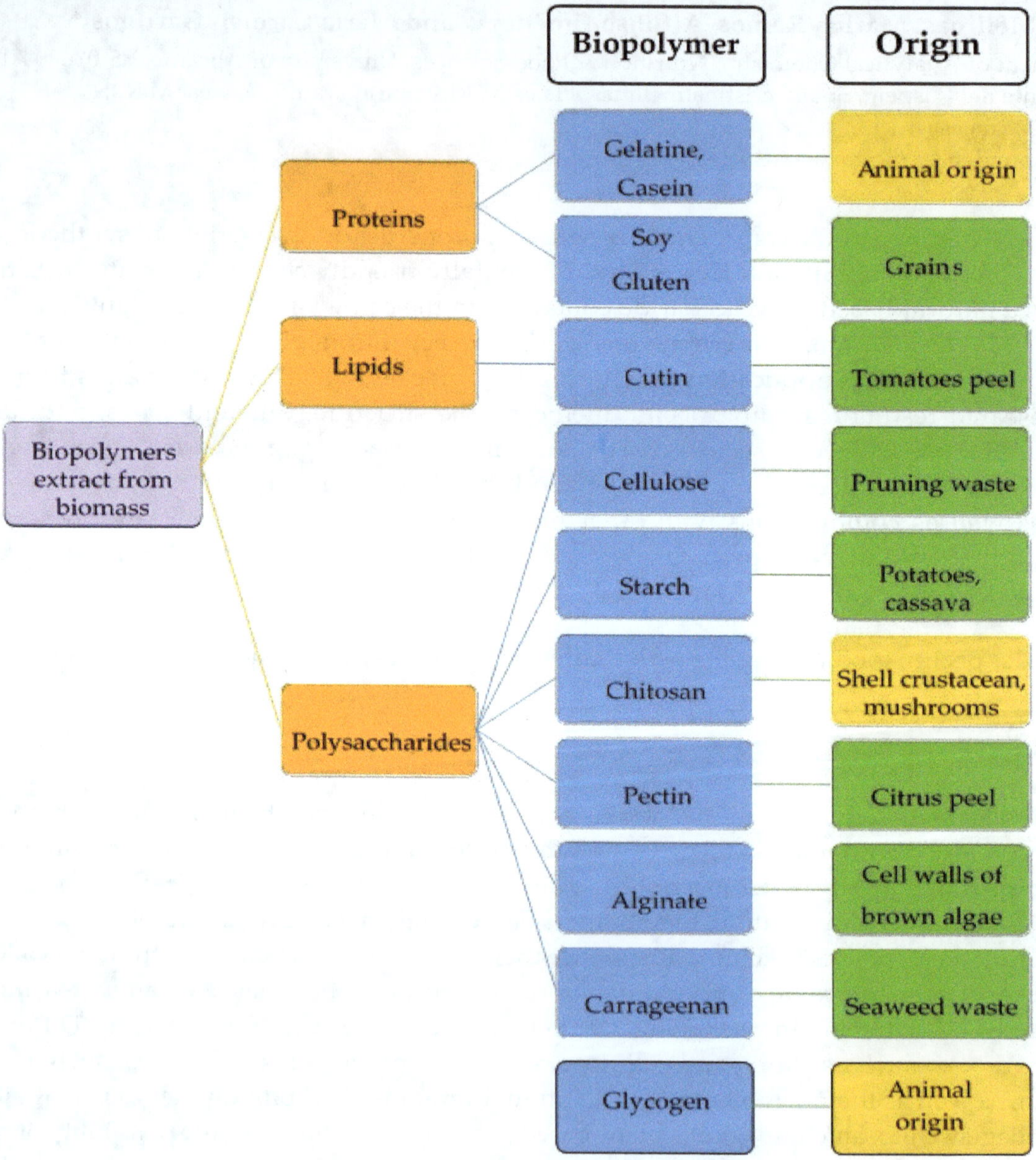

Figure 1. Different types of biopolymers obtained from animal and vegetable wastes.

Lipid-based polymers have been used in the last few years in food packaging or 3D printing materials. Their extraction from the natural sources is a necessary step to render isolated fatty acids to be further used in esterification reactions. For example, cutin extracted from tomato by-products [5] and lipids extracted from algae [6] have shown great potential to obtain specific fractions for the production of films with high barrier to water due to the repulsion caused by their high hydrophobic behaviour.

Polysaccharide-based polymers are the last group in Figure 1. These biopolymers are characterized by their biodegradable, biocompostable, sustainable and non-toxic characteristics. Additionally, polysaccharides are more thermally stable than other biopolymers, like lipids and proteins, since they are not irreversibly denatured via heating. However, their main disadvantages are the high sensitivity to moisture and low mechanical resistance [7]. In order to limit these problems, two different approaches have been proposed: the incorporation of different reinforcing additives to the polysaccharides matrices [8,9] and the combination with different polymers to obtain blends [10,11] or multilayer films [12,13]. The improvement in polysaccharides properties have permitted the extension of their use by the food industry in the last few years [14,15].

Pectin is one of the major structural polysaccharides present in many higher plant cells allowing primary cell wall extension and plant growth. It could be extracted and applied as an anionic biopolymer, soluble in water. A large number of recent articles have highlighted the advantages of using pectin over conventional polymers. Therefore, pectin is increasingly important for a multitude of food packaging applications, such as a thickening and gelling agent, colloidal stabilizer, texturizer, and emulsifier [16–18], a coating on fresh and cut fruits or vegetables [19–21] and as micro and nano-encapsulating agent for the controlled release of active principles with different functionalities [22]. Rodsamran et al. [16] reported that bioactive pectin films can retard soybean oil oxidation during 30 days of storage. Furthermore, Sucheta et al. [19] found that a pectin-corn flour-based coating significantly reduced the weight loss and decay per cent of tomatoes, delaying respiration with retention of biochemical quality of tomatoes. Additionally, polymeric blends of hydrocolloids obtained from chia seeds and apple pectin where developed with the aim to obtain antioxidant polymer blend films using the 2,2-diphenyl-1-picrylhydrazyl (DPPH) assay to estimate their antioxidant activity [10].

Pectin has been also extracted from waste biomass by using innovative methods, contributing to waste management in agriculture and food processing industries. Different pectin sources can be used, such as by-products of juice manufacturing as well as orange, mango, banana, lime and pomegranate peels and seeds. Therefore, this review aims to present and discuss the potential of pectin as a bio-based material in food packaging applications by its efficient extraction from waste biomass, while addressing a solution to the important environmental problems caused by the disposal of residues and by-products in the food sector.

2. Pectin

2.1. Pectin Structure

Pectin is a complex heteropolysaccharide and a major multifunctional component of the cell wall in many terrestrial plants. It is usually found in association with other compounds like cellulose, lignin or polyphenols present in the cell wall of plants [22]. Pectin is mainly composed of galacturonic acid units (Figure 2). The carboxyl groups of uronic acid residues can be present in different forms in the polymer structure, either free or as a salt form with sodium, calcium or other small counter-ions. In some cases, they can be also present as naturally-esterified groups, particularly with methanol, depending on the pectin source and/or the extraction method. Due to the presence of free carboxyl groups, pectin solutions exhibit acidic pH values. Galacturonic acid comprises approximately 70% of the pectin composition, depending on the plant species, and all the pectic polysaccharides contain galacturonic acid linked at the O-1 and the O-4 positions [23]. Pectin has a linear anionic backbone which regions showing no side chains known as "smooth regions" and regions with non-ionic side chains known as "hairy regions" [24].

Different pectin structural domains may be distinguished (Figure 2), influencing their properties depending on pectin proportions [23].

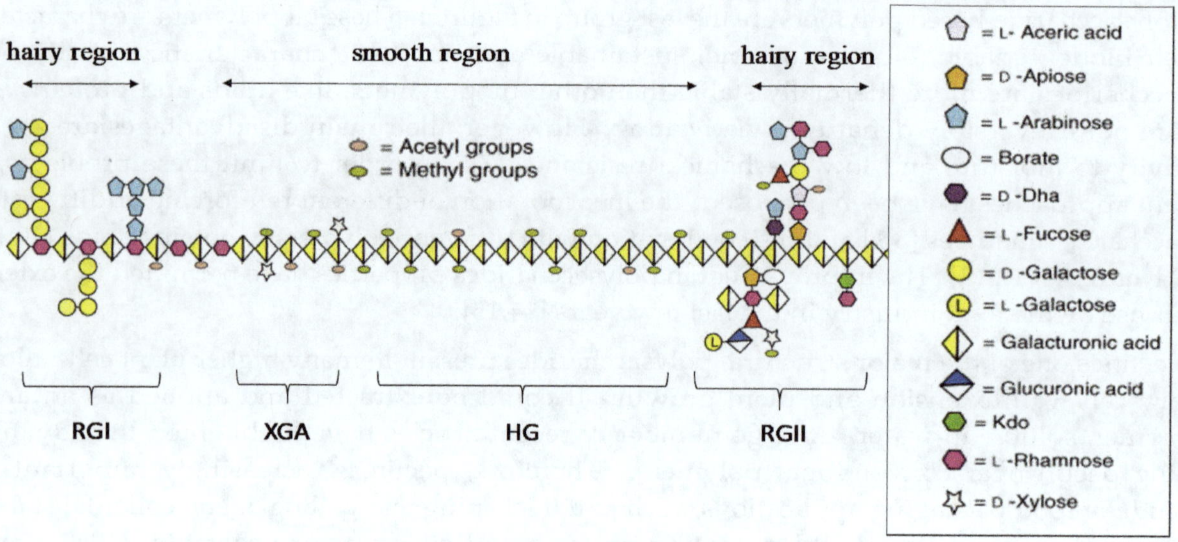

Figure 2. Schematic pectin structure, adapted from [23].

Homogalacturonan (HG): HG is the major domain of pectin in cell walls of plants, representing approximately 65% of the total pectin content. It is formed by galacturonic acid residues, linked by α (1→4) bonds, and their carboxyl groups are partially methyl esterified at position 6. Additionally, this domain may be acetylated at position 2 or 3 depending on the origin of the pectin. These domains are the main constituents of the above-mentioned "smooth regions" [24].

Rhamnogalacturonan I (RG-I) contains about 20%–35% of the total pectin and shows a more complex structure than HG. It contains repeated units of disaccharides consisting of L-rhamnose and galacturonic acid that can be also acetylated in the positions 2 or 3. It could have up to 100 units of (1,2)-α-L-Rha-(1,4)-α-D-GalA. In addition, large amount of L-rhamnose structures are substituted at O-4 by different neutral sugars such as L-galactose and L-arabinose [23].

Rhamnogalacturonan II (RG-II) contains about 10% of the total pectin and it is the structurally more complex component. Despite it is a relatively minor component in the pectin chain, RG-II plays a central role in the structure of plant cell walls. Small structure modifications of RG-II lead to reductions in the dimers formation and they can cause severe growth defects. So, dimerization of RG-II in the cell wall may be crucial for the normal growth and development of plants. This domain is composed of an HG backbone of (at least eight) 1,4-linked α-d-GalA residues decorated with side branches consisting of different types of sugars (rhamnose, fucose, xylose, galactose, apiose or aceric acid) in over 20 different types of linkages [23,25,26].

It is generally believed that the pectic polysaccharides are covalently bonded with high crosslinking densities since harsh chemical treatments or digestion by pectin-degrading enzymes are required to isolate HG, RG-I, and RG-II from each other. In addition, it has been reported that other components, such as xylogacturonan (XGA) and apiogalacturonan (AP), could replace the galacturonic acid units in some parts of the pectin chain [23]. The complexity of the pectin structures increases since it can be changed during the plant storage, extraction and processing, resulting in modifications of pectin functionalities and hindering its structural elucidation. It has been reported that the variations in chain lengths in each of the different domains are not the same, because HG and RGII have a highly homogeneous structure while RGI exhibits a wide heterogeneity in its composition [26].

2.2. *Type of Pectins*

The degree of esterification (DE) is an important parameter for the definition of the pectin applications and it is defined as the percentage of carboxyl groups esterified present in the structure of pectin. DE is often used to classify the different types of pectins (Figure 3). Depending on the DE different emulsifying, texturizing and gelling properties are observed. In general, when the DE increases, the water solubility decreases due to the hydrophobic nature of esters with long hydrocarbon chains. In contrast, when the DE increases the gelation rate also improves resulting in rapid gelation pectins [27]. Furthermore, the amount and composition of neutral sugars and the overall molecular weight of pectins have a great influence in their rheological properties [28].

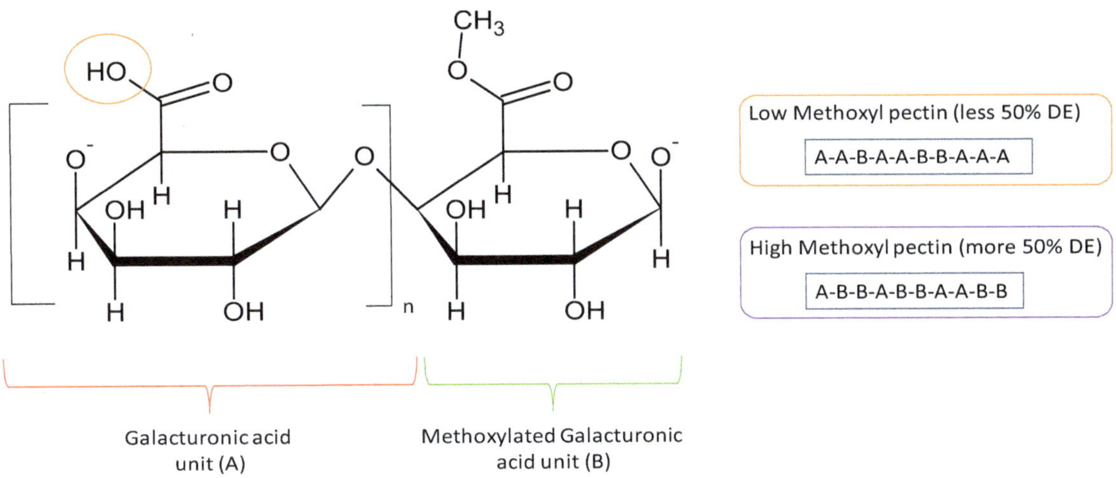

Figure 3. Structure of low and high methoxyl pectins.

High methoxyl pectin (HMP) shows DE higher than 50% and it is mainly used in the food industry by its thickening and gelling properties. It has been reported that HMP requires high amount of sugars for gelation and it is very sensitive to acidity [29]. HMP forms gels at low pH values and high concentrations of soluble solids due to the presence of hydrogen bonding and hydrophobic interactions between the pectin chains. Neutral sugars like sucrose play different roles in gelation through regulation of the hydrophobic interaction or directly binding to the polymer chains of HMP. Gels are formed when HG portions are cross-linked to form three dimensional crystalline networks in which water and other solutes are trapped [30]. The mechanism of formation of HMP gels is complex and it has been the subject of many investigations in the last decades. The glass transition theory has been proposed to explain the formation of HMP gels. Due to the high viscosity of molecules an arrest of the system kinetics occurs, giving as a result the formation of the gel due to the increased concentration of co-solutes and decrease in the water content. The transition from sol to gel behaviour is due to the combined effect of HMP and sucrose at pH 3, and occurs when excluded volume effects and attractive interactions are capable to give rise to an incipient three-dimensional network [31]. In addition, the effect of monovalent cations has been evaluated under different alkaline conditions (NaOH and KOH) at different pectin concentrations. It was suggested that HMP gel is formed through different mechanisms, such as de-esterification, self-aggregation, and entanglement under alkaline conditions. Na^+ or K^+ bind to dissociated carboxyl groups in HMP due to electronic attraction and this behaviour enables HMP molecules to move closer to each other, thereby improving gel network formation [32].

The emulsifying properties of HMP have been investigated by Jiang et al. in binary water-ethanol systems. They suggested that ethanol reduces the electrostatic repulsion and promotes pectin aggregation [33]. When HMP was mixed with the water-ethanol mixtures, the helix structure was broken, and electrostatic repulsion decreased. The compact and hydrophobic conformation enables pectin to adsorb better on the oil-water interface. The obtained emulsion showed good stability when using 21% of ethanol in the mixture.

Low methoxyl pectin (LMP) shows DE lower than 50% and it is generally formed by the de-esterification of HMP. Different agents can be used for the preparation of LMP from HMP, such as alkalis like sodium hydroxide or ammonia, enzymes (pectin methyl esterase) and concentrated acids [34].

LMP is widely used by the food industry to form low sugar-content jams as it does not require large amounts of sugar for gelation. It shows less sensitivity towards acidity and requires Ca^{2+} ions to form gels. The gelation mechanism in LMP is mediated by the formation of calcium bonds between two carboxyl groups from two chains in close contact [28]. Recently, Han et al. studied the effect of the calcium concentration, pH, soluble solids, and pectin concentration on the gel strength of LMP gels and they proposed different mechanisms of formation of pectin gels based on their rheological properties [35]. They observed that pH values close to the isoelectric point (pH = 3.50) and high calcium concentrations enhanced the storage modulus and gel strength by formation of calcium bridges at dissociated carboxyl groups. In addition, the sucrose content improved the gel strength because the neutral sugars provide hydroxyl groups to stabilize the gel and contribute to the formation of hydrogen bonds to immobilize free water. On the other hand, the formation of LMP gels under alkaline conditions was tested by Yang et al. [36]. They suggested that LMP can form relatively stable gels in the pH range of 3.5–9.5 using NaOH as pH regulator. In addition, they evaluated the effect of calcium concentration in the thermal and structural properties of the LMP gels and they concluded that the presence of calcium ions not only reduced the thermal stability but also the crystalline degree of LMP.

3. Sources of Pectin

Due to the high potential of pectin-based polymers, the extraction of pectin from biomass waste has been widely studied. Table 1 summarizes several published works based on the extraction of pectin from agro-waste sources.

Table 1. Different raw materials and extraction methods to obtain natural pectin.

Raw Material	Extraction Method	Conditions				Ref.
		T (°C)	Time (min)	LSR (mL/g)	Other Variables	
Eggplant peel waste	HAE	90	30	40	-	[37]
Orange peel waste	HAE	80	60	17.1	pH: 1.5	[38]
Orange peel waste	HAE	75	300	20	pH: 2.5	[39]
Orange peel waste	HAE	95	120	6	pH: 1.6	[40]
Pomelo peel waste	HAE	90	120	30	pH: 2	[41]
Pomegranate peel waste	HAE	70	120	10	-	[42]
Apple peel waste	HAE	85	120	25	-	[43]
Cashew apple pulp	HAE	100	120	5.15	-	[44]
Chamomile Waste	HAE	90	60	20	pH: 1.2	[45]
Durian rind waste	HAE	85	60	9	pH: 2.5	[46]
Hibiscus (*sabdariffa* L.)	HAE	100	30	20	pH: 2.5	[47]
Banana peel waste	HAE	86	360	50	pH: 2	[48]
Banana peel waste	HAE	90	30	20	pH: 1.5, 6	[49]
Mango peel waste	HAE	90	120	20	-	[50]
Jackfruit peel waste	HAE	138	9	17	-	[51]
Jackfruit peel waste	HAE	90	60	10	-	[52]
Jackfruit peel waste	HAE	90	60	20	-	[53]
Passion fruit rind	HAE	98	90	50	-	[54]
Tomato husk waste	HAE	100	15–25	30	-	[28]
Orange peel waste	HC	14.6–96	270	2.86	-	[55]
Artichoke (*Cynara scolymus* L.)	EAE	50	2880	15.4	pH: 5, Enzyme: 10.1 Ug^{-1}	[56]
Sisal Waste	EAE	50	1200	15	Enzyme: 88 Ug^{-1}, pH:5	[57]
Tobacco waste	MAE	-	4	20	550 W, pH: 1.8	[58]
Cocoa Pod Husk waste	SWE	121	30	27.5	103.4 bar	[59]
Custard apple peel waste	UAE	63	18	21	pH: 3	[60]
Mango peel waste	UAE	85	10	7.6	497.4 W/cm^2, pH: 2	[61]
Sisal Waste	UAE	-	60	15	450W, pH: 4	[57]
Passion fruit rind	UAE	-	10	20	135W	[62]
Jackfruit peel waste	UAE-MAE	86	29	48	-	[63]

HAE: Hydrothermal-assisted extraction; UAE: Ultrasound-assisted extraction; HC: Hydrodynamic cavitation; MAE: Microwave-assisted extraction; SWE: Subcritical water extraction; EAE: Enzyme-assisted extraction.

The peels of citrus fruits have been reported as the main source to obtain pectin at the industrial scale due to their good properties and high extraction yield. Hydrothermal extraction is the most usual method to obtain pectin from orange peels and it involves high temperatures (75–95 °C) and extraction times (60–300 min). Additionally, in all cases, the hydrothermal extraction of pectin takes place under acidic conditions using water as solvent. Pectin is very soluble in water and the acid medium decrease the presence of other compounds like polyphenols increasing extraction yields and helping to maintain the quality of the extracted pectin [38–41]. Other methods have been tested to reduce extraction times in citrus by-products. For example, microwave-assisted extraction (MAE) has been used in lime [64] and pomelo peels [65] reducing the extraction times to five and two minutes, respectively. However, high microwave powers (700–1100 W) were required to achieve these results. The hydrodynamic cavitation method was also used to obtain pectin derived from orange peel waste. Although a large decrease in the amount of solvent (2.86 mL/g of dry waste) was observed, long extraction times were also needed (270 min) [55].

The use of other sources to obtain pectin-based polymers in good grade and quality has been proposed in the last few years, such as eggplant peel [37], chamomile waste [45], cocoa pod husk [59,66], banana peel [49], mango peel [50,61,67] or tomato husk [28]. Tropical fruits have been also studied in the last years to obtain HMP. For example, passion fruit rind [54,62], durian rind [46] or jackfruit peels [51,52,63] have been proposed as interesting sources of pectins. Hydrothermal extraction is also the most used method in these types of wastes. Ultrasound-assisted extraction (UAE) has been also tested in passion fruit rind using 450 W and a water to dry sample ratio of 20 mL/g for 10 minutes. Results showed that the obtained pectin was mainly formed by homogalacturonans. Furthermore, their high degree of methylation indicated that the passion fruit pectin could be applied in gel forming products [62].

The use of innovative and sustainable extraction techniques is heading towards the study of hybrid techniques with the objective of combining their advantages, such as in the case of MAE and UAE. Pectin has been obtained from sisal waste by the combination of enzymatic and ultrasonic processes as an efficient strategy for the production of high-quality pectins since the enzymatic treatment disrupt the links between cellulose and xyloglucans in the cell wall of sisal and then the ultrasonic treatment produces mechanical destruction of the sisal structure to improve the release of pectin [57].

Finally, the introduction of new extraction techniques can be a great initial investment for companies since they offer the possibilities to get specific extractions of high added value purified compounds, although the costs of microwave or ultrasonic based equipment are higher than those of conventional extraction equipment, but in the long term, these devices are more profitable since the energy consumption, extraction time and the amount of expensive reagents used during pectin extraction are reduced [68,69].

4. Pectin-Based Materials for Food Packaging Applications

Pectin is a versatile compound that can be used to develop different materials in many food applications such as thickening and gelling agent, colloidal stabilizer, texturizer and emulsifier. These important applications are not limited to food processing, but also to packaging, coatings on fresh and cut fruits or vegetables and as microencapsulating agents (Table 2). Pectin is soluble in pure water and insoluble in organic solvents. Moreover, when dry pectin is mixed with water it tends to hydrate very rapidly, forming clumps. This behaviour is due to the formation of dry spheres of pectin contained in a highly hydrated outer coating. In order to eliminate these clumps, a vigorous and long agitation time is required [70]. In general terms, diluted pectin solutions present a Newtonian behaviour, but at high concentrations they show non-Newtonian behaviour, corresponding to pseudo-plastic characteristics. It was observed that the decrease in solubility and increase in viscosity contribute to increase the gelation capacity, i.e., the pectin concentration has a positive effect in gelation capacity and viscosity but a negative effect in solubility. Although it was stated that pectin properties are mainly dependent on structure, particularly DE [22,71], film forming, gelling and emulsifying properties should be also considered.

Table 2. Different types of pectin-based materials used in food packaging applications.

Type	Polymer Matrix	Additive	Application	Ref.
Film	LMP-bitter vetch protein	Transglutaminase	Drug delivery system	[72]
Film	LMP	Ascorbic acid	AO system	[73]
Film	HMP	Clove EO	AM system	[74]
Film	HMP	Marjoram EO	AO system	[75]
Film	HMP-Gluconaman	Tea extract	AO/AM system	[76]
Film	Pectin-Pullulan	AgNPs	AM system	[77]
Film	HM-Apple pectin	Chia seed hydrocolloid	AO system	[10]
Film	Chitosan-Starch-Pectin	Mint and rosemary oils Nisin	AO/AM system	[78]
Film	Fish gelatine-HMP	Hydroxytyrosol, dihydroxyphenylglycol	Preservation of beef meat	[79]
Film	HMP	Red cabbage extract	pH indicator	[80]
Film	Chitosan-HMP	Anthocyanin	pH indicator	[81]
Nanocomposite	Pectin	AgNPs, laponite	Coating polypropylene to improve barrier/AM properties	[82]
Nanocomposite	Pectin	Ag/AgCl-ZnONPs	AM system	[83]
Nanofiber	HMP	AgNPs	Reinforcement, AM	[84]
Nanofiber	LMP Polyethylene oxide	-	Reinforcement	[85]
Aerogel	Amidated pectin	TiO_2, NPs	AM under dark and UV illumination conditions	[86]
Hydrogel	LMP-Chitosan	Garlic and holy basil EOs	Incorporate to cellulose bag to improve AM properties	[87]
Oleogel	HMP	Camelia oil Tp-Palmitate	Drug delivery system	[88]
Emulsion	HMP	Clove EO	Bream fillets coating	[89]
Microemulsion	Chitosan-HMP	Cinnamaldehyde	AM system	[90]
Nanoemulsion	Food-grade pectin	Curcumin and garlic EOs	Coating chicken fillets	[91]
Nanoemulsion	HMP	Oregano, thyme, lemongrass, mandarin EOs	AM system	[92]
Nanoemulsion	HMP	Lemongrass EO	Addition in Cassava starch film to improve biodegradation properties	[93]
Multilayer emulsion	HMP-Chitosan	Astaxanthin	Release of hydrophobic carotenoids	[94]

EO: essential oil; AM: antimicrobial; AO: antioxidant.

4.1. Pectin-Based Films

Casting is the most used technique to obtain pectin-based films [10,73–75,79]. Pectin solutions (around 2–3 wt%) are mixed with the appropriate amount of plasticizer, commonly glycerol [95]. Then, the film forming solution is dried under controlled conditions of temperature and humidity forming a thin film. The incorporation of active agents, such as antimicrobial and/or antioxidant compounds, is performed after the incorporation of the plasticizer to obtain good compatibility between all components during the film processing [96–98]. Recently, Gouveira et al. [99] have reported the successful production of pectin-based films by using thermo-compression moulding of raw pectin with a natural deep eutectic solvent. The visual aspect of the obtained films was acceptable, since they were yellowish, visually homogenous, semi-transparent and without apparent pores, also showing high tensile strength and water resistance.

Pectin offers good compatibility with other biopolymers, such as proteins [73], lipids [100], other natural polysaccharides [101] or even synthetic biopolymers [82]. All these combinations represent alternatives when considering the final application of the obtained films. Both types of pectin (HMP and LMP) can form thin films under specific conditions. For example, LMP has been used as an appropriate matrix in new antioxidant systems with ascorbic acid as active additive by using casting as the processing method [72]. LMP was heated to 90 °C and then, glycerol and ascorbic acid were incorporated to the solution. Finally calcium chloride was added as the crosslinking agent to permit the formation of consistent and homogeneous pectin-based active films. In contrast, the addition of calcium ions is not necessary to develop films based on HMP, but low pH values and high sugar concentrations are needed to produce thin films. Nisar et al. [74] produced HMP films with antimicrobial properties incorporating clove essential oil by the casting method. Film forming solutions (3% *w/v*) were prepared by rehydrating pectin in sterile deionized water for 12 h at 20 °C. Glycerol was used as plasticizer at 30 wt % with magnetic stirring at 70 °C while pH was adjusted to 4.5. The clove essential oil with an emulsifier to improve the oil dispersion in the film aqueous solution were incorporated into the film forming solutions at different concentrations. A great integration of the clove essential oil into the polymer matrix was observed with a positive significant influence on the physico-chemical and functional properties, in particular barrier, mechanical, antioxidant and antimicrobial.

Marjoram, mint and rosemary essential oils are some of the active additives incorporated into the pectin matrix to get functionalities to these biopolymer materials. Almasi et al. [75] evaluated the effect on physico-chemical properties of marjoram essential oil in pectin films for food packaging applications. In order to prevent the degradation of the highly volatile essential oil and to control its release into food, it was incorporated using nanoemulsions and Pickering emulsions. Both types of emulsions combined the use of the essential oil and a low molecular weight surfactant with whey protein isolate or inulin as nanocarriers. Results obtained through X-ray diffraction (XRD), Fourier transformed infrared spectroscopy (FTIR) and field emission scanning electronic microscopy (FESEM) confirmed the high compatibility between pectin and both emulsions. The encapsulation of the essential oil through Pickering emulsions provided significantly slower releasing rates through the films when compared to nanoemulsions. For these reasons, authors concluded that the active pectin films containing Pickering emulsions showed the best potential to be used in active food packaging due to the slow release of the essential oil increasing food shelf-life. On the other hand, the synergic effect between mint and rosemary essential oils and nisin was investigated by Akhter et al. [78] using chitosan, starch and pectin blends. These authors concluded that the inclusion of rosemary essential oil and nisin improved the water barrier properties, tensile strength and thermal stability of the active biocomposites. Additionally, the combination of these compounds in a pectin matrix showed high antimicrobial action against some pathogenic strains (*Bacillus subtilis*, *Escherichia coli* and *Listeria monocytogenes*).

Furthermore, extracts derived from plants have been proposed to improve the functional properties of pectin. For example, tea extracts were incorporated into a HMP/Glucomannan blend. The influence of the addition of tea extracts at different concentrations (from 1% to 5% wt % on a dry basis) on the structural and physical properties of the blend, as well as on the antioxidant and antimicrobial activities were evaluated [76]. The authors found that concentrations of tea extract lower than 2 wt % improved all these properties but the effect was negative at high concentrations since some aggregation in the biopolymer macromolecules was observed. Red cabbage extract has been proposed for the development of a smart film based on HMP for meat and fish products. [80]. Red cabbage extract is rich in anthocyanins, showing the ability to change the colour of the biopolymer matrix at different pH values. It is known that the degradation of animal proteins produces an increase in pH due to the liberation of nitrogen compounds that can be monitored using a colorimetric sensor based on pectin and the red cabbage extract, offering innovative applications of pectin to the food industry [80]. These results showed the significant colour change in edible films when they are exposed to the headspace of meat and fish products at 21 °C and 4 °C, respectively.

The physical and functional properties of pectin-based films can be also modified by the combination of commercial pectin with corn flour and beetroot powder to minimize post-harvest decay, reducing ripening and improving sensorial properties of tomatoes [19]. In this study, results showed that pectin-based films protect from losses of polyphenols improving the antioxidant activity of these materials. In addition, other properties are modified due the presence of the edible coating. Pectin can modify the atmosphere around the fruit and/or vegetables, altering oxygen levels inside the fruit, retarding production of ethylene and, thus, limiting their physiological decay. In this work, pectin-based films showed low hydrophobicity to get optimum gas and water vapour permeability; reducing the ripening induced quality degradation in terms of texture and loss of bioactive compounds during storage.

Finally, the incorporation of nanoparticles to improve the physical and functional properties of pectin-based films has been evaluated. Biocomposites formed by a biopolymer matrix with metal or metal oxide nanoparticles are gaining importance in active food packaging since they could play a double role. On one hand, nanoparticles can act as nanofillers to enhance the mechanical and barrier properties of the biopolymer matrix and, on the other hand, they can interact directly with food due to their potential antimicrobial/antioxidant activity [102]. The effect of silver nanoparticles (AgNPs) has been tested in pectin, pullulan (a polysaccharide produced by fermentation by the *Aureobasidium pullulans* fungus), and pectin/pullulan blends [77]. Silver nanoparticles improved the mechanical properties of pullulan/AgNPs and pullulan/AgNPs/pectin composites while also showing high antimicrobial activity against foodborne pathogens, especially *Salmonella Typhimurium*, *Escherichia coli* and *Listeria monocytogenes*. AgNPs have been also proposed to develop nanocomposites based on pectin to be used as coatings for other polymer matrices with the aim to improve their barrier and mechanical properties as well as providing antimicrobial/antioxidant properties. Nanocomposites based on pectin with AgNPs and laponite have been evaluated to get a significant reduction in the oxygen transmission rate and water vapour transmission rate respect to neat polypropylene films taken as control [82]. The application of these new films showed antimicrobial activity against Gram-negative and Gram-positive bacteria, *Escherichia coli* and *Staphylococcus aureus*, respectively.

Other types of nanoparticles have been tested in pectin as the polymer matrix. Titanium oxide nanoparticles (TiO_2NPs) were incorporated at low concentrations (0–2 wt %) into biodegradable starch–pectin (3:1) films to improve their mechanical and barrier properties as well as their potential as antioxidant systems for food packaging applications [103]. In addition, visible and UV radiation was completely absorbed or scattered in these films by the addition of TiO_2NPs to get starch-pectin films with potential as UV screening packaging materials. On the other hand, the addition of halloysite nanotubes (HNT) offers great advantages to develop advanced food packaging materials. The effect of HNTs with salicylic acid [104], rosemary [97] and peppermint [105] essential oils in pectin films has been reported. HNTs showed high compatibility with pectin films improving their mechanical, thermal and moisture barrier properties [97,104]. The antimicrobial performance of these films was also improved due to the increase in the release rate of active compounds [97]. In fact, the antimicrobial activity of pectin-based films against Gram-negative *Escherichia coli* ATCC 25922, *Salmonella Typhimurium* ATCC 14028, *Pseudomonas aeruginosa* ATCC 10145, and Gram-positive *Staphylococcus aureus* ATCC 29213 was studied by the disk diffusion method, suggesting the effective antimicrobial properties of these functionalized films [104].

4.2. Emulsions and Gels

Pectins are widely used in the food industry as emulsifier and gelling agents. The ability of pectins to form gels under specific conditions has been used to obtain aerogels [86,106], hydrogels [87,107] or oleogels [88,108]. In particular, hydrogels are the most popular gel compositions used in food packaging, since they are able to absorb large amounts of water or other biological fluids inside their structure. For example, Torpol et al. studied the encapsulation of two different antimicrobial compounds: garlic and holy basil essential oils in chitosan-pectin hydrogel beads [87]. The entrapment of essential oils in the matrix structure was successful and it showed antimicrobial capacity against *Bacillus cereus*, *Clostridium perfringens*, *Escherichia coli*, *Pseudomonas fluorescens*, *Listeria monocytogenes* and *Staphylococcus aureus*, but not against *Lactobacillus plantarum* and *Salmonella Typhimurium*. Hydrogel coatings have been also proposed to reduce the deterioration of fresh fruit, meat or fish, as they can provide a semi-permeable protection to gases and water vapour and some other environmental factors that could damage food. By promoting food perspiration, these films also help to reduce enzymatic browning and water loss. Furthermore, this protection may also be enhanced by the addition of other ingredients, such as minerals, antioxidants, nutrients, vitamins or probiotics [109]. On the other hand, when the extraction of solvents at their supercritical state is produced in hydrogels or alcogels, the resultant material is called aerogels. Due to their unique properties, such as high porosity, high specific surface area, low relative density and thermal conductivity, these pectin-based biopolymers represent an innovative approach as advanced materials for food packaging since they can be used as internal layers, oxygen scavengers or drug delivery systems [110,111]. Recently, pectin-based aerogels have been developed for the storage of temperature-sensitive food. In this regard, TiO_2 nanoparticles were incorporated into the pectin matrix to improve the mechanical, thermal and antimicrobial properties of pectin when compared to control films [86].

As it has been mentioned above, pectin can be used as nanoemulsions, which are kinetically stable, but thermodynamically unstable, systems whose production requires emulsifiers to stabilize the dispersed phase [93]. Different types of essential oils have been encapsulated using nanoemulsions to delay and control the release processes. Several authors have studied the use of nanoemulsions in pectin matrices with different essential oils extracted from curcumin [91], lemongrass [92,93] and oregano [92]. Although essential oils are especially interesting in food packaging applications due to their antioxidant and/or antimicrobial properties, Mendes et al. [93] incorporated pectin nanoemulsions with the lemongrass essential oil into cassava starch film to improve the biodegradation rate of these formulations. Results obtained for the film with nanoemulsions showed a suitable degradation in vegetal compost, ensuring their complete biodegradation in a short time increasing their potential application in the food industry.

5. Conclusions

Researchers and scientists have achieved great success in the development of new systems based on pectins, as a natural bio-based biopolymer that can be obtained from agro-waste products, contributing to the implementation of the circular economy concept by improving waste management. This review article has considered the latest results obtained by researchers on the extraction, functionalization and potential applications in the food industry (including packaging), such as the production of films, emulsions and gels. However, due to the variability in the pectin structure the final application of pectin matrices is very diverse but very promising in many fields related to food packaging, particularly when active formulations are searched. Further studies on pectin matrices and optimization of polymer processes will be needed to better control the resulting pectin-based products.

Author Contributions: Conceptualization: C.M., M.R., A.J. and M.C.G.; methodology: C.M., M.R., A.J. and M.C.G.; formal analysis, discussion and supervision: C.M., M.R., A.J. and M.C.G. All authors have read and agreed to the published version of the manuscript.

Acknowledgments: Authors would like to thank the Spanish Ministry of Science, Innovation and Universities for their support through the project referenced MAT2017-84909-C2-1-R.

References

1. Jha, A.; Kumar, A. Biobased technologies for the efficient extraction of biopolymers from waste biomass. *Bioprocess Biosyst. Eng.* **2019**, *42*, 1893–1901. [CrossRef] [PubMed]
2. Martău, G.A.; Mihai, M.; Vodnar, D.C. The Use of Chitosan, Alginate, and Pectin in the Biomedical and Food Sector—Biocompatibility, Bioadhesiveness, and Biodegradability. *Polymers* **2019**, *11*, 1837. [CrossRef] [PubMed]
3. Adhikari, B.B.; Chae, M.; Bressler, D.C. Utilization of slaughterhouse waste in value-added applications: Recent advances in the development of wood adhesives. *Polymers* **2018**, *10*, 176. [CrossRef]
4. Nishinari, K.; Fang, Y.; Guo, S.; Phillips, G.O. Soy proteins: A review on composition, aggregation and emulsification. *Food Hydrocoll.* **2014**, *39*, 301–318. [CrossRef]

5. Benítez, J.J.; Castillo, P.M.; del Río, J.C.; León-Camacho, M.; Domínguez, E.; Heredia, A.; Guzmán-Puyol, S.; Athanassiou, A.; Heredia-Guerrero, J.A. Valorization of Tomato Processing by-Products: Fatty Acid Extraction and Production of Bio-Based Materials. *Materials* **2018**, *11*, 2211. [CrossRef]
6. Tran, D.-T.; Lee, H.R.; Jung, S.; Park, M.S.; Yang, J.-W. Lipid-extracted algal biomass based biocomposites fabrication with poly(vinyl alcohol). *Algal Res.* **2018**, *31*, 525–533. [CrossRef]
7. Damm, T.; Commandeur, U.; Fischer, R.; Usadel, B.; Klose, H. Improving the utilization of lignocellulosic biomass by polysaccharide modification. *Process Biochem.* **2016**, *51*, 288–296. [CrossRef]
8. Valdés, A.; Mellinas, A.C.; Ramos, M.; Garrigós, M.C.; Jiménez, A. Natural additives and agricultural wastes in biopolymer formulations for food packaging. *Front. Chem.* **2014**, *2*. [CrossRef]
9. Shankar, S.; Tanomrod, N.; Rawdkuen, S.; Rhim, J.-W. Preparation of pectin/silver nanoparticles composite films with UV-light barrier and properties. *Int. J. Biol. Macromol.* **2016**, *92*, 842–849. [CrossRef] [PubMed]
10. da Silva, I.S.V.; de Sousa, R.M.F.; de Oliveira, A.; de Oliveira, W.J.; Motta, L.A.C.; Pasquini, D.; Otaguro, H.; da Silva, I.S.V.; de Sousa, R.M.F.; de Oliveira, A.; et al. Polymeric blends of hydrocolloid from chia seeds/apple pectin with potential antioxidant for food packaging applications. *Carbohydr. Polym.* **2018**, *202*, 203–210. [CrossRef]
11. Correa, J.P.; Molina, V.; Sanchez, M.; Kainz, C.; Eisenberg, P.; Massani, M.B. Improving ham shelf life with a polyhydroxybutyrate/polycaprolactone biodegradable film activated with nisin. *Food Packag. Shelf Life* **2017**, *11*, 31–39. [CrossRef]
12. Xia, C.; Wang, W.; Wang, L.; Liu, H.; Xiao, J. Multilayer zein/gelatin films with tunable water barrier property and prolonged antioxidant activity. *Food Packag. Shelf Life* **2019**, *19*, 76–85. [CrossRef]
13. Wang, H.; Gong, X.; Miao, Y.; Guo, X.; Liu, C.; Fan, Y.-Y.; Zhang, J.; Niu, B.; Li, W. Preparation and characterization of multilayer films composed of chitosan, sodium alginate and carboxymethyl chitosan-ZnO nanoparticles. *Food Chem.* **2019**, *283*, 397–403. [CrossRef] [PubMed]
14. Mellinas, C.; Valdés, A.; Ramos, M.; Burgos, N.; Del Carmen Garrigós, M.; Jiménez, A. Active edible films: Current state and future trends. *J. Appl. Polym. Sci.* **2016**, *133*. [CrossRef]
15. Abdul Khalil, H.P.S.; Chong, E.W.N.; Owolabi, F.A.T.; Asniza, M.; Tye, Y.Y.; Rizal, S.; Nurul Fazita, M.R.; Mohamad Haafiz, M.K.; Nurmiati, Z.; Paridah, M.T. Enhancement of basic properties of polysaccharide-based composites with organic and inorganic fillers: A review. *J. Appl. Polym. Sci.* **2019**, *136*. [CrossRef]
16. Rodsamran, P.; Sothornvit, R. Lime peel pectin integrated with coconut water and lime peel extract as a new bioactive film sachet to retard soybean oil oxidation. *Food Hydrocoll.* **2019**, *97*. [CrossRef]
17. Sun, X.; Cameron, R.G.; Bai, J. Effect of spray-drying temperature on physicochemical, antioxidant and antimicrobial properties of pectin/sodium alginate microencapsulated carvacrol. *Food Hydrocoll.* **2020**, *100*. [CrossRef]
18. de Oliveira Alves Sena, E.; Oliveira da Silva, P.S.; de Aragão Batista, M.C.; Alonzo Sargent, S.; Ganassali de Oliveira Junior, L.F.; Almeida Castro Pagani, A.; Gutierrez Carnelossi, M.A. Calcium application via hydrocooling and edible coating for the conservation and quality of cashew apples. *Sci. Hortic* **2019**, *256*. [CrossRef]
19. Sucheta; Chaturvedi, K.; Sharma, N.; Yadav, S.K. Composite edible coatings from commercial pectin, corn flour and beetroot powder minimize post-harvest decay, reduces ripening and improves sensory liking of tomatoes. *Int. J. Biol. Macromol.* **2019**, *133*, 284–293. [CrossRef]
20. Pizato, S.; Chevalier, R.C.; Dos Santos, M.F.; Da Costa, T.S.; Arévalo Pinedo, R.; Cortez Vega, W.R. Evaluation of the shelf-life extension of fresh-cut pineapple (Smooth cayenne) by application of different edible coatings. *Br. Food J.* **2019**, *121*, 1592–1604. [CrossRef]
21. Jiang, Y.; Li, F.; Li, D.; Sun-Waterhouse, D.; Huang, Q. Zein/Pectin Nanoparticle-Stabilized Sesame Oil Pickering Emulsions: Sustainable Bioactive Carriers and Healthy Alternatives to Sesame Paste. *Food Bioprocess Technol.* **2019**, *12*, 1982–1992. [CrossRef]
22. Noreen, A.; Nazli, Z.-H.; Akram, J.; Rasul, I.; Mansha, A.; Yaqoob, N.; Iqbal, R.; Tabasum, S.; Zuber, M.; Zia, K.M. Pectins functionalized biomaterials; a new viable approach for biomedical applications: A review. *Int. J. Biol. Macromol.* **2017**, *101*, 254–272. [CrossRef] [PubMed]
23. Mohnen, D. Pectin structure and biosynthesis. *Curr. Opin. Plant Biol.* **2008**, *11*, 266–277. [CrossRef] [PubMed]
24. Naqash, F.; Masoodi, F.A.; Rather, S.A.; Wani, S.M.; Gani, A. Emerging concepts in the nutraceutical and functional properties of pectin—A Review. *Carbohydr. Polym.* **2017**, *168*, 227–239. [CrossRef]

25. O'Neill, M.A.; Ishii, T.; Albersheim, P.; Darvill, A.G. RHAMNOGALACTURONAN II: Structure and Function of a Borate Cross-Linked Cell Wall Pectic Polysaccharide. *Annu. Rev. Plant Biol.* **2004**, *55*, 109–139. [CrossRef]
26. Voragen, A.G.J.; Coenen, G.J.; Verhoef, R.P.; Schols, H.A. Pectin, a versatile polysaccharide present in plant cell walls. *Struct. Chem.* **2009**, *20*, 263–275. [CrossRef]
27. Sañudo Barajas, J.A.; Ayón, M.; Velez, R.; Verdugo-Perales, M.; Lagarda, J.; Allende, R. *Pectins: From the Gelling Properties to the Biological Activity*; Nova Publishers: Hauppauge, NY, USA, 2014; pp. 203–224.
28. Morales-Contreras, B.E.; Rosas-Flores, W.; Contreras-Esquivel, J.C.; Wicker, L.; Morales-Castro, J. Pectin from Husk Tomato (Physalis ixocarpa Brot.): Rheological behavior at different extraction conditions. *Carbohydr. Polym.* **2018**, *179*, 282–289. [CrossRef]
29. Giacomazza, D.; Bulone, D.; San Biagio, P.L.; Marino, R.; Lapasin, R. The role of sucrose concentration in self-assembly kinetics of high methoxyl pectin. *Int. J. Biol. Macromol.* **2018**, *112*, 1183–1190. [CrossRef]
30. do Nascimento, G.E.; Simas-Tosin, F.F.; Iacomini, M.; Gorin, P.A.J.; Cordeiro, L.M.C. Rheological behavior of high methoxyl pectin from the pulp of tamarillo fruit (Solanum betaceum). *Carbohydr. Polym.* **2016**, *139*, 125–130. [CrossRef]
31. Giacomazza, D.; Bulone, D.; San Biagio, P.L.; Lapasin, R. The complex mechanism of HM pectin self-assembly: A rheological investigation. *Carbohydr. Polym.* **2016**, *146*, 181–186. [CrossRef]
32. Wang, H.; Wan, L.; Chen, D.; Guo, X.; Liu, F.; Pan, S. Unexpected gelation behavior of citrus pectin induced by monovalent cations under alkaline conditions. *Carbohydr. Polym.* **2019**, *212*, 51–58. [CrossRef] [PubMed]
33. Jiang, W.; Qi, J.-R.; Huang, Y.; Zhang, Y.; Yang, X.-Q. Emulsifying properties of high methoxyl pectins in binary systems of water-ethanol. *Carbohydr. Polym.* **2020**, *229*, 115420. [CrossRef] [PubMed]
34. Fishman, M.L.; Chau, H.K.; Qi, P.X.; Hotchkiss, A.T.; Garcia, R.A.; Cooke, P.H. Characterization of the global structure of low methoxyl pectin in solution. *Food Hydrocoll.* **2015**, *46*, 153–159. [CrossRef]
35. Han, W.; Meng, Y.; Hu, C.; Dong, G.; Qu, Y.; Deng, H.; Guo, Y. Mathematical model of Ca^{2+} concentration, pH, pectin concentration and soluble solids (sucrose) on the gelation of low methoxyl pectin. *Food Hydrocoll.* **2017**, *66*, 37–48. [CrossRef]
36. Yang, X.; Nisar, T.; Liang, D.; Hou, Y.; Sun, L.; Guo, Y. Low methoxyl pectin gelation under alkaline conditions and its rheological properties: Using NaOH as a pH regulator. *Food Hydrocoll.* **2018**, *79*, 560–571. [CrossRef]
37. Kazemi, M.; Khodaiyan, F.; Hosseini, S.S.; Najari, Z. An integrated valorization of industrial waste of eggplant: Simultaneous recovery of pectin, phenolics and sequential production of pullulan. *Waste Manag.* **2019**, *100*, 101–111. [CrossRef]
38. Senit, J.J.; Velasco, D.; Gomez Manrique, A.; Sanchez-Barba, M.; Toledo, J.M.; Santos, V.E.; Garcia-Ochoa, F.; Yustos, P.; Ladero, M. Orange peel waste upstream integrated processing to terpenes, phenolics, pectin and monosaccharides: Optimization approaches. *Ind. Crops Prod.* **2019**, *134*, 370–381. [CrossRef]
39. Hilali, S.; Fabiano-Tixier, A.S.; Ruiz, K.; Hejjaj, A.; Ait Nouh, F.; Idlimam, A.; Bily, A.; Mandi, L.; Chemat, F. Green Extraction of Essential Oils, Polyphenols, and Pectins from Orange Peel Employing Solar Energy: Toward a Zero-Waste Biorefinery. *ACS Sustain. Chem. Eng.* **2019**, *7*, 11815–11822. [CrossRef]
40. Tovar, A.K.; Godínez, L.A.; Espejel, F.; Ramírez-Zamora, R.-M.; Robles, I. Optimization of the integral valorization process for orange peel waste using a design of experiments approach: Production of high-quality pectin and activated carbon. *Waste Manag.* **2019**, *85*, 202–213. [CrossRef]
41. Roy, M.C.; Alam, M.; Saeid, A.; Das, B.C.; Mia, M.B.; Rahman, M.A.; Eun, J.B.; Ahmed, M. Extraction and characterization of pectin from pomelo peel and its impact on nutritional properties of carrot jam during storage. *J. Food Process. Preserv.* **2018**, *42*, 1–9. [CrossRef]
42. Shakhmatov, E.G.; Makarova, E.N.; Belyy, V.A. Structural studies of biologically active pectin-containing polysaccharides of pomegranate Punica granatum. *Int. J. Biol. Macromol.* **2019**, *122*, 29–36. [CrossRef] [PubMed]
43. Cho, E.-H.; Jung, H.-T.; Lee, B.-H.; Kim, H.-S.; Rhee, J.-K.; Yoo, S.-H. Green process development for apple-peel pectin production by organic acid extraction. *Carbohydr. Polym.* **2019**, *204*, 97–103. [CrossRef] [PubMed]
44. Tamiello-Rosa, C.S.; Cantu-Jungles, T.M.; Iacomini, M.; Cordeiro, L.M.C. Pectins from cashew apple fruit (Anacardium occidentale): Extraction and chemical characterization. *Carbohydr. Res.* **2019**, *483*, 107752. [CrossRef] [PubMed]
45. Slavov, A.; Yantcheva, N.; Vasileva, I. Chamomile Wastes (Matricaria chamomilla): New Source of Polysaccharides. *Waste Biomass Valor.* **2018**, *10*, 1–12. [CrossRef]

46. Hasem, N.H.; Mohamad Fuzi, S.F.Z.; Kormin, F.; Abu Bakar, M.F.; Sabran, S.F. Extraction and partial characterization of durian rind pectin. *IOP Conf. Ser. Earth Environ. Sci.* **2019**, *269*, 012019. [CrossRef]
47. Esparza-Merino, R.M.; Macías-Rodríguez, M.E.; Cabrera-Díaz, E.; Valencia-Botín, A.J.; Estrada-Girón, Y. Utilization of by-products of Hibiscus sabdariffa L. as alternative sources for the extraction of high-quality pectin. *Food Sci. Biotechnol.* **2019**, *28*, 1003–1011. [CrossRef]
48. Marenda, F.R.B.; Colodel, C.; Canteri, M.H.G.; de Olivera Müller, C.M.; Amante, E.R.; de Oliveira Petkowicz, C.L.; de Mello Castanho Amboni, R.D. Investigation of cell wall polysaccharides from flour made with waste peel from unripe banana (Musa sapientum) biomass. *J. Sci. Food Agric.* **2019**, *99*, 4363–4372. [CrossRef]
49. Maneerat, N.; Tangsuphoom, N.; Nitithamyong, A. Effect of extraction condition on properties of pectin from banana peels and its function as fat replacer in salad cream. *J. Food Sci. Technol.* **2017**, *54*, 386–397. [CrossRef]
50. Banerjee, J.; Singh, R.; Vijayaraghavan, R.; MacFarlane, D.; Patti, A.F.; Arora, A. A hydrocolloid based biorefinery approach to the valorisation of mango peel waste. *Food Hydrocoll.* **2018**, *77*, 142–151. [CrossRef]
51. Li, W.J.; Fan, Z.G.; Wu, Y.Y.; Jiang, Z.G.; Shi, R.C. Eco-friendly extraction and physicochemical properties of pectin from jackfruit peel waste with subcritical water. *J. Sci. Food Agric.* **2019**, *99*. [CrossRef]
52. Sundarraj, A.A.; Thottiam Vasudevan, R.; Sriramulu, G. Optimized extraction and characterization of pectin from jackfruit (Artocarpus integer) wastes using response surface methodology. *Int. J. Biol. Macromol.* **2018**, *106*, 698–703. [CrossRef] [PubMed]
53. Fazio, A.; La Torre, C.; Dalena, F.; Plastina, P. Screening of glucan and pectin contents in broad bean (Vicia faba L.) pods during maturation. *Eur. Food Res. Technol.* **2019**, *246*, 333–347. [CrossRef]
54. Inayati, I.; Puspita, R.I.; Fajrin, V.L. Extraction of pectin from passion fruit rind (Passiflora edulis var. flavicarpa Degener) for edible coating. *AIP Conf. Proc.* **2018**, *1931*. [CrossRef]
55. Meneguzzo, F.; Brunetti, C.; Fidalgo, A.; Ciriminna, R.; Delisi, R.; Albanese, L.; Zabini, F.; Gori, A.; dos Santos Nascimento, L.B.; De Carlo, A.; et al. Real-Scale Integral Valorization of Waste Orange Peel via Hydrodynamic Cavitation. *Processes* **2019**, *7*, 581. [CrossRef]
56. Sabater, C.; Corzo, N.; Olano, A.; Montilla, A. Enzymatic extraction of pectin from artichoke (Cynara scolymus L.) by-products using Celluclast®1.5L. *Carbohydr. Polym.* **2018**, *190*, 43–49. [CrossRef] [PubMed]
57. Yang, Y.; Wang, Z.; Hu, D.; Xiao, K.; Wu, J.-Y. Efficient extraction of pectin from sisal waste by combined enzymatic and ultrasonic process. *Food Hydrocoll.* **2018**, *79*, 189–196. [CrossRef]
58. Zhang, M.; Zeng, G.; Pan, Y.; Qi, N. Difference research of pectins extracted from tobacco waste by heat reflux extraction and microwave-assisted extraction. *Biocatal. Agric. Biotechnol.* **2018**, *15*, 359–363. [CrossRef]
59. Muñoz-Almagro, N.; Valadez-Carmona, L.; Mendiola, J.A.; Ibáñez, E.; Villamiel, M. Structural characterisation of pectin obtained from cacao pod husk. Comparison of conventional and subcritical water extraction. *Carbohydr. Polym.* **2019**, *217*, 69–78. [CrossRef]
60. Shivamathi, C.S.; Moorthy, I.G.; Kumar, R.V.; Soosai, M.R.; Maran, J.P.; Kumar, R.S.; Varalakshmi, P. Optimization of ultrasound assisted extraction of pectin from custard apple peel: Potential and new source. *Carbohydr. Polym.* **2019**, *225*, 115240. [CrossRef]
61. Guandalini, B.B.V.; Rodrigues, N.P.; Marczak, L.D.F. Sequential extraction of phenolics and pectin from mango peel assisted by ultrasound. *Food Res. Int.* **2019**, *119*, 455–461. [CrossRef]
62. de Souza, C.G.; Rodrigues, T.H.S.; e Silva, L.M.A.; Ribeiro, P.R.V.; de Brito, E.S. Sequential extraction of flavonoids and pectin from yellow passion fruit rind using pressurized solvent or ultrasound. *J. Sci. Food Agric.* **2018**, *98*, 1362–1368. [CrossRef] [PubMed]
63. Xu, S.-Y.; Liu, J.-P.; Huang, X.; Du, L.-P.; Shi, F.-L.; Dong, R.; Huang, X.-T.; Zheng, K.; Liu, Y.; Cheong, K.-L. Ultrasonic-microwave assisted extraction, characterization and biological activity of pectin from jackfruit peel. *LWT* **2018**, *90*, 577–582. [CrossRef]
64. Rodsamran, P.; Sothornvit, R. Microwave heating extraction of pectin from lime peel: Characterization and properties compared with the conventional heating method. *Food Chem.* **2019**, *278*, 364–372. [CrossRef] [PubMed]
65. Wandee, Y.; Uttapap, D.; Mischnick, P. Yield and structural composition of pomelo peel pectins extracted under acidic and alkaline conditions. *Food Hydrocoll.* **2019**, *87*, 237–244. [CrossRef]
66. Lu, F.; Rodriguez-Garcia, J.; Van Damme, I.; Westwood, N.J.; Shaw, L.; Robinson, J.S.; Warren, G.; Chatzifragkou, A.; McQueen Mason, S.; Gomez, L.; et al. Valorisation strategies for cocoa pod husk and its fractions. *Curr. Opin. Green Sustain. Chem.* **2018**, *14*, 80–88. [CrossRef]

67. Maran, J.P.; Swathi, K.; Jeevitha, P.; Jayalakshmi, J.; Ashvini, G. Microwave-assisted extraction of pectic polysaccharide from waste mango peel. *Carbohydr. Polym.* **2015**, *123*, 67–71. [CrossRef]
68. Wang, W.; Ma, X.; Jiang, P.; Hu, L.; Zhi, Z.; Chen, J.; Ding, T.; Ye, X.; Liu, D. Characterization of pectin from grapefruit peel: A comparison of ultrasound-assisted and conventional heating extractions. *Food Hydrocoll.* **2016**, *61*, 730–739. [CrossRef]
69. Albuquerque, B.R.; Prieto, M.A.; Vazquez, J.A.; Barreiro, M.F.; Barros, L.; Ferreira, I.C.F.R. Recovery of bioactive compounds from Arbutus unedo L. fruits: Comparative optimization study of maceration/microwave/ultrasound extraction techniques. *Food Res. Int.* **2018**, *109*, 455–471. [CrossRef]
70. Raj, S. A Review on Pectin: Chemistry due to General Properties of Pectin and its Pharmaceutical Uses. *Sci. Rep.* **2012**, *1*, 550. [CrossRef]
71. Alistair, M.; Stephen, G.O.P. *Food Polysaccharides and Their Applications*; CRC Press: Boca Raton, FL, USA, 2006.
72. Porta, R.; Di Pierro, P.; Sabbah, M.; Regalado-Gonzales, C.; Mariniello, L.; Kadivar, M.; Arabestani, A. Blend films of pectin and bitter vetch (Vicia ervilia) proteins: Properties and effect of transglutaminase. *Innov. Food Sci. Emerg. Technol.* **2016**, *36*, 245–251. [CrossRef]
73. Chiarappa, G.; De'Nobili, M.D.; Rojas, A.M.; Abrami, M.; Lapasin, R.; Grassi, G.; Ferreira, J.A.; Gudiño, E.; de Oliveira, P.; Grassi, M. Mathematical modeling of L-(+)-ascorbic acid delivery from pectin films (packaging) to agar hydrogels (food). *J. Food Eng.* **2018**, *234*, 73–81. [CrossRef]
74. Nisar, T.; Wang, Z.-C.; Yang, X.; Tian, Y.; Iqbal, M.; Guo, Y. Characterization of citrus pectin films integrated with clove bud essential oil: Physical, thermal, barrier, antioxidant and antibacterial properties. *Int. J. Biol. Macromol.* **2018**, *106*, 670–680. [CrossRef] [PubMed]
75. Almasi, H.; Azizi, S.; Amjadi, S. Development and characterization of pectin films activated by nanoemulsion and Pickering emulsion stabilized marjoram (*Origanum majorana* L.) essential oil. *Food Hydrocoll.* **2020**, *99*, 105338. [CrossRef]
76. Lei, Y.; Wu, H.; Jiao, C.; Jiang, Y.; Liu, R.; Xiao, D.; Lu, J.; Zhang, Z.; Shen, G.; Li, S. Investigation of the structural and physical properties, antioxidant and antimicrobial activity of pectin-konjac glucomannan composite edible films incorporated with tea polyphenol. *Food Hydrocoll.* **2019**, *94*, 128–135. [CrossRef]
77. Lee, J.H.; Jeong, D.; Kanmani, P. Study on physical and mechanical properties of the biopolymer/silver based active nanocomposite films with antimicrobial activity. *Carbohydr. Polym.* **2019**, *224*, 115159. [CrossRef] [PubMed]
78. Akhter, R.; Masoodi, F.A.; Wani, T.A.; Rather, S.A. Functional characterization of biopolymer based composite film: Incorporation of natural essential oils and antimicrobial agents. *Int. J. Biol. Macromol.* **2019**, *137*, 1245–1255. [CrossRef] [PubMed]
79. Bermúdez-Oria, A.; Rodríguez-Gutiérrez, G.; Rubio-Senent, F.; Fernández-Prior, Á.; Fernández-Bolaños, J. Effect of edible pectin-fish gelatin films containing the olive antioxidants hydroxytyrosol and 3,4-dihydroxyphenylglycol on beef meat during refrigerated storage. *Meat Sci.* **2019**, *148*, 213–218. [CrossRef]
80. Dudnyk, I.; Janeček, E.-R.; Vaucher-Joset, J.; Stellacci, F. Edible sensors for meat and seafood freshness. *Sens. Actuators B Chem.* **2018**, *259*, 1108–1112. [CrossRef]
81. Maciel, V.B.V.; Yoshida, C.M.P.; Franco, T.T. Chitosan/pectin polyelectrolyte complex as a pH indicator. *Carbohydr. Polym.* **2015**, *132*, 537–545. [CrossRef]
82. Vishnuvarthanan, M.; Rajeswari, N. Food packaging: Pectin—laponite—Ag nanoparticle bionanocomposite coated on polypropylene shows low O2 transmission, low Ag migration and high antimicrobial activity. *Environ. Chem. Lett.* **2019**, *17*, 439–445. [CrossRef]
83. Yu, N.; Peng, H.; Qiu, L.; Wang, R.; Jiang, C.; Cai, T.; Sun, Y.; Li, Y.; Xiong, H. New pectin-induced green fabrication of Ag@AgCl/ZnO nanocomposites for visible-light triggered antibacterial activity. *Int. J. Biol. Macromol.* **2019**, *141*, 207–217. [CrossRef] [PubMed]
84. Li, K.; Cui, S.; Hu, J.; Zhou, Y.; Liu, Y. Crosslinked pectin nanofibers with well-dispersed Ag nanoparticles: Preparation and characterization. *Carbohydr. Polym.* **2018**, *199*, 68–74. [CrossRef] [PubMed]
85. McCune, D.; Guo, X.; Shi, T.; Stealey, S.; Antrobus, R.; Kaltchev, M.; Chen, J.; Kumpaty, S.; Hua, X.; Ren, W.; et al. Electrospinning pectin-based nanofibers: A parametric and cross-linker study. *Appl. Nanosci.* **2018**, *8*, 33–40. [CrossRef]

86. Nešić, A.; Gordić, M.; Davidović, S.; Radovanović, Ž.; Nedeljković, J.; Smirnova, I.; Gurikov, P. Pectin-based nanocomposite aerogels for potential insulated food packaging application. *Carbohydr. Polym.* **2018**, *195*, 128–135. [CrossRef] [PubMed]
87. Torpol, K.; Sriwattana, S.; Sangsuwan, J.; Wiriyacharee, P.; Prinyawiwatkul, W. Optimising chitosan–pectin hydrogel beads containing combined garlic and holy basil essential oils and their application as antimicrobial inhibitor. *Int. J. Food Sci. Technol.* **2019**, *54*, 2064–2074. [CrossRef]
88. Luo, S.-Z.; Hu, X.-F.; Jia, Y.-J.; Pan, L.-H.; Zheng, Z.; Zhao, Y.-Y.; Mu, D.-D.; Zhong, X.-Y.; Jiang, S.-T. Camellia oil-based oleogels structuring with tea polyphenol-palmitate particles and citrus pectin by emulsion-templated method: Preparation, characterization and potential application. *Food Hydrocoll.* **2019**, *95*, 76–87. [CrossRef]
89. Nisar, T.; Yang, X.; Alim, A.; Iqbal, M.; Wang, Z.-C.; Guo, Y. Physicochemical responses and microbiological changes of bream (Megalobrama ambycephala) to pectin based coatings enriched with clove essential oil during refrigeration. *Int. J. Biol. Macromol.* **2019**, *124*, 1156–1166. [CrossRef]
90. Gong, C.; Lee, M.C.; Godec, M.; Zhang, Z.; Abbaspourrad, A. Ultrasonic encapsulation of cinnamon flavor to impart heat stability for baking applications. *Food Hydrocoll.* **2020**, *99*, 105316. [CrossRef]
91. Abdou, E.S.; Galhoum, G.F.; Mohamed, E.N. Curcumin loaded nanoemulsions/pectin coatings for refrigerated chicken fillets. *Food Hydrocoll.* **2018**, *83*, 445–453. [CrossRef]
92. Guerra-Rosas, M.I.; Morales-Castro, J.; Cubero-Márquez, M.A.; Salvia-Trujillo, L.; Martín-Belloso, O. Antimicrobial activity of nanoemulsions containing essential oils and high methoxyl pectin during long-term storage. *Food Control* **2017**, *77*, 131–138. [CrossRef]
93. Mendes, J.F.; Norcino, L.B.; Martins, H.H.A.; Manrich, A.; Otoni, C.G.; Carvalho, E.E.N.; Piccoli, R.H.; Oliveira, J.E.; Pinheiro, A.C.M.; Mattoso, L.H.C. Correlating emulsion characteristics with the properties of active starch films loaded with lemongrass essential oil. *Food Hydrocoll.* **2020**, *100*, 105428. [CrossRef]
94. Liu, C.; Tan, Y.; Xu, Y.; McCleiments, D.J.; Wang, D. Formation, characterization, and application of chitosan/pectin-stabilized multilayer emulsions as astaxanthin delivery systems. *Int. J. Biol. Macromol.* **2019**, *140*, 985–997. [CrossRef] [PubMed]
95. Sganzerla, W.G.; Paes, B.B.; Azevedo, M.S.; Ferrareze, J.P.; da Rosa, C.G.; Nunes, M.R.; Veeck, A.P.L. Bioactive and biodegradable film packaging incorporated with acca sellowiana extracts: physicochemical and antioxidant characterization. *Chem. Eng. Trans.* **2019**, *75*, 445–450.
96. Spatafora Salazar, A.S.; Sáenz Cavazos, P.A.; Mújica Paz, H.; Valdez Fragoso, A. External factors and nanoparticles effect on water vapor permeability of pectin-based films. *J. Food Eng.* **2019**, *245*, 73–79. [CrossRef]
97. Gorrasi, G. Dispersion of halloysite loaded with natural antimicrobials into pectins: Characterization and controlled release analysis. *Carbohydr. Polym.* **2015**, *127*, 47–53. [CrossRef] [PubMed]
98. Bernhardt, D.C.; Pérez, C.D.; Fissore, E.N.; De'Nobili, M.D.; Rojas, A.M. Pectin-based composite film: Effect of corn husk fiber concentration on their properties. *Carbohydr. Polym.* **2017**, *164*, 13–22. [CrossRef] [PubMed]
99. Gouveia, T.I.A.; Biernacki, K.; Castro, M.C.R.; Gonçalves, M.P.; Souza, H.K.S. A new approach to develop biodegradable films based on thermoplastic pectin. *Food Hydrocoll.* **2019**, *97*, 105175. [CrossRef]
100. Manrich, A.; Moreira, F.K.V.; Otoni, C.G.; Lorevice, M.V.; Martins, M.A.; Mattoso, L.H.C. Hydrophobic edible films made up of tomato cutin and pectin. *Carbohydr. Polym.* **2017**, *164*, 83–91. [CrossRef]
101. Gao, H.-X.; He, Z.; Sun, Q.; He, Q.; Zeng, W.-C. A functional polysaccharide film forming by pectin, chitosan, and tea polyphenols. *Carbohydr. Polym.* **2019**, *215*, 1–7. [CrossRef]
102. Ramos, M.; Fortunati, E.; Peltzer, M.; Kenny, J.M.; Garrigós, M.C. Characterization and disintegrability under composting conditions of PLA-based nanocomposite films with thymol and silver nanoparticles. *Polym. Degrad. Stab.* **2016**, *132*, 2–10. [CrossRef]
103. Dash, K.K.; Ali, N.A.; Das, D.; Mohanta, D. Thorough evaluation of sweet potato starch and lemon-waste pectin based-edible films with nano-titania inclusions for food packaging applications. *Int. J. Biol. Macromol.* **2019**, *139*, 449–458. [CrossRef]
104. Makaremi, M.; Pasbakhsh, P.; Cavallaro, G.; Lazzara, G.; Aw, Y.K.; Lee, S.M.; Milioto, S. Effect of Morphology and Size of Halloysite Nanotubes on Functional Pectin Bionanocomposites for Food Packaging Applications. *ACS Appl. Mater. Interfaces* **2017**, *9*, 17476–17488. [CrossRef] [PubMed]
105. Biddeci, G.; Cavallaro, G.; Di Blasi, F.; Lazzara, G.; Massaro, M.; Milioto, S.; Parisi, F.; Riela, S.; Spinelli, G. Halloysite nanotubes loaded with peppermint essential oil as filler for functional biopolymer film. *Carbohydr. Polym.* **2016**, *152*, 548–557. [CrossRef]

106. Groult, S.; Budtova, T. Thermal conductivity/structure correlations in thermal super-insulating pectin aerogels. *Carbohydr. Polym.* **2018**, *196*, 73–81. [CrossRef] [PubMed]
107. Mehrali, M.; Thakur, A.; Kadumudi, F.B.; Pierchala, M.K.; Cordova, J.A.V.; Shahbazi, M.-A.; Mehrali, M.; Pennisi, C.P.; Orive, G.; Gaharwar, A.K.; et al. Pectin Methacrylate (PEMA) and Gelatin-Based Hydrogels for Cell Delivery: Converting Waste Materials into Biomaterials. *ACS Appl. Mater. Interfaces* **2019**, *11*, 12283–12297. [CrossRef]
108. Wijaya, W.; Sun, Q.-Q.; Vermeir, L.; Dewettinck, K.; Patel, A.R.; Van der Meeren, P. pH and protein to polysaccharide ratio control the structural properties and viscoelastic network of HIPE-templated biopolymeric oleogels. *Food Struct.* **2019**, *21*, 100112. [CrossRef]
109. Sanchís, E.; Ghidelli, C.; Sheth, C.C.; Mateos, M.; Palou, L.; Pérez-Gago, M.B. Integration of antimicrobial pectin-based edible coating and active modified atmosphere packaging to preserve the quality and microbial safety of fresh-cut persimmon (*Diospyros kaki* Thunb. cv. Rojo Brillante). *J. Sci. Food Agric.* **2017**, *97*, 252–260. [CrossRef]
110. Ahmadzadeh, S.; Nasirpour, A.; Keramat, J.; Desobry, S. Powerful Solution to Mitigate the Temperature Variation Effect: Development of Novel Superinsulating Materials. *Food Packag. Preserv.* **2018**, 137–176.
111. de Oliveira, J.P.; Bruni, G.P.; el Halal, S.L.M.; Bertoldi, F.C.; Dias, A.R.G.; Zavareze, E.d.R. Cellulose nanocrystals from rice and oat husks and their application in aerogels for food packaging. *Int. J. Biol. Macromol.* **2019**, *124*, 175–184. [CrossRef]

4

Hybrid Biocomposites based on Poly(Lactic Acid) and Silica Aerogel for Food Packaging Applications

Alejandro Aragón-Gutierrez [1], Marina P. Arrieta [2,3,*], Mar López-González [1], Marta Fernández-García [1,4] and Daniel López [1,4,*]

1 Instituto de Ciencia y Tecnología de Polímeros, ICTP-CSIC, Juan de la Cierva 3, 28006 Madrid, Spain; alejandro.aragon.gutierrez@gmail.com (A.A.-G.); mar@ictp.csic.es (M.L.-G.); martafg@ictp.csic.es (M.F.-G.)

2 Departamento de Ingeniería Química y del Medio Ambiente, Escuela Politécnica Superior de Ingenieros Industriales, Universidad Politécnica de Madrid (ETSII-UPM), Calle José Gutiérrez Abascal 2, 28006 Madrid, Spain

3 Grupo de Investigación: Polímeros, Caracterización y Aplicaciones (POLCA), 28006 Madrid, Spain

4 Interdisciplinary Platform for Sustainable Plastics towards a Circular Economy-Spanish National Council (SusPlast-CSIC), 28006 Madrid, Spain

* Correspondence: m.arrieta@upm.es (M.P.A.); daniel.l.g@ictp.csic.es (D.L.).

Abstract: Bionanocomposites based on poly (lactic acid) (PLA) and silica aerogel (SiA) were developed by means of melt extrusion process. PLA-SiA composite films were plasticized with 15 wt.% of acetyl (tributyl citrate) (ATBC) to facilitate the PLA processability as well as to attain flexible polymeric formulations for films for food packaging purposes. Meanwhile, SiA was added in four different proportions (0.5, 1, 3 and 5 wt.%) to evaluate the ability of SiA to improve the thermal, mechanical, and barrier performance of the bionanocomposites. The mechanical performance, thermal stability as well as the barrier properties against different gases (carbon dioxide, nitrogen, and oxygen) of the bionanocomposites were evaluated. It was observed that the addition of 3 wt.% of SiA to the plasticized PLA-ATBC matrix showed simultaneously an improvement on the thermal stability as well as the mechanical and barrier performance of films. Finally, PLA-SiA film formulations were disintegrated in compost at the lab-scale level. The combination of ATBC and SiA sped up the disintegration of PLA matrix. Thus, the bionanocomposites produced here show great potential as sustainable polymeric formulations with interest in the food packaging sector.

Keywords: biocomposites; biodegradable polymers; plasticizers; food packaging; poly (lactic acid); citrate esters; silica aerogel

1. Introduction

As a result of environmental awareness, the use of the petrochemical resources to produce short term plastics products as well as the disposal of large amounts of waste from daily use polymers, such as food packaging, is among one of the foremost concerns in current times. Therefore, the use of both biobased and biodegradable polymers for food packaging applications emerges as a promising solution to this rapidly rising waste problem with the main objective to reduce the consumption of non-renewable resources and prevent the accumulation of plastics wastes [1,2]. Among all biobased and biodegradable thermoplastic polymers currently used in the packaging field (i.e., poly (lactic acid) (PLA), poly (hydroxyalcanoates) (PHAs), thermoplastic starch (TPS), and poly (butylene adipate co-terephthalate) (PBAT)), PLA is the most widely used and investigated for short-term applications [2–4]. In this sense, PLA presents a series of promising properties interesting for film applications intended for agricultural mulch films or food packaging, such as high transparency, excellent printability, availability in the market, economically competitive, ease of processing by the already existing technologies in the plastic processing industry (i.e., melt-blending, extrusion, injection molding and film forming), and it presents inherent environmentally friendly characteristics (i.e., it is biobased and biodegradable) [2,3,5]. In this sense, it is synthesized by the ring opening polymerization of lactide (LA) which in turn is coming from the fermentation of renewable agricultural crops (corn, sugar-cane and other polysaccharides). Moreover, it is compostable showing a high rate of disintegration disappearing completely in less than one month [6–9]. Thus, PLA based products can be wasted along with organic waste instead of being discarded with the rest of plastic materials [10]. This is why PLA represents a sustainable alternative to replace non-compostable conventional polymers derived from petroleum. However, PLA presents some drawbacks with respect to their petrochemical counterparts that restrict its application as flexible film for food packaging such as its low thermal stability, limited stretchability, poor barrier properties, low crystallization rate, and degree [3].

Blending approaches are the most widely used to improve PLA properties for industrial applications. Therefore, the poor ductility of PLA is usually enhanced with the addition of a plasticizer, although it has been reported that the addition of a plasticizer to the PLA matrix increases gas permeability, which may be detrimental for food packaging applications [11–13]. Several monomers and oligomers have been investigated as plasticizers for PLA, but citrate esters are probably the most commonly used and investigated since they are accepted for food contact, available at the market and economical competitive [14–16]. Among all citrate esters used as plasticizers in the food packaging industry, triethyl citrate (TEC), tributyl citrate (TBC), acetyl triethyl citrate (ATEC), and acetyl tributyl citrate (ATBC) are noteworthy [16–18]. In this sense, ATBC has shown good thermal processability as well as effective plasticization effect, and therefore is one of the most recommended and widely used plasticizers for PLA and other biopolyesters [12,14,19]. To obtain the required flexibility for food packaging applications, citrate esters should be added in amounts between 10 and 20 wt.% [7,12,14,20,21]. Additionally, because of its low toxicity, citrate esters are accepted for food contact [22]. However, PLA plasticization will further reduce the well-known PLA poor barrier performance [21,23]. Thus, different strategies have been used to improve plasticized PLA barrier properties including multilayer approaches, surface film coating, co-extrusion with other biopolymers [21,23], and increasing its crystallization degree by blending, co-polimerization [24,25] as well as the development of composites and/or nanocomposites [12,14,26,27].

Composites have been positioned as a promising option in the provision of excellent barrier properties for food packaging applications. Inorganic materials such as silica exhibits excellent thermal stability and higher hardness [28]. In this sense, silica based particles are low-price particles widely used at industrial level since the incorporation of well dispersed silica particles into the polymeric matrix was proven to be an effective way to improve the thermomechanical performance of biopolymers [28,29]. Aerogels are ultra-light weight open-celled mesoporous materials whose structure is composed of an interconnected network of channels made of thin ligaments [28]. For the development of the present work, silica aerogels (SiA) have been chosen as reinforcement for the plasticized polymer matrix. SiA are three-dimensional highly porous materials with nanoporous networks, high specific surface and extremely low density, while polymer-silica aerogel based composites tend to exhibit enhanced mechanical and thermal properties [30–32]. SiA reinforced materials has also been proposed for food contact packaging applications. For instance, Chen et al. developed films based on poly vinyl alcohol (PVA) incorporated with SiA for food packaging with improved thermal insulation and oxygen barrier performance [32]. Ventaka Prasad et al. have used SiA particles to improve the mechanical properties of PLA/sisal composites and they conclude that the improvement of the interfacial adhesion between the polymeric matrix and sisal fibers due to the SiA particles presence increased the tensile and flexural strength of the final materials [28].

Thus, considering that among the main drawbacks of PLA to be used in food packaging are its rigidity, its low gas barrier properties, and its low flexibility, the main objective of the present work is the development of hybrid biocomposites based on plasticized PLA with 15 wt.% of a frequently used PLA plasticizer, such as ATBC, and further reinforced with different amounts of silica aerogels (SiA) particles (0.5, 1, 3, and 5 wt.%) to obtain biocomposite films with enhanced barrier, thermal as well as mechanical properties. Additionally, the disintegration behavior under composting medium at laboratory scale level of the SiA nanoreinforced plasticized PLA formulations was studied to corroborate their compostable character.

2. Materials and Methods

2.1. Materials

Poly (lactic acid) was kindly provided by ERCROS S.A. (ErcrosBio® LL703 density 1.25 g cm^{-3}) (Barcelona, Spain) M_n = 23,250 Da [7]. Acetyl tri-n-butyl citrate (ATBC, M = 402 g mol^{-1}, 98% purity) was purchased from Sigma-Aldrich (Madrid, Spain). Silica aerogel (SiA) powder was kindly provided by JIOS Aerogel (Seoul, Korea) (particle size 1–20 μm, pore diameter 20 nm and surface area 600–800 m/g^2).

2.2. Film Samples Preparation

PLA and plasticized PLA biocomposites were prepared in a twin conical co-rotating screws (Thermo Scientific MiniLab Haake Rheomex CTW5) (Waltham, MA, USA) with a capacity of 7 cm^3. PLA undergoes hydrolytic degradation, which is influenced by ambient moisture and temperature [33], in order to avoid the presence of moisture PLA pellets were previously dried overnight at 80 °C [9,33–35]. The same procedure was applied to SiA, while ATBC was dried at 80 °C during 2 h [12]. The appropriate amounts of PLA and SiA were firstly processed in the Minilab mixer (Waltham, MA, USA) operating at a rotation speed of 100 rpm and at 190 °C. Once PLA had reached the melt state (after 2 min), ATBC was added in 15 wt.% to avoid ATBC thermal degradation, following an already optimized process [21]. Therefore, the total time of biocomposite formulations mixing was 3 min. Films were then processed from the obtained extruded blends by compression molding process at 200 °C in a hot press (Collin P-200-P) (GmbH, Ebersberg, Germany). Each pelletized blend formulation (1 g) was kept between the hot plates firstly at atmospheric pressure and at 200 °C until reaching the melt state (one minute). Subsequently, the hot press was submitted to an already optimized pressure cycles for PLA-ATBC based formulations, 1 MPa for one minute, 5 MPa for one minute, and 5 MPa for two minutes [7]. Six formulations were obtained and were named as shown in Table 1.

Table 1. Biocomposite film formulations.

Formulation	PLA (wt.%)	ATBC (wt.%)	SiA (wt.%)
PLA	100	0	0
PLA-ATBC	85	15	0
PLA-ATBC-SiA0.5%	84.5	15	0.5
PLA-ATBC-SiA1%	84	15	1
PLA-ATBC-SiA3%	82	15	3
PLA-ATBC-SiA5%	80	15	5

2.3. Thermal Characterization

2.3.1. Differential Scanning Calorimetry (DSC)

DSC was used to determine glass transition temperature (T_g), cold crystallization temperature (T_{cc}), melting temperature (T_m) and degree of crystallinity (χ_c). Experiments were conducted in a DSC calorimeter Q-2000 (New Castle, DE, USA). The DSC runs were performed in a nitrogen atmosphere. Samples (around 3.5 mg) were subjected to a cycle program consisting in a first heating run from 25 °C to 200 °C (rate = 10 °C min^{-1}) followed by a cooling process from 200 °C to −50 °C and finally a second heating up to 200 °C at a rate of 10 °C min^{-1}. The DSC thermal parameters: glass transition temperature

(T_g), the melting temperature (T_m), and cold crystallization temperature (T_{cc}), were obtained from the first DSC heating scan. The degree of crystallinity (χ_c) of PLA, PLA-ATBC and its plasticized biocomposites was calculated by using Equation (1):

$$\chi_c = 100 \times \left[\frac{\Delta H_m - \Delta H_{cc}}{\Delta H_m^c} \right] \times \frac{1}{W_{PLA}} \quad (1)$$

where ΔH_m and ΔH_{cc} are the melting enthalpy and the cold crystallization enthalpy of the sample, respectively. ΔH_m^c is the melting heat of purely crystalline PLA taken as 93 J/g [36].

2.3.2. Thermogravimetric Analysis (TGA)

Thermogravimetric dynamic measurements were run in a thermogravimetric analyzer (TGA Q-500) (TA Instruments, New Castle, DE, USA) at a constant heating rate of 10 °C/min. Temperature was raised from 30 to 700 °C under nitrogen atmosphere. The onset degradation temperature (T_0) was calculated at 5% weight loss [7]; meanwhile, the first derivative of the TGA curves (DTG) was used to calculate the maximum degradation rate temperatures (T_{max}) for each stage.

2.4. Mechanical Characterization

Mechanical properties of the samples were obtained by using a universal testing system (3366 Instron dynamometer) (INSTRON, Norwood, MA, USA), using a 100 N cell and 10 mm min^{-1} of speed. Dimensions of the halterio type probes, were 35 mm length, 2 mm width and a thickness at around 0.2 mm. Separation between the clamps was 20 mm. Results to obtain the Young modulus (E), tensile strength (TS), and the average percentage elongation at break ($\varepsilon_{B\%}$) were calculated from the obtained stress–strain curves as the average of at least five measurements. Significance in the mechanical properties differences were statistically determined by means of one-way analysis of variance (ANOVA) with OriginPro 8.6 software (OriginLab corporation Northampton, MA, USA). Means comparison was performed employing a Tukey's test, at 95% of the confidence level, in order to identify which formulations were significantly different from other formulations.

2.5. Scanning Electron Microscopy (SEM)

SEM micrographs of the crio-fractured cross-section of the films were obtained by means of a Philips XL30 SEM (Phillips, Eindhoven, The Netherlands). PLA and plasticized PLA formulations were previously coated under vacuum conditions with a palladium/gold layer previous to their SEM observation with the main objective to enhance their electrical conductivity.

2.6. Permeability to Gases

Permeation measurements of O_2, CO_2, and N_2 through the films were carried out in a lab-made permeator, described elsewhere by Tiemblo et al. [37]. In brief, biocomposite films were placed in the center of a cell to separate the upstream chamber (the high pressure chamber) from the downstream chamber (low-pressure chamber). The system was thermostatically controlled in a water bath at 30 °C. Then, the permeation device was brought to a high vacuum with an Alcatel turbomolecular pump. The vacuum was maintained overnight to remove the last traces of gas and solvent in the membrane until low pressure was reached, which is about 10^{-6} mbar. Subsequently, the measurements were performed, and thus the gas in a reservoir at the pressure of interest was allowed to flow into the downstream chamber, and it was further monitored as a function of time with an MKS 628B Baratron (MKS Instruments Inc., Andover, MA, USA), which is an absolute pressure transducer which worked in the 0–10 mbar pressure range, connected to a computer. A Gometric transducer of 0–10 bar range measured the gas pressures in the upstream chamber. The permeability (*P*) and diffusion (*D*) coefficients were obtained from the plotted curves measuring the pressure increases in the downstream chamber at intervals of 1 s. Before each measurement, the inlet of air was measured as a function of time and it was then subtracted from the obtained curves [38]. Thus, the pressure of permeant versus time was plotted. Biocomposite films with thickness between 100 to 200 μm were used and three independent experiments were carried out for each gas. The volume V of the downstream semicell is given in cm^3, the thickness (L) in cm, the diffusion area (A) of the membrane is in cm^2, and the pressure in cm Hg, and thus the permeability coefficient (P) of each gas (O_2, CO_2 and N_2) in the membranes is reported in barrer (1 barrer = 10^{-10} cm^3(STP)cm/(cm^2 s cmHg)) following Equation (2):

$$P = \frac{273}{76}\left(\frac{VL}{ATp_0}\right)\left(\frac{dp(t)}{dt}\right), \quad (2)$$

where p_0 is the upstream gas pressure, p is downstream gas pressure, and T is the absolute temperature. The p against t isotherms present a transitory state followed by a straight line ($t \rightarrow \infty$), which corresponds to the conditions in the steady-state. The intersection of this straight line with the abscissa axis of the plot represents the time lag θ. This time lag θ is related by means of the Equation (3) [38] to the apparent gas diffusion coefficient (D) in the:

$$D = \frac{L^2}{6\theta} \quad (3)$$

The diffusion coefficient (D) is given in $cm^2\ s^{-1}$. The apparent solubility coefficient (S) is given in (cm^3(STP)/(cm^3 cm Hg)), following Equation (4):

$$S = \frac{P}{D} \quad (4)$$

2.7. Disintegrability under Composting Conditions

Disintegration under composting conditions at a laboratory scale level was performed based on the UNE-EN ISO-20200 standard [39]. The solid synthetic waste (SSW) was elaborated by mixing 10% of compost, 30% rabbit food, 40% sawdust, 10% starch, 5% sugar, 1% urea, and 4% corn oil. Water was then added to the SSW in 45:55 SSW:water proportions. All formulations were cut in 1.5 × 1.5 cm^2 and buried inside a polyester mesh to permit taking away the disintegrated film materials from the reactors containing the compost soil, but at the same time allowing the access of moisture and furthermore the access of microorganisms [40]. The materials were further subjected to aerobic disintegration conditions in composting containers at 58 °C in an oven for a period of 35 days. Water was added periodically to the composting containers maintain the required humidity. Each film formulation was recovered from the composting containers at 1, 4, 10, 21, 28, and 35 days. Each recovered sample was cleaned using distilled water and then dried at 40 °C in an oven for around 24 h and further re-weighted. The disintegration degree of each formulation was calculated by normalizing the sample weight at each disintegration time to the initial weight. Additionally, photographs were taken to all recovered disintegrated films from the composting containers in order to qualitatively follow the disintegration process. The degree of disintegration was calculated following Equation (5):

$$GD = \frac{m_i - m_r}{m_i} \times 100 \quad (5)$$

where GD is the degree of disintegration, m_i is the initial dry mass of the tested formulation, and m_r is the final dry mass of the recovered sample.

3. Results

The processing technology used for the development of SiA reinforced plasticized PLA formulations allowed us to obtain transparent films with thickness ranging between 100 μm and 200 μm. The visual appearance of the PLA-ATBC-SiA3% film formulation is showed in Figure 1a as an example. The films showed good transparency after the addition of the plasticizer (ATBC) in agreement with already reported works [7,12,14,21], even after the addition of silica aerogel suggesting good dispersion of this filler into the plasticized PLA matrix. These findings were confirmed by SEM. Thus, morphological characteristics of the cross cryo-fracture sections of film samples were evaluated by SEM and the images are shown in Figure 1. As expected, PLA presented a smooth and uniform surface where a rigid fracture typical of amorphous polymer [9] could be observed (Figure 1b), while plasticized samples presented plastic deformation, showing more ductile fracture patterns [12] (Figure 1c–h), in accordance with previous works [12,21,41]. In the case of silica aerogel based composites, the fracture surface resulted in more roughness (Figure 1d–h) as SiA concentration increased in the formulation. Small white spots on the cross-section of the plasticized PLA-ATBC-SiA films can be observed, and they became denser with increasing SiA content. Similar findings have been observed in PVA based films loaded with SiA [32]. At higher magnifications, it could be observed that silica aerogel particles appeared well dispersed as crystalline aggregates in the polymeric matrix, as can be seen in PLA-ATBC-SiA1% (Figure 1h). This fact influenced that the films were still transparent after the addition of the silica aerogel particles (see Figure 1a). In fact, the well dispersion of silica aerogel particles suggests that no phase separation had taken place during the extrusion process.

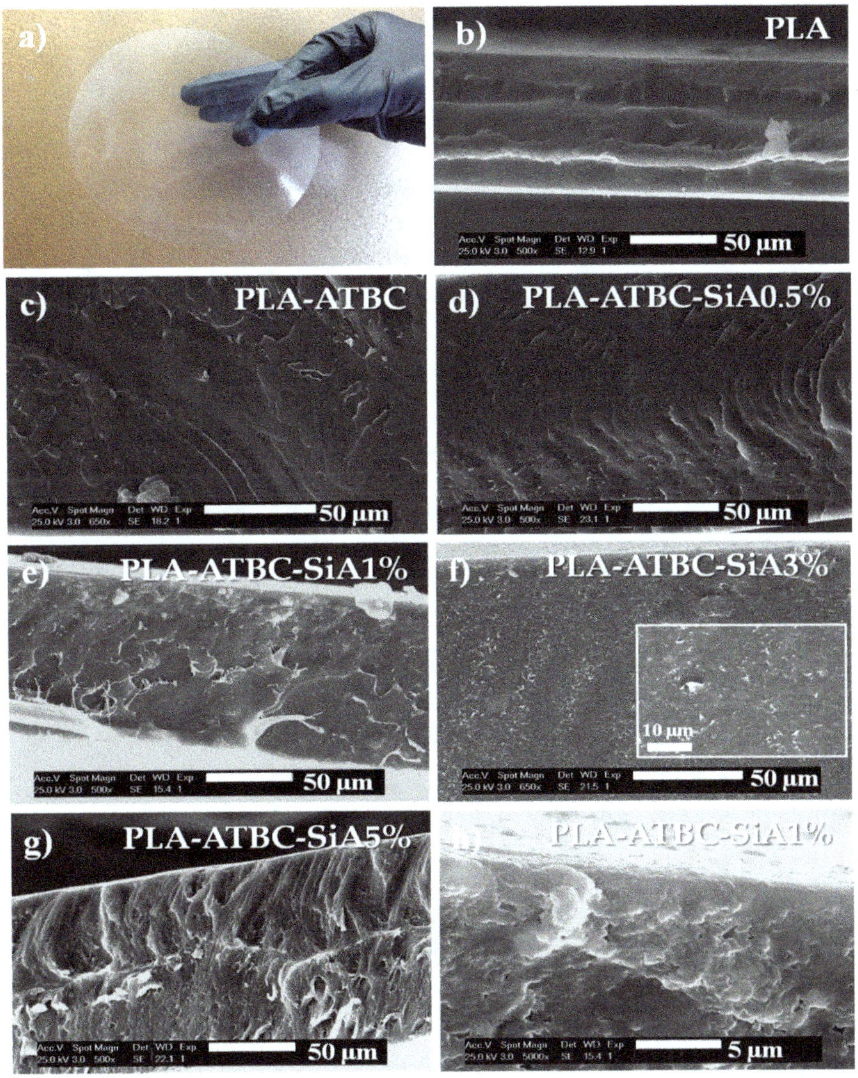

Figure 1. (**a**) Visual appearance of PLA-ATBC-SiA3% composite film. Cryo-fractured SEM micrographs of: (**b**) PLA, (**c**) PLA-ATBC, (**d**) PLA-ATBC-SiA0.5% aerogel, (**e**) PLA-ATBC-SiA1%, (**f**) PLA-ATBC-SiA3% (zoom image at 2000×), (**g**) PLA-ATBC-SiA5%, and (**h**) PLA-ATBC-SiA1% at higher magnification (5000×).

The thermal stability of all formulations was studied by thermogravimetric analysis. The main thermal parameters obtained from TGA and DTG curves are summarized in Table 2, while DTG curves are shown in Figure 2a. Neat PLA film decomposed in a single degradation step process with an onset degradation temperature (T_0) of 322 °C, while the maximum degradation rate temperature (T_{max}) was centered at 360 °C. The presence of ATBC plasticizer decreased the onset degradation temperature (T_0) of the polymeric material. A slightly growth of the residual weight percentage at the end of the TGA analysis was evident in those formulation with a higher amount of SiA, 3 wt.% and 5 wt.%.

Table 2. TGA and DTG parameters of PLA, PLA-ATBC, and hybrid plasticized biocomposite films.

Film Formulations	T_0 (°C)	$T_{max\ I}$ (°C)	$T_{max\ II}$ (°C)
PLA	322	-	360
PLA-ATBC	205	217	361
PLA-ATBC-SiA0.5%	215	212	358
PLA-ATBC- SiA1%	204	215	360
PLA-ATBC- SiA3%	236	216	361
PLA-ATBC- SiA5%	249	213	348

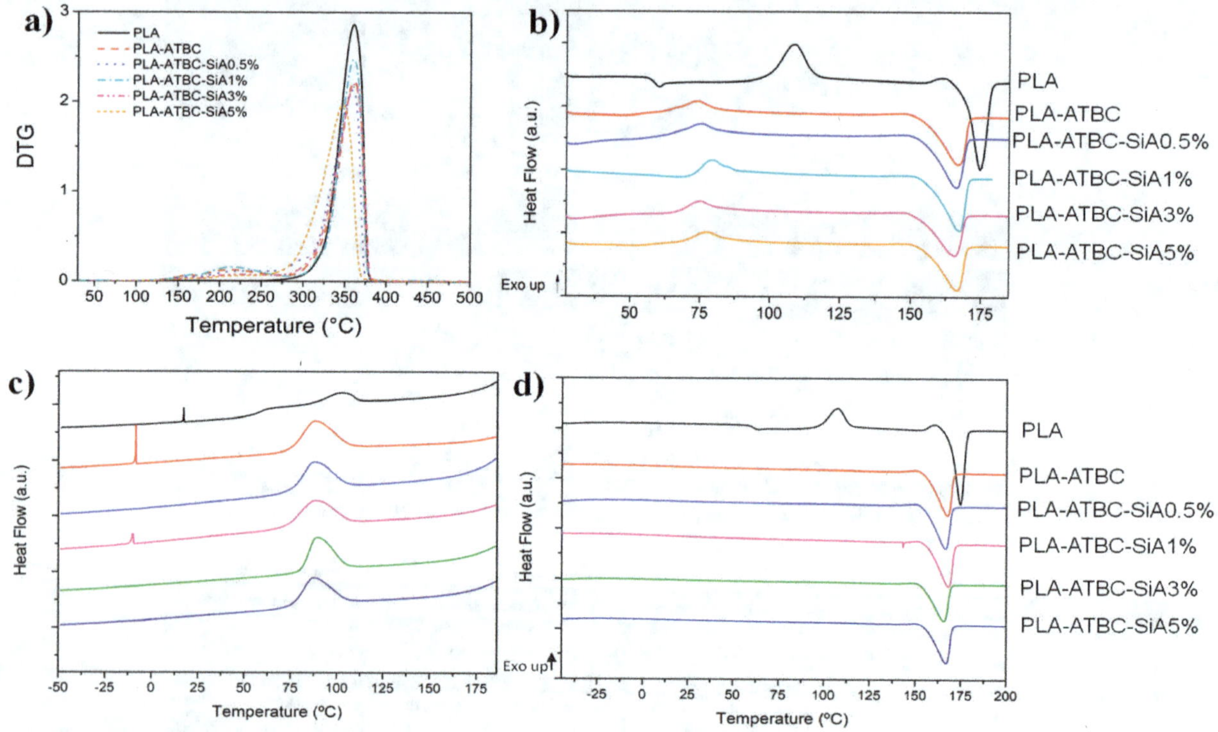

Figure 2. (**a**) DTG curves, (**b**) first heating DSC curves, (**c**) cooling DSC curves, and (**d**) second heating DSC curves of PLA, PLA-ATBC, and hybrid plasticized biocomposite films.

The DTG curves of Figure 2a showed that PLA-ATBC showed a degradation peak prior to the main degradation of the polymeric matrix, which has been already observed ascribed with ATBC plasticizer vaporization [21]. Conversely, the addition of silica aerogel increases the onset degradation temperature (T_0) with respect to the plasticized PLA-ATBC films suggestive of some stabilization of the PLA-ATBC continuous phase. Regarding the T_{max} values corresponding to the PLA-ATBC and PLA-ATBC reinforced with aerogel, no significant changes were not observed in either of the two degradation stages, that of plasticizer and that of the polymeric matrix. In conclusion, reinforcing plasticized PLA with silica aerogel resulted in little changes in their thermal stability. Nevertheless, it should be highlighted that not significant modifications on the TGA thermal parameters were observed at the processing as well as the intended use temperatures and thus confirming the good thermal stability of the developed biocomposites.

Figure 2 shows the DSC thermograms of all samples during the first heating scan (Figure 2b), the cooling scan (Figure 2c), and the second heating scan (Figure 2d). Meanwhile, Table 3 summarizes the corresponding values of the main thermal parameters. As expected, the addition of ATBC had a clear effect on the plasticization of the PLA polymeric matrix, since it induced a decrease in the T_g in all plasticized films. This reduction in the T_g has been ascribed to the ability of citrate plasticizers to increase the mobility of the PLA polymer chains, caused by an increase in the free volume between them [14,21,41]. Although, from the obtained DSC curves in blend formulations, the two parallel baselines, before and after the T_g, are not well defined, it is evident that the typical T_g of PLA could not be seen at 58 °C, suggesting that, in PLA, reinforced films further decreased the T_g, with respect to the PLA-ATBC sample. The midpoint between intersections can be estimated and these values are summarized in Table 3. The disappearance of PLA T_g at 59 °C confirms the good interaction of the silica aerogel microparticles with the polymeric matrix as well as with the plasticizer in the blend.

Table 3. Thermal parameters of PLA, PLA-ATBC and hybrid plasticized biocomposite films obtained from DSC during the first heating scan.

Film Formulations	T_g (°C)	T_{cc} (°C)	ΔH_{cc} (J g^{-1})	T_m (°C) First Heating Scan	ΔH_m (J g^{-1})	χ_c (%)
PLA	59	109	30.2	175	52.0	22.1
PLA-ATBC	50	74	13.4	166	54.0	51.3
PLA-ATBC-SiA0.5%	49	75	11.9	166	51.9	50.9
PLA-ATBC-SiA1%	47	79	13.7	167	53.0	50.2
PLA-ATBC-SiA3%	48	75	10.9	166	48.3	49.0
PLA-ATBC-SiA5%	48	78	12.6	167	49.2	49.1

Film Formulations	Cooling Scan T_g (°C)	T_{cc} (°C)	ΔH_{cc} (J g^{-1})	Second Heating Scan T_{cc} (°C)	ΔH_{cc} (J g^{-1})	T_m (°C)	ΔH_m (J g^{-1})	χ_c (%)
PLA	63	109	21.0	107	4.0	175	53.4	30.5
PLA-ATBC	-	88	28.5	-	0.6	167	46.2	58.4
PLA-ATBC-SiA0.5%	-	88	23.2	-	-	167	46.2	58.7
PLA-ATBC-SiA1%	-	89	24.3	-	-	168	46.8	59.9
PLA-ATBC-SiA3%	-	89	24.3	-	-	166	40.6	53.2
PLA-ATBC-SiA5%	-	87	25.4	-	-	167	40.3	54.2

The cold crystallization in neat PLA suggests that the processing conditions did not induce the crystallization of PLA. Nevertheless, in the case of plasticized systems, it could be observed a clear decrease not only in the enthalpy, but also in the cold crystallization temperature with respect to the neat PLA, suggesting that ATBC favors PLA crystallization during processing. Indeed, the degree of crystallinity was higher for all plasticized formulations (Table 3). It has been observed that ATBC plasticizer interacts well with PLA matrix, and thus enhancing the PLA crystallization rate [7,16]. Although no significant variations in the cold crystallization temperature between the plasticized PLA-ATBC films and composites reinforced with silica aerogel were observed, it seems that the combination of silica aerogel with ATBC plasticizer in the PLA based blend formulations lead to a slightly increase of the T_{cc} value. This synergic effect on the crystallization of PLA plasticized with ATBC and a potential nucleating agent has already been reported in PLA-ATBC reinforced with talc [20] and nanocellulose [12].

It is remarkable that neat PLA presents a small exothermic peak right before melting, which disappears with the presence of the plasticizer. This peak has been related with restrictions in the polymer chains due to the transition of the disorder to order PLA crystallites [42]—thus suggesting that a great part of the polymeric system was in an amorphous state as a consequence of the fast DSC cooling rate applied after the compression molding process [7]. As expected, ATBC reduced the melting temperature of PLA confirming again the good interaction between the ATBC plasticizer and the polymeric matrix, in accordance with previous works [7,21,41]. The melting temperatures of the composites resulted in being almost invariant with respect to the PLA-ATBC film.

In plasticized formulations, the addition of ATBC plasticizer to the PLA polymer matrix induced an increase in the degree of crystallinity, related to the fact that the ATBC chains can interact with the polymeric chains of PLA, which result in increased polymeric chains mobility, which influences the nucleation and thus promotes a better packaging of segments [23]. However, the degree of crystallinity was reduced increasing the silica aerogel amount in the formulation, suggesting that higher silica aerogel contents restrict the polymeric chains mobility in the system, which is required for its regular packing into crystal lattices, because of enlarged interaction sites between PLA matrix and the inorganic silica phase, ultimately retarding the crystallization.

PLA is well known for its slow nucleation and crystallization rates. In the present work, the cooling DSC conditions conducted at slow rate allowed PLA crystallization (Figure 2c) in all formulations, while T_g values could not be observed during the cooling probably due to the low amount of sample used. Thus, all samples presented higher crystallinity degrees during the second DSC heating (Figure 2d and Table 3) than in the first DSC heating (Figure 2b and Table 3). The second DSC heating allowed better comparison between samples as a consequence of the elimination of the PLA-based formulations thermal history (Figure 2d). Neat PLA was the only film that showed the cold crystallization in the second DSC heating. This result shows that PLA was not capable to crystallize completely, even at the low cooling rate used during DSC cooling scan. During the second DSC heating (Figure 2d), all the plasticized formulations did not show the cold crystallization and only the melting temperature is clearly observed, showing complete crystallized materials, highlighting the ability of ATBC plasticizer to crystallize PLA during the cooling.

The mechanical behavior of the different plasticized PLA-based materials was assessed and the tensile test results are reported in Figure 3. The tensile test results for neat PLA film are also reported for comparison. Figure 3a shows the obtained results of the Young's Modulus (E) for all tested samples. The high Young's modulus values corresponding to the neat PLA are symptomatic of the high stiffness and brittleness that typically present the neat polymer, with elastic modulus values approaching 2.5 GPa. As expected, the addition of ATBC caused a significant ($p < 0.05$) reduction in the Young's modulus up to 60%, since plasticizers induce ductile fracture. In general, biocomposites presented comparable values of elastic modulus, showing not significant ($p > 0.05$) changes with respect to the plasticized PLA-ATBC film, with the exception of PLA-ATBC-SiA3% which showed some ($p > 0.05$) improvement of the Young's modulus with respect to PLA-ATBC suggesting a reinforcing effect.

In the same way as for the Young's modulus, the tensile strength decreased significantly ($p < 0.05$) with the presence of the plasticizer (Figure 3b). The incorporation of silica aerogel slightly reduced the tensile strength of PLA-ATBC film ($p > 0.05$). The presence of ATBC in the plasticized PLA and its biocomposites produced a reduction in the tensile strength up to 55% when compared to neat PLA ($p < 0.05$).

Regarding the elongation at break (Figure 3c), neat PLA did not show an elastic behavior, presenting a value of elongation at break of 4% demonstrating the well-known poor flexibility of PLA. A noticeable significant ($p < 0.05$) improvement in elongation at break values was evidenced for PLA plasticized with ATBC and its biocomposite films, in good accordance with the reduction in the T_g values in these formulations as well as with already reported works [7,12,21]. The elongation at break reached values between 140% and 185% in the plasticized samples, as a result of the increase in the polymer chains mobility. The best improvement in the elongation at break of the biocomposites was observed for the biocomposite loaded with the less amount of SiA (PLA-ATBC-SiA0.5%). In this sense, although not statistically significant ($p > 0.05$), it can be observed that the addition of low amounts of SiA, which is 0.5 wt% and 1 wt%, caused a somewhat plasticizing effect.

Similar findings have already been observed in PLA based polyurethanes reinforced with nano hydroxyapatite in which less reinforcing material of 0.5 and 1 wt.% produced a slightly plasticizing effect, while higher amounts of 3 wt.% produced a reinforcing effect [25]. In fact, the less elongation at break value was observed for PLA-ATBC-SiA3% biocomposite formulation ($p < 0.05$), in agreement with the highest Young's modulus, but still showing enough stretchability for the intended use. In this sense, it should be highlighted that films intended for food packaging applications require high flexibility at room temperature and the elongation at break values obtained here for PLA-ATBC and PLA based biocomposites are comparable in terms of flexibility to frequently used traditional petrol-based and non-biodegradable polymers. One of the most used conventional plastic materials at the industrial level for food packaging applications is low density polyethylene (LDPE), with tensile strength values between 10 and 20 MPa and elongation at break between 80 and 500% [43–45]. Therefore, the biocomposite films developed here offer possibilities for their processing at industrial level and their use in the food packaging sector.

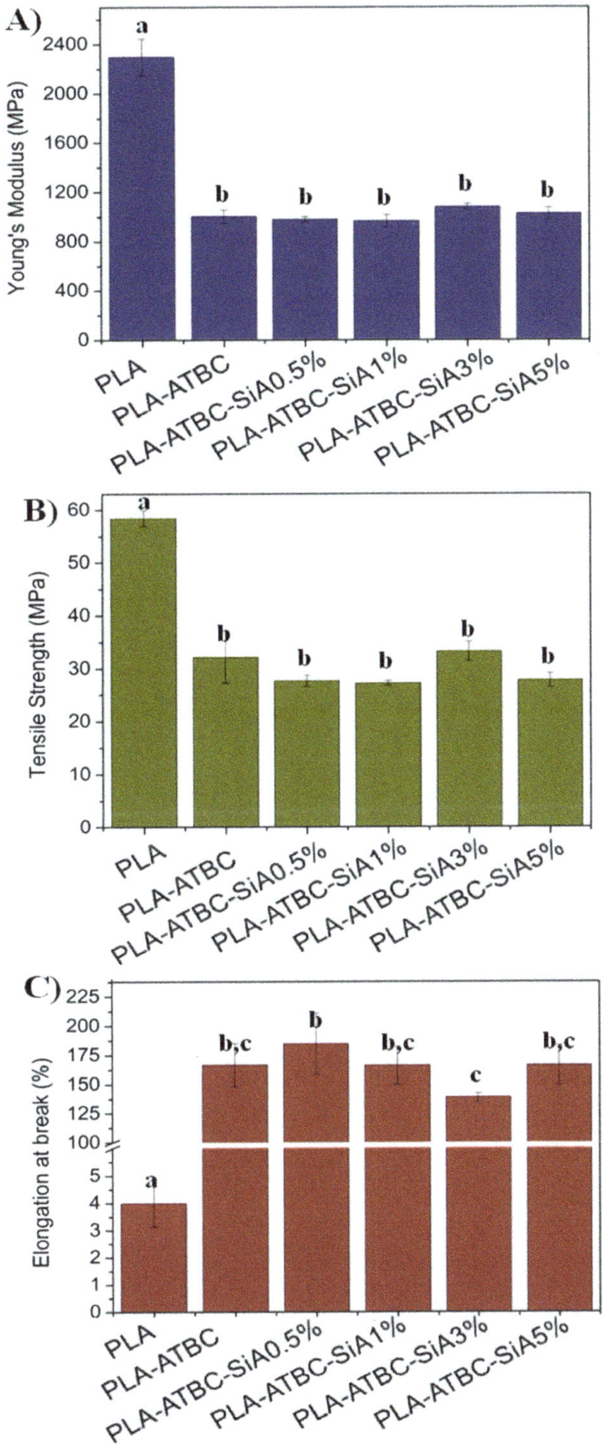

Figure 3. Tensile test results: (**A**) Young Modulus, (**B**) Tensile strength and (**C**) elongation at break. ([a–c] Different letters on the bars within the same image indicate significant differences between formulations ($p < 0.05$)).

Besides containing and protecting packed food from physical damage, food packaging materials should maintain foodstuff quality and extend their shelf-life by controlling the mass transfer processes [46]. Considering that plastic, and particularly bioplastics, are relatively permeable to small molecules [47], the polymer barrier performance in food packaging materials based on biopolymers is of fundamental importance. Table 4 shows the permeability, diffusion, and solubility coefficients values obtained for all formulations by using the time-lag method for CO_2, N_2 and O_2. Meanwhile, for a better comparison of the results of plasticized samples, the diffusivity and permeability parameters for PLA, PLA-ATBC and the biocomposites as a function of the silica aerogel content are represented in Figure 4a,b, respectively.

Table 4. Comparative values at 30 °C of permeability (P), diffusion (D) and apparent solubility (S) coefficients of carbon dioxide, nitrogen and oxygen in PLA, PLA-ATBC and plasticized biocomposite films.

Film Formulation	Film Thickness (μm)	Gas	P (Barrer)	$D \times 10^8$ ($cm^2\ s^{-1}$)	$S \times 10^3$ (cm^3 (STP) cm^{-3} (cm Hg)$^{-1}$)
PLA	99	CO_2	1.03	0.31	33.1
		N_2	0.06	0.44	1.3
		O_2	0.32	1.83	1.7
PLA-ATBC	185	CO_2	3.21	1.66	19.4
		N_2	0.22	1.93	1.2
		O_2	0.84	7.36	1.1
PLA-ATBC-SiA0.5%	185	CO_2	10.91	4.62	23.6
		N_2	0.47	10.66	0.4
		O_2	1.72	17.62	1.0
PLA-ATBC-SiA1%	168	CO_2	11.21	4.95	22.7
		N_2	0.44	9.17	0.5
		O_2	1.64	17.06	1.0
PLA-ATBC-SiA3%	175	CO_2	3.40	1.33	25.6
		N_2	0.16	1.78	0.9
		O_2	0.75	5.42	1.4
PLA-ATBC-SiA5%	197	CO_2	3.56	1.38	25.9
		N_2	0.20	1.46	1.4
		O_2	0.76	5.40	1.4

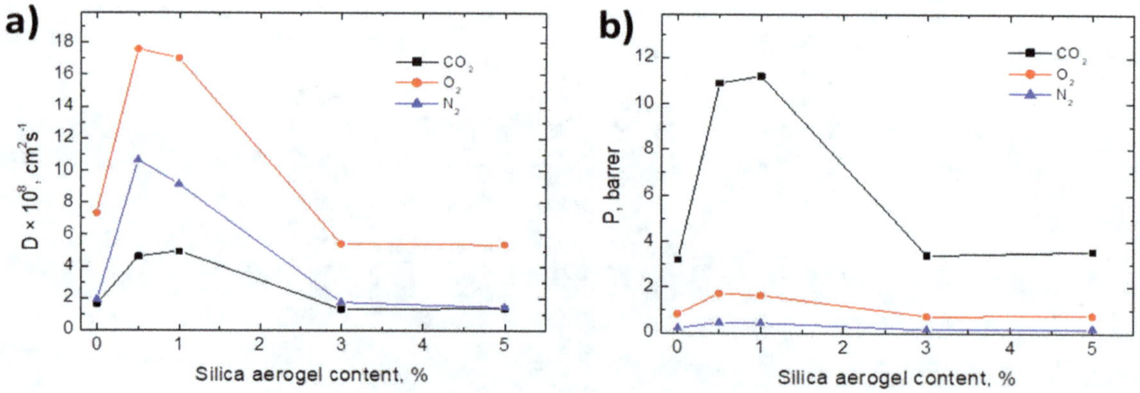

Figure 4. (**a**) Diffusivity and (**b**) permeability as a function of the aerogel content.

As it was expected and can be seen in Table 4, the permeability followed the trends $P_{CO2} > P_{O2} > P_{N2}$ in all film samples and the values of apparent solubility coefficients were higher for more condensable gases, so $S_{CO2} > S_{O2} > S_{N2}$. As it was already commented, the addition of ATBC plasticizer increased the free volume leading to an increase in the mobility of the polymer chains. Thus, gases diffused easily through the films and consequently the permeability and diffusion coefficients increased with the addition of ATBC from pristine PLA, in agreement with previously developed PLA-ATBC based materials [12,21]. It could be observed that, for biocomposites with 0.5 wt.% and 1 wt.% of silica aerogel contents, the diffusion coefficients increased for all gases tested. Therefore, in these cases, the permeabilities were improved with the addition of SiA fillers up to 1 wt.%. Nevertheless, higher aerogel contents of 3 wt.% and 5 wt.% in the composite materials provoked a decrease in diffusivity values, which were slightly lower compared to PLA-ATBC formulation. Regarding the permeability, comparable values were obtained between PLA-ATBC film sample and those biocomposites loaded with 3 wt.% and 5 wt.% of silica aerogel.

It is noteworthy that N_2 has lower diffusivity and permeability compared to CO_2 and O_2. In the case of neat PLA film, N_2 permeability is about 0.06 Barrer, 18 times smaller than CO_2 and 6 times smaller than O_2. This N_2 permeation was too low to be accurately measured in our experimental system. In relation to the apparent solubility coefficient, no significant differences were observed between the plasticized biocomposites film formulations. In all cases, CO_2 is the gas that presented a higher solubility due to its bigger condensability.

As expected, the increase in the free volume in plasticized systems led to an increment in diffusion coefficient and, thus, in the permeability to gases. By adding low contents of silica aerogel, 0.5 wt.% and 1 wt.%, the diffusivity through the films becomes significantly larger. Considering the good compatibility between the additives and the polymeric matrix (as can be seen in Figure 1), this increase in the diffusion coefficients suggest that a low content of silica aerogel caused a further plasticizing effect, in agreement with the higher elongation at break showed for these biocomposites. It should be also considered that crystalline regions act as an impermeable barrier for the diffusion and adsorption and, as commented before, the addition of silica aerogel decreases the degree of crystallinity, increasing in this way the permeability to gases. On the other hand, for high aerogel contents, 3 wt.% and 5% wt.%, the negative effect of the decrease in the crystallinity degree would have been compensated by a more tortuous pathway for gas diffusion. Thus, in these cases, the result of both effects was the decrease of the diffusion coefficients to values slightly lower than for PLA-ATBC film. The reduction of the diffusion coefficient caused the consequent decrease in the permeability of composites with 3 wt.% and 5 wt.% of aerogel content, reaching permeability values comparable to that PLA-ATBC film. Similarly, Chen et al. need at least 3 wt.% of SiA content to improve PVA barrier performance [32]. This behavior has been related with the good dispersion and good adherence of fillers into the polymeric matrix, which hinders micro-Brownian motions and thus minimizes the formation of non-permanent holes that would promote gas diffusivity through the film [48]. It should be mentioned that it is known that the PLA permeability to gases decreases as temperature decreases [47]. Thus, it is expected that, in foreseeable conditions of the intended use, such as at fridge storage temperature (i.e., 4 °C), the gas permeability values of biocomposites will decrease further.

Biodegradable polymers are able to decompose into simple molecules found in the environment (i.e., carbon dioxide, methane, water, inorganic compounds, or biomass), mediated by the enzymatic action of the microorganisms, in a determined period of time and includes the formation of humus soil that can be further used as soil conditioner [40,49,50]. Thus, finally, the PLA and plasticized PLA formulations were disintegrated under composting conditions performed at a laboratory scale level. The compostable character was qualitatively checked by photographs taken. Figure 5a shows the visual aspect of the recovered film samples at different testing days of exposition to the compost medium. Meanwhile, the degree of disintegration as a function of time was quantitatively followed as it is showed in Figure 5b.

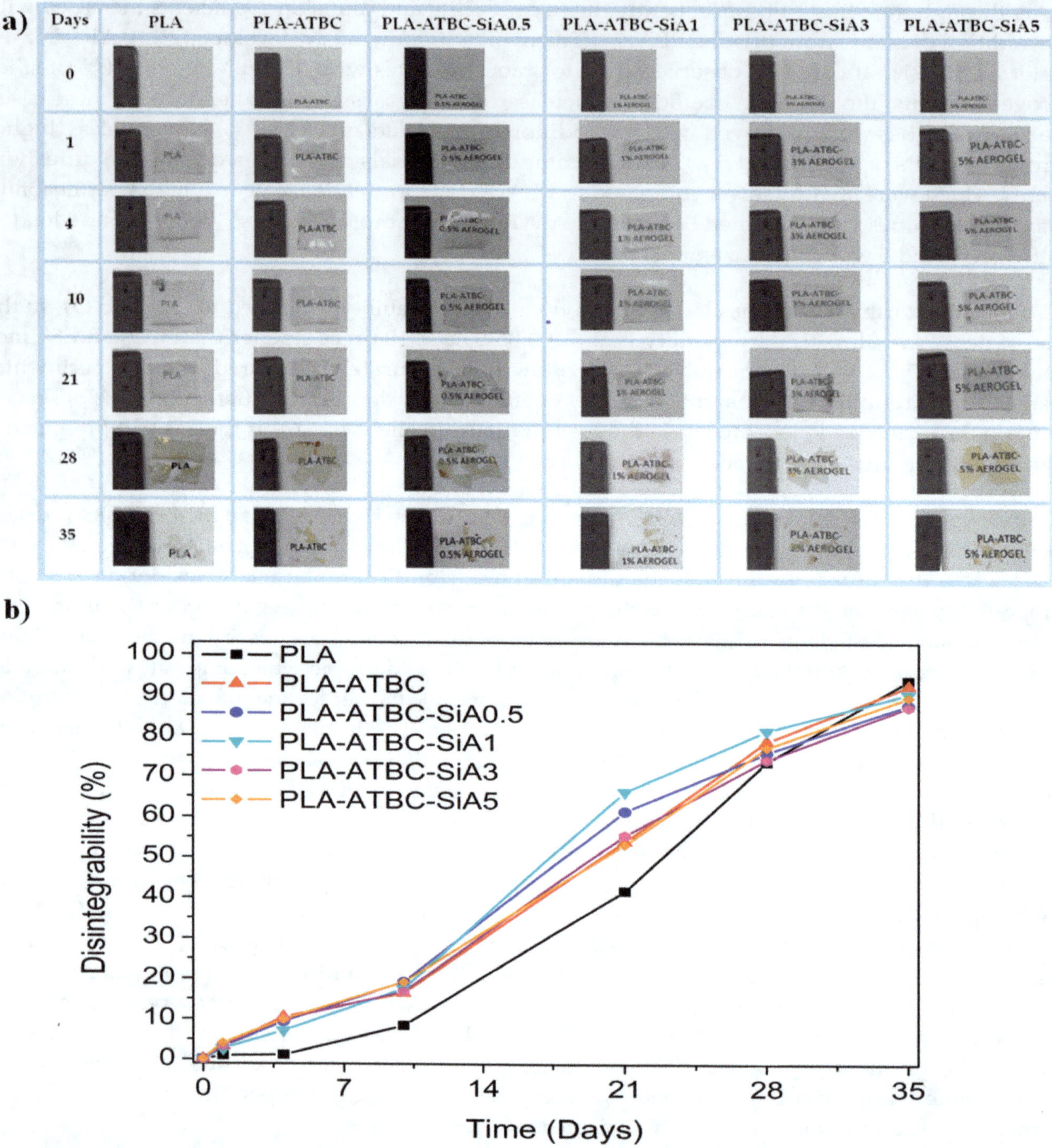

Figure 5. (**a**) visual aspect of film samples at different degradation times, (**b**) degree of disintegration under composting conditions as a function of time.

It was observed that all PLA and PLA-ATBC based samples changed their visual appearance during the first week under composting conditions—becoming opaque. The changes of the refraction index of the polymeric materials during composting has already been observed and related with the beginning of the degradation process of the polymer matrices as a result of the water absorption and/or the presence of products generated during the hydrolytic processes [40,51]. The presence of silica aerogel speeds up the disintegration process. The fact that the plasticized PLA based formulations were still visible after 35 days compared to previous works [12,40] is related to the higher thickness of the films developed here as well as to the higher degree of crystallinity promoted by silica aerogel, which could be somewhat enhanced at the temperature of the composting container during the test. An increase in crystallinity of the samples entails a decrease in their degradation rate since the disintegration process starts in the amorphous regions of the polymeric matrix and due to the ordered structure located in the crystalline fractions can retain the microorganisms' action [52]. Conversely, the presence of ATBC in the formulations leads to a clear increase in the disintegration rate. In fact, at the temperature of the composting test, the plasticized materials are over its T_g, and thus the presence of a plasticizer increases the polymer chain mobility, facilitating the water diffusion and thus the hydrolysis of polymeric matrix, which consequently further accelerates the disintegration process mediated by the enzymatic attack of microorganisms [8,9,40].

4. Conclusions

Silica aerogels were used to reinforce plasticized PLA with ATBC. The incorporation of the silica aerogel in 0.5, 1, 3, and 5 wt.% resulted in a small improvement in the thermal stability of the biocomposites. The presence of a plasticizer decreased, as expected, the glass transition temperature due to the ability of ATBC to increase the mobility of the polymer chains. The resulting higher chain mobility promoted an increment of the degree of crystallinity of PLA as well as the stretchability. On the contrary, it could be observed that the addition of silica aerogel had a clear tendency to decrease the degree of crystallinity with respect to PLA-ATBC samples, which is at the detriment of the enhancement of barrier properties of the composites. Nevertheless, silica aerogel in the amounts up to 3 wt.% was able to generate a more tortuous path which decreased the permeability of the biocomposite films (PLA-ATBC-SiA3% and PLA-ATBC-SiA5%), slightly improving the barrier performance of PLA-ATBC. Although the plasticizer decreased the Young's modulus and tensile strength, the elongation at break reached values up to 166% in plasticized PLA formulations. In general, no significant modifications were observed in the mechanical behavior of the plasticized biocomposites with the addition of silica aerogel. Nevertheless, in the case of the addition of silica aerogel in 3 wt.%, some improvement of the mechanical resistance was observed which was, however, accompanied with a reduction of the elongation at break, but being stretchable enough (around 140%) for the intended use. PLA-ATBC and its biocomposite films presented a higher disintegration rate than neat PLA due to the presence of the plasticizer, which promote the hydrolysis of polymer chains in the compost medium.

In summary, plasticized PLA loaded with 3 wt.% of silica aerogel formulation, PLA-ATBC-SiA3%, showed the best performance for biodegradable films for food packaging applications with high transparency, improved elongation at break, moderate barrier properties, and being compostable in around 40 days.

Author Contributions: Conceptualization, M.P.A. and D.L.; methodology, A.A.-G., M.P.A. and D.L.; validation, A.A.-G., M.P.A., M.L.-G., M.F.-G. and D.L.; formal analysis, A.A.-G., M.P.A., M.L.-G. and D.L.; investigation, A.A.-G., M.P.A., M.L.-G., M.F.-G. and D.L.; resources, D.L.; data curation, A.A.-G., M.P.A. and M.L.-G.; writing—original draft preparation, A.A.-G. and M.P.A.; writing—review and editing, A.A.-G., M.P.A., M.L.-G., M.F.-G. and D.L.; visualization, A.A.-G., M.P.A., M.L.-G., M.F.-G. and D.L.; supervision, M.P.A. and D.L.; project administration, D.L.; funding acquisition, D.L. All authors have read and agreed to the published version of the manuscript.

Acknowledgments: The authors thank Ercros S.A. for kindly providing PLA pellets.

References

1. Guo, X.; Xiang, D.; Duan, G.; Mou, P. A review of mechanochemistry applications in waste management. *Waste Manag.* **2010**, *30*, 4–10. [CrossRef]
2. Arrieta, M.; Samper, M.; Aldas, M.; López, J. On the use of PLA-PHB blends for sustainable food packaging applications. *Materials* **2017**, *10*, 1008. [CrossRef]
3. Auras, R.; Harte, B.; Selke, S. An overview of polylactides as packaging materials. *Macromol. Biosci.* **2004**, *4*, 835–864. [CrossRef]

4. Armentano, I.; Bitinis, N.; Fortunati, E.; Mattioli, S.; Rescignano, N.; Verdejo, R.; Lopez-Manchado, M.A.; Kenny, J.M. Multifunctional nanostructured pla materials for packaging and tissue engineering. *Prog. Polym. Sci.* **2013**, *38*, 1720–1747. [CrossRef]
5. Lim, L.T.; Auras, R.; Rubino, M. Processing technologies for poly(lactic acid). *Prog. Polym. Sci. (Oxf.)* **2008**, *33*, 820–852. [CrossRef]
6. Arrieta, M.P.; López, J.; López, D.; Kenny, J.M.; Peponi, L. Development of flexible materials based on plasticized electrospun pla–phb blends: Structural, thermal, mechanical and disintegration properties. *Eur. Polym. J.* **2015**, *73*, 433–446. [CrossRef]
7. Arrieta, M.P.; Peponi, L.; López, D.; Fernández-García, M. Recovery of yerba mate (*Ilex paraguariensis*) residue for the development of PLA-based bionanocomposite films. *Ind. Crop. Prod.* **2018**, *111*, 317–328. [CrossRef]
8. Kale, G.; Kijchavengkul, T.; Auras, R.; Rubino, M.; Selke, S.E.; Singh, S.P. Compostability of bioplastic packaging materials: An overview. *Macromol. Biosci.* **2007**, *7*, 255–277. [CrossRef]
9. Fortunati, E.; Armentano, I.; Iannoni, A.; Barbale, M.; Zaccheo, S.; Scavone, M.; Visai, L.; Kenny, J. *New Multifunctional Poly(Lactide Acid) Composites: Mechanical, Antibacterial, and Degradation Properties*; Wiley: Hoboken, NJ, USA, 2012; Volume 124.
10. Gonçalves de Moura, I.; Vasconcelos de Sá, A.; Lemos Machado Abreu, A.S.; Alves Machado, A.V. 7—Bioplastics from agro-wastes for food packaging applications a2—Grumezescu, alexandru mihai. In *Food Packaging*; Academic Press: Cambridge, MA, USA, 2017; pp. 223–263.
11. Burgos, N.; Martino, V.P.; Jiménez, A. Characterization and ageing study of poly (lactic acid) films plasticized with oligomeric lactic acid. *Polym. Degrad. Stab.* **2013**, *98*, 651–658. [CrossRef]
12. Arrieta, M.P.; Fortunati, E.; Dominici, F.; López, J.; Kenny, J.M. Bionanocomposite films based on plasticized PLA–PHB/cellulose nanocrystal blends. *Carbohydr. Polym.* **2015**, *121*, 265–275. [CrossRef]
13. Ferri, J.M.; Samper, M.D.; García-Sanoguera, D.; Reig, M.J.; Fenollar, O.; Balart, R. Plasticizing effect of biobased epoxidized fatty acid esters on mechanical and thermal properties of poly(lactic acid). *J. Mater. Sci.* **2016**, *51*, 5356–5366. [CrossRef]
14. Courgneau, C.; Domenek, S.; Guinault, A.; Avérous, L.; Ducruet, V. Analysis of the structure-properties relationships of different multiphase systems based on plasticized poly (lactic acid). *J. Polym. Environ.* **2011**, *19*, 362–371. [CrossRef]
15. Arrieta, M.P.; López, J.; López, D.; Kenny, J.M.; Peponi, L. Effect of chitosan and catechin addition on the structural, thermal, mechanical and disintegration properties of plasticized electrospun PLA-PHB biocomposites. *Polym. Degrad. Stab.* **2016**, *132*, 145–156. [CrossRef]
16. Labrecque, L.; Kumar, R.A.; Davé, V.; Gross, R.A.; McCarthy, S. Citrate esters as plasticizers for poly(lactic acid). *J. Appl. Polym. Sci.* **1997**, *66*, 1507–1513. [CrossRef]
17. Liu, H.; Zhang, J. Research progress in toughening modification of poly(lactic acid). *J. Polym. Sci. Part B Polym. Phys.* **2011**, *49*, 1051–1083. [CrossRef]
18. Arrieta, M.P.; López de Dicastillo, C.; Garrido, L.; Roa, K.; Galotto, M.J. Electrospun pva fibers loaded with antioxidant fillers extracted from durvillaea antarctica algae and their effect on plasticized PLA bionanocomposites. *Eur. Polym. J.* **2018**, *103*, 145–157. [CrossRef]
19. Aliotta, L.; Vannozzi, A.; Panariello, L.; Gigante, V.; Coltelli, M.-B.; Lazzeri, A. Sustainable micro and nano additives for controlling the migration of a biobased plasticizer from pla-based flexible films. *Polymers* **2020**, *12*, 1366. [CrossRef]
20. Courgneau, C.; Ducruet, V.; Avérous, L.; Grenet, J.; Domenek, S. Nonisothermal crystallization kinetics of poly(lactide)—Effect of plasticizers and nucleating agent. *Polym. Eng. Sci.* **2013**, *53*, 1085–1098. [CrossRef]
21. Arrieta, M.P.; Samper, M.D.; López, J.; Jiménez, A. Combined effect of poly(hydroxybutyrate) and plasticizers on polylactic acid properties for film intended for food packaging. *J. Polym. Environ.* **2014**, *22*, 460–470. [CrossRef]
22. European Food Safety Authority. Scientific opinion on flavouring group evaluation 10, revision 3 (fge.10rev3): Aliphatic primary and secondary saturated and unsaturated alcohols, aldehydes, acetals, carboxylic acids and esters containing an additional oxygenated functional group and lactones from chemical groups 9, 13 and 30. *EFSA J.* **2012**, *10*, 25–63.
23. Arrieta, M.P.; López, J.; Hernández, A.; Rayón, E. Ternary PLA–PHB–limonene blends intended for biodegradable food packaging applications. *Eur. Polym. J.* **2014**, *50*, 255–270. [CrossRef]

24. Arrieta, M.P.; Sessini, V.; Peponi, L. Biodegradable poly(ester-urethane) incorporated with catechin with shape memory and antioxidant activity for food packaging. *Eur. Polym. J.* **2017**, *94*, 111–124. [CrossRef]
25. Navarro-Baena, I.; Arrieta, M.P.; Sonseca, A.; Torre, L.; López, D.; Giménez, E.; Kenny, J.M.; Peponi, L. Biodegradable nanocomposites based on poly(ester-urethane) and nanosized hydroxyapatite: Plastificant and reinforcement effects. *Polym. Degrad. Stab.* **2015**, *121*, 171–179. [CrossRef]
26. Bao, L.; Dorgan, J.R.; Knauss, D.; Hait, S.; Oliveira, N.S.; Maruccho, I.M. Gas permeation properties of poly(lactic acid) revisited. *J. Membr. Sci.* **2006**, *285*, 166–172. [CrossRef]
27. Burgos, N.; Armentano, I.; Fortunati, E.; Dominici, F.; Luzi, F.; Fiori, S.; Cristofaro, F.; Visai, L.; Jiménez, A.; Kenny, J.M. Functional properties of plasticized bio-based poly(lactic acid)_poly(hydroxybutyrate) (PLA_PHB) films for active food packaging. *Food Bioprocess Technol.* **2017**, *10*, 770–780. [CrossRef]
28. Venkata Prasad, C.; Sudhakara, P.; Prabhakar, M.N.; Ur Rehman Shah, A.; Song, J.I. An investigation on the effect of silica aerogel content on thermal and mechanical properties of sisal/pla nano composites. *Polym. Compos.* **2018**, *39*, 835–840. [CrossRef]
29. Salgado, C.; Arrieta, M.P.; Peponi, L.; López, D.; Fernández-García, M. Photo-crosslinkable polyurethanes reinforced with coumarin modified silica nanoparticles for photo-responsive coatings. *Prog. Org. Coat.* **2018**, *123*, 63–74. [CrossRef]
30. Choi, H.; Parale, V.G.; Lee, K.-Y.; Nah, H.-Y.; Driss, Z.; Driss, D.; Bouabidi, A.; Euchy, S.; Park, H.-H. Polypropylene/silica aerogel composite incorporating a conformal coating of methyltrimethoxysilane-based aerogel. *J. Nanosci. Nanotechnol.* **2019**, *19*, 1376–1381. [CrossRef]
31. Zolfaghari, S.; Paydayesh, A.; Jafari, M. Mechanical and thermal properties of polypropylene/silica aerogel composites. *J. Macromol. Sci. Part B* **2019**, *58*, 305–316. [CrossRef]
32. Chen, C.; Ding, R.; Yang, S.; Wang, J.; Chen, W.; Zong, L.; Xie, J. Development of thermal insulation packaging film based on poly(vinyl alcohol) incorporated with silica aerogel for food packaging application. *LWT* **2020**, *129*, 109568. [CrossRef]
33. Ndazi, B.S.; Karlsson, S. *Characterization of Hydrolytic Degradation of Polylactic Acid/Rice Hulls Composites in Water at Different Temperatures*; University of Dar es Salaam: Dar es Salaam, Tanzania, 2011.
34. Bitinis, N.; Verdejo, R.; Bras, J.; Fortunati, E.; Kenny, J.M.; Torre, L.; López-Manchado, M.A. Poly(lactic acid)/natural rubber/cellulose nanocrystal bionanocomposites part i. Processing and morphology. *Carbohydr. Polym.* **2013**, *96*, 611–620. [CrossRef]
35. Murariu, M.; Doumbia, A.; Bonnaud, L.; Dechief, A.L.; Paint, Y.; Ferreira, M.; Campagne, C.; Devaux, E.; Dubois, P. High-performance polylactide/ZnO nanocomposites designed for films and fibers with special end-use properties. *Biomacromolecules* **2011**, *12*, 1762–1771. [CrossRef]
36. Turner Ii, J.F.; Riga, A.; O'Connor, A.; Zhang, J.; Collis, J. Characterization of drawn and undrawn poly-l-lactide films by differential scanning calorimetry. *J. Therm. Anal. Calorim.* **2004**, *75*, 257–268. [CrossRef]
37. Tiemblo, P.; Guzman, J.; Riande, E.; Salvador, E.; Peinado, C. Gas-transport properties in crosslinked polymers. I. Aliphatic polyurethane–acrylate-based adhesives. *J. Polym. Sci. Part B Polym. Phys.* **2001**, *39*, 786–795. [CrossRef]
38. Vargas, J.; Santiago, A.A.; Cruz-Morales, J.A.; Tlenkopatchev, M.A.; de Lys, T.; López-González, M.; Riande, E. Gas transport properties of hydrogenated and fluorinated polynorbornene dicarboximides. *Macromol. Chem. Phys.* **2013**, *214*, 2607–2615. [CrossRef]
39. UNE-EN ISO. *ISO 20200:2015. Plastics. Determination of the Degree of Disintegration of Plastic Materials under Simulated Composting Conditions in a Laboratory-Scale Test*; ISO: Geneva, Switzerland, 2016.
40. Arrieta, M.P.; López, J.; Rayón, E.; Jiménez, A. Disintegrability under composting conditions of plasticized pla–phb blends. *Polym. Degrad. Stab.* **2014**, *108*, 307–318. [CrossRef]
41. García-Arroyo, P.; Arrieta, M.P.; Garcia-Garcia, D.; Cuervo-Rodríguez, R.; Fombuena, V.; Mancheño, M.J.; Segura, J.L. Plasticized poly(lactic acid) reinforced with antioxidant covalent organic frameworks (COFs) as novel nanofillers designed for non-migrating active packaging applications. *Polymer* **2020**, *196*, 122466. [CrossRef]
42. Muller, J.; Jiménez, A.; González-Martínez, C.; Chiralt, A. Influence of plasticizers on thermal properties and crystallization behaviour of poly (lactic acid) films obtained by compression moulding. *Polym. Int.* **2016**, *65*, 970–978. [CrossRef]

43. Hauser, C.; Peñaloza, A.; Guarda, A.; Galotto, M.J.; Bruna, J.E.; Rodríguez, F.J. Development of an active packaging film based on a methylcellulose coating containing murta (*Ugni molinae turcz*) leaf extract. *Food Bioprocess Technol.* **2016**, *9*, 298–307. [CrossRef]
44. Arrieta, M.P.; Garrido, L.; Faba, S.; Guarda, A.; Galotto, M.J.; Dicastillo, C.L.d. *Cucumis metuliferus* fruit extract loaded acetate cellulose coatings for antioxidant active packaging. *Polymers* **2020**, *12*, 1248. [CrossRef]
45. Briassoulis, D. Analysis of the mechanical and degradation performances of optimised agricultural biodegradable films. *Polym. Degrad. Stab.* **2007**, *92*, 1115–1132. [CrossRef]
46. Arrieta, M.P.; Peponi, L.; López, D.; López, J.; Kenny, J.M. An overview of nanoparticles role in the improvement of barrier properties of bioplastics for food packaging applications. In *Food Packaging*; Elsevier: Amsterdam, The Netherlands, 2017; pp. 391–424.
47. Auras, R.A. Chapter 19—Solubility of gases and vapors in polylactide polymers. In *Thermodynamics, Solubility and Environmental Issues*; Letcher, T.M., Ed.; Elsevier: Amsterdam, The Netherlands, 2007; pp. 343–368.
48. Sereda, L.; López-González, M.; Yuan Visconte, L.L.; Nunes, R.C.R.; Guimarães Furtado, C.R.; Riande, E. Influence of silica and black rice husk ash fillers on the diffusivity and solubility of gases in silicone rubbers. *Polymer* **2003**, *44*, 3085–3093. [CrossRef]
49. Jiang, L.; Zhang, J. 7—Biodegradable and biobased polymers a2—Kutz, myer. In *Applied Plastics Engineering Handbooki*, 2nd ed.; William Andrew Publishing, Myer Kutz Associates. Inc.: Delmar, NY, USA, 2017; pp. 127–143.
50. Kale, G.; Auras, R.; Singh, S.P. Comparison of the degradability of poly (lactide) packages in composting and ambient exposure conditions. *Packag. Technol. Sci. Int. J.* **2007**, *20*, 49–70. [CrossRef]
51. Fortunati, E.; Puglia, D.; Santulli, C.; Sarasini, F.; Kenny, J. Biodegradation of phormium tenax/poly (lactic acid) composites. *J. Appl. Polym. Sci.* **2012**, *125*, E562–E572. [CrossRef]
52. Tuominen, J.; Kylmä, J.; Kapanen, A.; Venelampi, O.; Itävaara, M.; Seppälä, J. Biodegradation of lactic acid based polymers under controlled composting conditions and evaluation of the ecotoxicological impact. *Biomacromolecules* **2002**, *3*, 445–455. [CrossRef]

5

Antioxidant/Antibacterial Electrospun Nanocoatings Applied onto PLA Films

Bogdanel Silvestru Munteanu [1,*], Liviu Sacarescu [2], Ana-Lavinia Vasiliu [2], Gabriela Elena Hitruc [2], Gina M Pricope [3], Morten Sivertsvik [4,*], Jan Thomas Rosnes [4] and Cornelia Vasile [2]

1 Faculty of Physics, Alexandru Ioan Cuza University, 11 Carol I bvd, 700506 Iasi, Romania
2 "P. Poni" Institute of Macromolecular Chemistry, Romanian Academy, 41A Grigore GhicaVoda Alley, 700487 Iasi, Romania; livius@icmpp.ro (L.S.); vasiliu.lavinia@icmpp.ro (A.-L.V.); gabihit@icmpp.ro (G.E.H.); cvasile@icmpp.ro (C.V.)
3 Veterinary and the Food Safety Laboratory, Food Safety Department, 700489 Iasi, Romania; ginacornelia@yahoo.com
4 Nofima AS, Deptartment of Processing Technology, Muninbakken 9-13, Tromsø 9291, Norway; thomas.rosnes@nofima.no
* Correspondence: muntb@uaic.ro (B.S.M.); Morten.Sivertsvik@nofima.no (M.S.)

Abstract: Polylactic acid (PLA) films were coated by coaxial electrospinning with essential and vegetable oils (clove and argan oils) and encapsulated into chitosan, in order to combine the biodegradability and mechanical properties of PLA substrates with the antimicrobial and antioxidant properties of the chitosan–oil nanocoatings. It has been established that the morphology of the electrospun nanocoatings mainly depend on the average molecular weight (MW) of chitosan. Oil beads, encapsulated into the main chitosan nanofibers, were obtained using high-MW chitosan (Chit-H). Oil encapsulated in chitosan naoparticles resulted when low-MW chitosan (Chit-L) was used. The coating layer, with a thickness of 100 ± 20 nm, had greater roughness for the samples containing Chit-H compared with the samples containing Chit-L. The coated PLA films had higher antibacterial activity when the nanocoating contained clove oil rather than when argan oil was used, for both types of chitosan. Nanocoatings containing Chit-H had higher antibacterial activity compared with those containing Chit-L, for both types of oil tested, due to the larger surface area of the rougher nanoscaled morphology of the coating layer that contained Chit-L. The chitosan–clove oil combination had higher antioxidant activity compared to the simple chitosan nanocoating, which confirmed their synergistic activities. The low activity of systems containing argan oil was explained by big differences between their chemical composition and viscosity.

Keywords: electrospinning; nanocoating; chitosan; vegetable oil; essential oil; cold-press oil; antimicrobial; antioxidant

---

1. Introduction

Due to increased demand for food packaging, efforts are being made to increase the storage- and shelf-life of food products by developing active antimicrobial packaging that does not require the use of synthetic additives [1]. The antibacterial agent may be incorporated into the packaging material [2] or coated onto it [3,4].

Various natural antibacterial agents have been encapsulated into different matrices to create coatings onto base packaging materials, such as cinnamaldehyde and carvacrol encapsulated in soy protein isolates [5]; nisin and natamycin in polyvinylchloride lacquer [6]; sorbic acid in polyvinyl acetate [7]; essential and vegetable oils, such as garlic oil and rosemary oleoresin, in soy protein isolate [8]; and oregano essential oil in ethylene-vinyl alcohol copolymer film [9].

Due to its well-known antibacterial activity against various pathogens such as *Klebsiella pneumoniae*, *Escherichia coli*, *Staphylococcus aureus*, and *Pseudomonas aeruginosa* [10], chitosan has been extensively studied to obtain antibacterial coatings [11]. Chitosan was also used to encapsulate another antibacterial agent, nisin [12], which, in combination with chitosan, could improve the antimicrobial activity of the coatings [13]. The biodegradable chitosan is non-toxic and confers antibiofilm, antioxidant, and antifungal properties to the coated layer [14].

Various procedures have been proposed for preparation of the antimicrobial coating: Spreading with a thin-layer chromatography applicator [5] or a lab bench coater [7]; using a bar coater [15] or a brush [16]; spraying (nebulization) [17]; lamination [18]; gravure printing [9]; dip-coating [19]; and translated, at industrial scale, by coupling with a wet-coating station and roll-to-roll system [20].

In this work, electrospinning/electrospraying is used as the coating technique. The properties of the coatings, containing the essential and cold-pressed oils, depend on the way they are distributed onto the base polymer (substrate) and also on the way they make contact with the food. In this respect, the nanoscale dimensions and high area-to-volume ratio of the electrospun/electrosprayed nanostructures (nanofibers and nanoparticles), help to improve the contact between the coatings and the packed food, while the encapsulation of the essential and vegetable oils, into the polymeric nanostructures, helps to maintain their antibacterial and antioxidant activities over a longer time period. Therefore, electrospinning/electrospraying is an effective and convenient method to obtain nanocoatings [21] due to several advantages offered by this technique, such as:

(a) By electrospinning, very thin nanofibers with a high porosity and area-to-volume ratio can be obtained.

(b) The thickness of the coated (deposited) layer can be easily controlled by changing the deposition time or the flow rate of the electrospinning solutions. Thus, it is possible to obtain a very thin coating (nanocoating) with a very small quantity of materials, which in many cases is enough to obtain the desired antibacterial and antioxidant [4] or antifungal activity [22]. In comparison, with the coating thickness obtained by other methods, such as solvent casting or dip-coating with special film applicators (3 μm [23], 2 ÷ 9 μm [6], 2 ÷ 3 μm [24]), the very thin layer (100 nm or even less) has the added advantage of requiring a low amount of the coating material.

(c) Plasticizers are often added to the chitosan coating layer [12] to overcome the brittleness exhibited during the package deformation, and to improve the flexibility and processability. As a coating layer, the nanofibers can exhibit better ductility than the corresponding bulk material [25] due to low-nanofiber crystallinity, which resulted from rapid solidification of the ultrafine electrospun jets [26]. Thus, it is expected that the nanosized electrospun coating layers will exhibit the needed flexibility, either in the form of nanofibers or nanoparticles, without the plasticizer addition.

In the present work, the polylactic acid (PLA) films were coated by electrospinning with bio-formulations containing chitosan and one of the two vegetable oils (i.e., clove and argan oils), which were chosen due to their different compositions. According to the literature data, acylglycerols constitute 99% of the argan oil composition, while the unsaponifiable matter contains tocopherols, squalene, sterols, and phenols as the main antioxidant compounds [27,28]. Eugenol (~80%) is the main volatile component of clove oil, which is responsible for its strong antioxidant and antimicrobial activities [29]. The different compositions of the two types of oils are expected to result in different antimicrobial and antioxidant activities of the coated PLA films, and also in different rheological properties, which, in turn, will determine differences in the morphologies of the electrospun coated layers at nanoscale level.

PLA films were coated with active bio-formulations with the intention to combine the biodegradability and mechanical properties of PLA substrate polymer with the antimicrobial and antioxidant activities and biological functions of clove and argan oils, in combination with the chitosan. The oils were encapsulated in chitosan by coaxial electrospinning so as to prolong their functions as antimicrobial and antioxidant agents over a long time, to prevent their degradation and loss during the processing conditions, and also to prevent their rapid diffusion and migration into the packaged food.

2. Materials and Methods

2.1. Materials

Chitosan with two different average molecular weights (MW) was purchased from Sigma-Aldrich (Schnelldorf, Germany): Highly viscous chitosan (Chit-H), MW = 310,000–375,000 g/mol, deacetylation degree DD ≈ 77% [21], with a dynamic viscosity in 1% acetic acid (20 °C) > 400 mPa·s; and low viscous chitosan (Chit-L), MW = 50,000–190,000, DD ≈ 87% [21], with viscosity in 1% acetic acid of 20–300 cP. The range values for MW (average molecular weight) were taken from producer specifications.

Glacial acetic acid (analytical purity) and chloroform were obtained from Chemical Company (Iasi, Romania).

Vegetable oils (argan and clove oils) with antioxidant and antimicrobial activities were chosen and were purchased from the Fares SA (Orastie, Romania) and Herbavit SA (Oradea, Romania) companies, respectively. The phenol content of the oils were 1.16 mg GAE/g DW (gallic acid equivalents/dry weight) for the clove oil and 0.02 mg GAE/g DW for the argan oil.

Two types of PLA films were coated: (a) Hot-pressed films obtained from PLA 2002D pellets (from NatureWorks LLC, Minnetonka, MN, USA), and (b) commercial NATIVIA® NTSS 40 μm PLA foils (from Taghleef Industries, Newark, DE, USA). The hot-pressed PLA substrates with a thickness of 0.30 ± 0.05 mm were obtained using a Carver press at 175 °C (2 min pre-melting and 2 min pressing at 240 bar). The NATIVIA® commercial PLA foils (Taghleef Industries, Newark, DE, USA) were used in order to test the applicability of this coating method for commercial products that currently exist in the market.

2.2. Preparation of the Nanostructured Coatings

The electrospinning system, used for coating the PLA foils (Figure 1), consisted of a high voltage supply (HV), a rotating metallic-plate collector, and two syringes with a coaxial needle oriented perpendicular to the metal plate. The PLA substrate film was placed on the metallic collector. The high direct voltage (0 to 30 kV) was applied between the metal plate and the metallic needle.

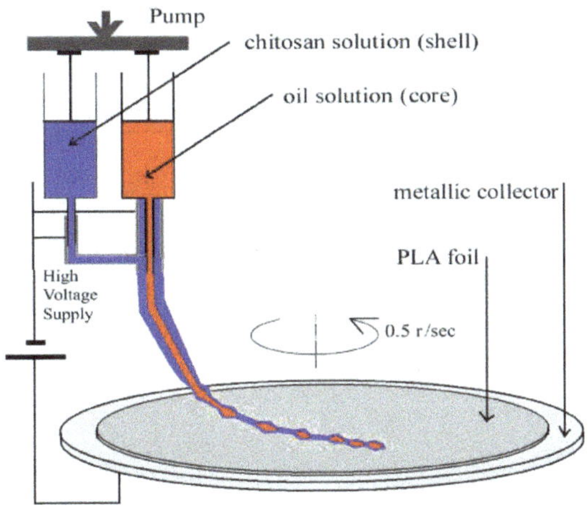

Figure 1. The coaxial electrospinning set-up used for coating the polylactic acid (PLA) foils. The PLA foil is placed directly onto the metallic collector.

The low surface energy of polymeric substrates often leads to poor adhesion of the coatings to the substrates. To improve the adhesion, the surface is usually activated or functionalized by corona [9] or plasma [30] treatment. Before getting coated with the chitosan–oil formulations by electrospinning, the PLA substrates were treated by exposure to cold plasma (nitrogen gas discharge atmosphere, 1.3 MHz frequency, 100 W power, 0.4 mbar pressure). Plasma treatment, and the consequent air exposure, leads to the implementation of some functionalities on the surface, such as groups and free radicals containing nitrogen and oxygen, which are able to interact with functional groups of the components of the nanocoating that were deposited by electrospinning [31].

After plasma treatment, the coaxial electrospinning was applied to obtain oil-loaded nanostructures, by encapsulating the clove or argan oil into the chitosan nanostructures, simultaneous with being coated onto the plasma-treated PLA substrates. The coaxial system consisted of chitosan supplied to the outer nozzle and oil solutions supplied to the inner nozzle.

For the hot-pressed PLA film substrates, the samples were designated the codes HC, HA, LC, and LA where H and L signifies high- and low-viscous chitosan, respectively, and C and A signifies clove and argan oil, respectively.

Correspondingly, for the commercial NATIVIA® NTSS substrate, the samples were labelled as HC-NATIVIA and HA-NATIVIA. Additionally, H-NATIVIA was designated the commercial NATIVIA® NTSS 40 μm, coated only with Chit-H, and PLA-NATIVIA was designated the uncoated PLA commercial foil (Table 1).

Table 1. PLA samples uncoated and coated with chitosan and clove oil or argan oil, using the electrospinning technique.

Code	Sample Description
	Uncoated Samples
PLA Hot-Pressed	Hot-pressed PLA films obtained from PLA 2002D pellets (NatureWorks LLC)
PLA-NATIVIA	Commercial NATIVIA® NTSS 40 μm PLA foils (from Taghleef Industries)
	Coated Samples: PLA Hot-Pressed Substrate
PLA-H	PLA hot-pressed film coated with Chit-H
HC	PLA hot-pressed film coated with Chit-H/clove oil
HA	PLA hot-pressed film coated with Chit-H/argan oil
LC	PLA hot-pressed film coated with Chit-L/clove oil
LA	PLA hot-pressed film coated with Chit-L/argan oil
	Coated Samples: PLA-NATIVIA Substrate
HC-NATIVIA	Commercial PLA-NATIVIA® NTSS 40 μm coated with Chit-H/clove oil
HA-NATIVIA	Commercial PLA-NATIVIA® NTSS 40 μm coated with Chit-H/argan oil
H-NATIVIA	Commercial PLA-NATIVIA® NTSS 40 μm coated with Chit-H

Both of the chitosan and oil solutions had the same concentration of 1.5 wt%. Chitosan was dissolved in 9:1 glacial acetic acid and water, respectively, argan oil was dissolved in chloroform, and clove oil was dissolved in glacial acetic acid. Electrospinning parameters (for all samples) were 1 kV/cm electrical field strength, 1.2 μL/min feed rate (for both the inner and outer needle), and 30 min deposition time. To be sure of complete solvent removal, the electrospun nanostructures were placed for two days in a vacuum desiccator (Binder GmbH, Tuttlingen, Germany). The absence of the acetic acid in the deposited mesh in was confirmed by IR spectroscopy (using a Bruker VERTEX 70 spectrometer, Ettlingen, Germany).

2.3. Investigation Methods

Scanning electron microscopy (SEM/EDX) analyses were carried out using a QUANTA 200 scanning electronic microscope (Thermo Fisher Scientific, Waltham, MA, USA) with an integrated (EDX system, (Thermo Fisher Scientific Waltham, Waltham, MA, USA), and a GENESIS XM 2i EDAX with SUTW detector (FEI Company, Eindhoven, The Netherlands).

Atomic force microscopy (AFM) investigations were done with a Solver-Pro-M type instrument (Solver Inc., Moscow, Russia) under ambient conditions, using standard tips of Si_3N_4 (10 nm curvature radius). The root mean square roughness was calculated from the total image sample after a second-order flatness treatment of the raw data. NT-MDT Nova v.1.26.0.1443 software (NT-MDT Spectrum Instruments, Moscow, Russia) was used for the acquisition and analysis of the images.

Transmission electron microscope (TEM) investigations were conducted with a Hitachi High-Tech HT7700 microscope (Hitachi High-Tech GLOBAL, Tokyo, Japan) high contrast mode at 100 kV accelerating voltage) on electrospun grids (300 mesh holey carbon coated copper grids). The grids were placed onto the PLA substrate during the electrospinning process.

Antimicrobial activity was determined, using two kinds of tests, in order to verify the differences between the antimicrobial activities of the clove and argan oils.

In the first test the suspension and culture medium were prepared as is described in the ISO 22196 (ISO 22196: Plastics—measurement of antibacterial activity on plastics and other non-porous surfaces (2011)). The test strains were incubated for 24 h at 37 °C. The antimicrobial testing method consisted of placing a drop of a suspension, of either *Escherichia coli* (ATCC 8739) or *Staphylococcus aureus* (ATCC 6538), onto the surface of the tested material (with a surface area of 5 cm × 5 cm). Each tested sample was prepared in a separate sterile vessel, with the test surface facing upwards. Then 0.4 mL of the test inoculum was dropped onto the tested surface. The test inoculum was covered with a piece of film (with a surface area 4 cm × 4 cm), which did not have anti-bacterial properties (neutral film), and was pressed mildly onto the film so that the test inoculum spread to the periphery. Half of the untreated specimens were processed instantly after inoculation, by adding 10 mL of Soybean Casein Lecithin Polysorbate Medium (SCDLP) broth (ISO 22196) to the vessel containing the test specimen. After this, the sum of viable bacterial cells was assessed. This obtained amount was used to evaluate the recovery rate of the bacteria from the investigated test specimens. After the inoculation of the specimen and the application of the cover film, the lid of the vessel was changed. After a 24 h incubation period, at 35 °C, the bacterial suspension was released from the coverslip-test sample and the sum of the viable bacterial cells that had survived was determined. The log values were the mean of 3 parallel samples (for each film type). The log reduction was the difference between the growth obtained with a neutral film (Stomacher bag) and the growth obtained with an active treated film. Sometimes a reduction, rather than growth, was observed with the neutral film, because of the low pH of the free lactic acid containing the PLA.

The second test was performed as it is described in the ISO 16649-2/2007: Microbiology of alimentary and animal products. There are several phases in the procedure for testing antimicrobial activity against *Escherichia coli*, *Salmonella typhymurium*, and *Listeria monocytogenes*: Sterilization of samples; inoculation with ATCC culture bacteria; 24 and 48 h (at 44 °C) inoculation and incubation, respectively; and the identification of target germs. Sterilization of the samples was performed for 20 min at 0.5 bar and at 110 °C, in an autoclave. Formation of the ATCC cultures was performed by seeding the average pre-enrichment, followed by incubation for 24 h at 37 °C; counting the number of colonies in the 0.1 mL culture after the culture medium separation; and then seeding 0.1 mL of bacterial culture ATCC, using sterile swab samples taken from the surface. The following standardized bacteriological procedures were employed in order to identify the target germs: For *Escherichia coli*, the SR ISO 16649, horizontal method for beta-glucuronidase-positive *Escherichia coli* quantification, was used. Then colonies were counted at 44 °C using 5-bromo-4-chloro-3-indolyl beta-D-glucuronide according to Minerals Modified Glutamate Broth (Cat. 1365), which produce blue or green–blue colonies on agar glucuronide; for *Listeria monocytogenes*, SR EN ISO 11290 was used; and for *Salmonella* sp., SR EN ISO 6579/2003/AC/2004/AC/2006, Amd.1:2007 was used.

The antioxidant activity of PLA (NATIVIA) coated films was measured by DPPH (1,1-diphenyl-2-picryl-hydrazyl) free radical assay [32]. The samples (coated films) were immersed in the DPPH solution and the decrease of the absorbance at 517 nm, which reflected the amount of DPPH radicals in the solution, was measured by means of a Cary 60-UV-Vis Spectrophotometer (Mettler Toledo, Columbus, OH, USA). The measurements were done at different immersion times (after 0, 15, 40 and 60 h immersion time). The DPPH solution (in the absence of the coated films) was used as a control. The radical scavenging activity (RSA) of the samples was expressed by the relative decrease in DPPH absorbance at 517 nm (RSA = $1 - A_{SAMPLE}(t)/A_{DPPH}(t)$ where $A_{DPPH}(t)$ and $A_{SAMPLE}(t)$ are the absorbance of the control and sample at the time of measurement). For testing, a 4 cm^2 area of the coated film was cut into pieces and immersed in 1.5 mL of DPPH solution in methanol with a concentration of 8 micrograms/mL. Before immersion and testing, the samples were stored for seven days at room temperature. Between measurements the solutions were stored in the dark at 24 °C.

3. Results and Discussion

3.1. Transmission Electron Microscope (TEM) Results

The TEM images show a hybrid fiber and particle morphology for the HA and HC samples (Figure 2a,b), and particle morphology for the LA and LC samples (Figure 2c,d).

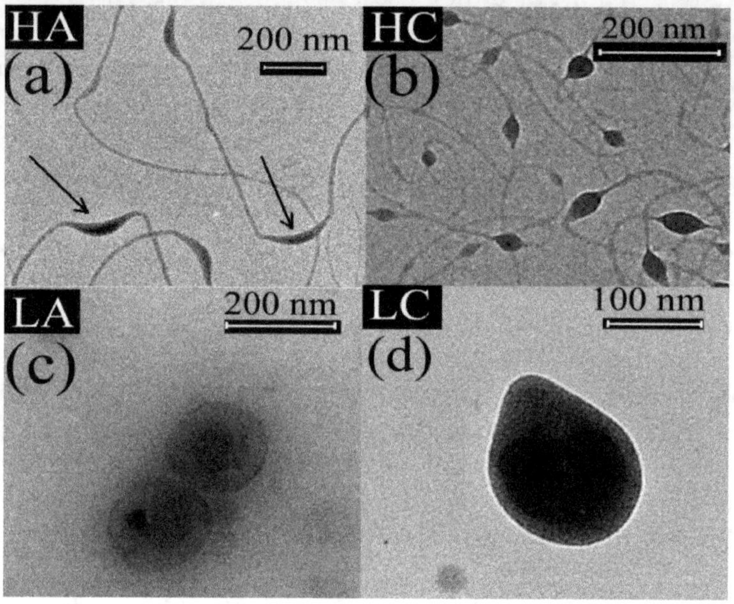

Figure 2. Transmission electron microscope (TEM) images of the studied samples: (**a**) HA; (**b**) HC; (**c**) LA; and (**d**) LC samples.

It is well known that sufficient chain entanglements should occur in the electrospun solution [33] in order to obtain nanofibers by electrospinning. Thus, the presence of fibers for the HA and HC samples could be explained by the high MW of chitosan, which resulted in better macromolecular entanglements in the electrospun solution. Due to the low MW, the chain entanglements in LA and LC samples were not enough [34] to sustain the fiber formation, which resulted in the spraying of beads (which ultimately may form a rather compact film). Thus, the only spinnable (nanofiber mesh forming) samples are those containing Chit-H.

The TEM images also show two different encapsulation types, of the vegetable oils into the chitosan, depending on the MW of the chitosan used for the shell solution: The LA and LC samples (Figure 2c,d), with low-MW chitosan, had a particle morphology whereby isolated oil beads were encapsulated by chitosan; whereas the HA and HC samples (Figure 2a,b), with high-MW chitosan, had oil encapsulated as scattered beads along the main chitosan fiber.

The beads-into-fiber structure of the HA and HC samples can be explained by the different molecular structure and viscosities of the shell and core solutions used in the coaxial system, that is, a highly viscous (>400 mPa·s) electrospinnable macromolecular (Chit-H) shell solution and a sprayable oil core solution with very low viscosity (1–2 mPa·s). It is known that the similarity between the rheological properties of the core and shell solutions control the viscous dragging exerted by the shell solution on the core solution [35]. If the viscosity of the core solution is too low, compared with the shell solution, the viscous drag exerted by the shell solution is unable to overcome the cohesive forces of the core solution [36]. This was the case for the solutions used for the HA and HC samples. The entanglements provided by the highly viscous shell Chit-H solution were enough to prevent fiber rupture, which can be caused by low viscosity of the core oil solution, but, due to the low viscosity, the core oil solution was unable to follow the stretching of the macromolecular shell solution [37], thus taking the form of beads into the chitosan fibers [38]. It can also be observed that the HA and HC samples had a similar fiber diameter (10–20 nm), which supports the assertion that the HA and HC fibers were mostly made of Chit-H and the oil was encapsulated into the beads along the fibers. Similar morphologies, with beads-into-fiber structures, were obtained by other authors for proteins [39] or dexamethasone [40] encapsulated along the polymeric nanofibers as beads, which can act as depots for sustained drug release [40]. Other authors reported that coaxial electrospining employed to encapsulate sunflower oil into PEO produced beaded fibers, with the oil encapsulated into the beads [36]. The authors also reported that the oil was confined or encapsulated solely inside these beads [36]. This is in good agreement with our findings, that is, the fibers contained mainly chitosan and the oil was distributed as beads along the main chitosan fiber.

From the TEM images, another observation can be made regarding the morphology of the HA and HC samples: The beads distributed along the Chit-H fibers were much more elongated for the HA sample (Figure 2a) compared with the HC sample (Figure 2b). Because a higher viscosity of the core fluid increases the viscous drag, exerted by the shell solution onto the low molecular core fluid [36], the more elongated argan oil beads can be explained by the higher viscosity of the argan oil compared with the clove oil. Considering the rapid evaporation of the solvent during the jet stretching [41], it is more relevant to compare the viscosities of pure oils instead of the oil solutions. In our case, argan oil was ten times more viscous (~0.6 poise [42]) than the clove oil (~0.06 poise [43]), which explains the more elongated argan oil beads.

3.2. Atomic Force Microscopy (AFM) Results

The particle and fiber diameters evaluated from AFM images (Figure 3) are all in the nanometer scale with similar values as those revealed by TEM images (Table 2).

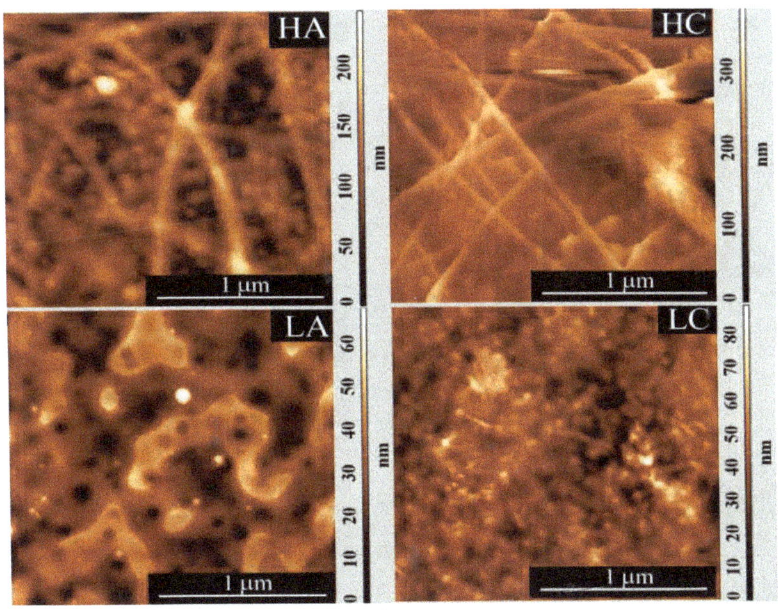

Figure 3. Atomic force microscopy (AFM) images of the studied samples: (**a**) HA; (**b**) HC; (**c**) LA; and (**d**) LC samples.

Table 2. Low and high values (range) of the particle and fiber diameter and height distributions evaluated from AFM images.

Sample	Range of the Particle Diameter Distribution (nm)	Range of the Fiber Diameter Distribution (nm)	Range of the AFM Height Distribution (nm)
HA	20–140	20–140	150–500
HC	10–80	30–140	150–600
LA	10–60	-	10–55
LC	20–80	-	10–80

It is interesting to compare the width (range) of the particle and fiber diameter distributions, determined from the AFM images, with the width of the AFM height distribution (Table 2, Figures 4 and 5). The width of the AFM height distribution was similar to the width of the particle AFM diameter distributions for the LA and LC samples, but was much wider than the particle and fiber diameter distributions for the HA and HC samples.

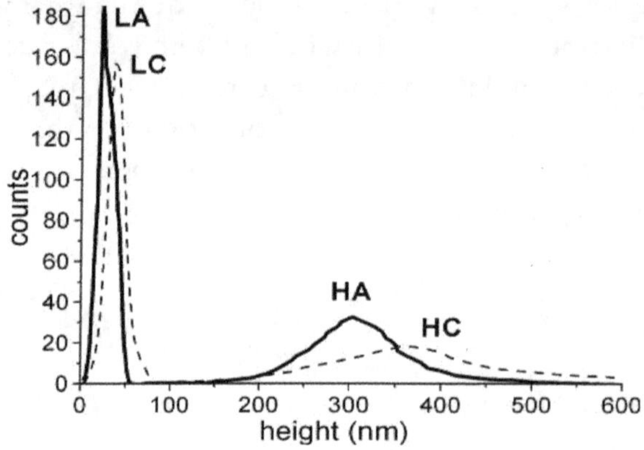

Figure 4. AFM height distributions for the studied samples: (**a**) HA; (**b**) HC; (**c**) LA; and (**d**) LC samples.

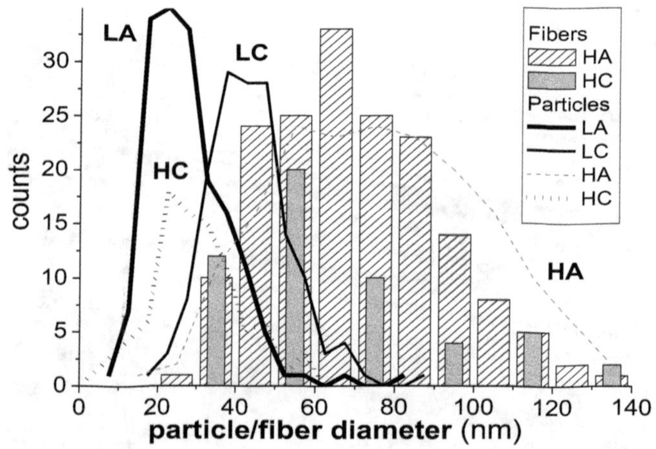

Figure 5. Distributions of the particle (lines) and fiber (bars) diameters, in the coated layers, determined from AFM.

This is because, compared with the HA and HC samples, the LA and LC samples had a more compact particle arrangement, with smaller voids between them, which narrowed the height distribution. This was also shown by other authors [44,45], who correlated the roughness (or the width of the height distribution) of nanoparticle layers with the experimentally measured nanoparticle sizes, and also with the way the particles are packed [44]. For a closely (or orderly) packed spherical structure, with spheres tangent to one another (particle-near-particle), the roughness is low and the AFM height distribution approaches the particle diameter distribution (roughness approaches the average diameter of the particles). But when the same particles are randomly packed in a more porous arrangement, the roughness increases [45]. Thus, it can be concluded that, for the LA and LC samples, the structures were more compact with lower porosity (smaller voids between particles) and, consequently, with narrower height distribution than the HA and HC samples.

3.3. Scanning Electron Microscopy (SEM) Results

The SEM images for the HC-NATIVIA coated sample show the presence of nanofibers with diameters around 60–70 nm (Figure 6a). In contrast with the NATIVIA® PLA commercial foils, the hot-pressed PLA films experienced sample damage due to high electron acceleration during the SEM analysis, which lowered the resolution of the SEM images. Based on the presence of nitrogen, the SEM-EDX analysis was used to investigate the presence of chitosan on the PLA substrate. The mapping of nitrogen distribution, with an average nitrogen content of 5–6%, revealed a uniform coating on the surface at a micrometric scale.

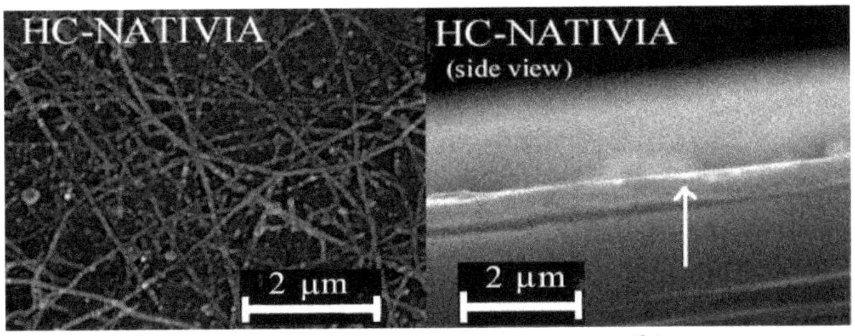

(**a**) (**b**)

Figure 6. Scanning electron microscopy (SEM) images of the HC-NATIVIA sample: (**a**) Morphology of the coated surface; and (**b**) side view showing the coated layer.

Because of their smooth and glossy finish, the PLA-NATIVIA® films were used to evidence the coated layer. The side view of the coated PLA-NATIVIA® foil (Figure 6b) shows the coating layer with a thickness of 100 ± 20 nm. Considering that the entire output of the electrospinning syringe was deposited onto the PLA-NATIVIA® foil (10 μg/cm^2), based on chitosan and oil density, we would need to have the coated layer at a thickness of around 250 nm for maximum compactness of the chitosan (with no porosity of the coated layer). These results, therefore, show that only a part of the syringe output was deposited onto the PLA substrate (about 4 μg/cm^2 assuming no porosity of the chitosan mesh). Considering porosity (void percent) of ~80% for the chitosan mesh [46], results suggest that the amount of the antimicrobial oil deposited onto the PLA substrate was at least 0.4 μg/cm^2. For meat packaging applications, the contact between the antibacterial coated layer and the meat packaged is realized by means of the moist layer, which exists between the packaging foil and the meat. If we consider a moist layer with a thickness of approximately 1 mm is in contact with the meat, the oil concentration throughout the moist layer would be ~4 μg/mL, which is well above the minimum inhibitory concentration of clove oil against *E. coli* (0.1 mg/mL) [47].

Therefore, nanostructured chitosan morphologies were obtained by electrospinning, with clove and argan oils encapsulated as beads into the chitosan nanofibers or nanoparticles. The nanofibres were predominant in the samples obtained from the electro-spinnable Chit-H. Nanoparticles were obtained when Chit-L was used.

3.4. Antibacterial Tests

The results of the antibacterial tests are presented in Table 3 (ISO 22196) and the Table 4 (ISO 16649-2/2007).

Table 3. Results of the antibacterial tests performed according to ISO 22196:2007 (E) (only the upper face was treated).

Sample	Log Reduction of the Number of Viable Bacterial Cells	
	E. coli	*S. Aureus*
	PLA Hot-Pressed	
PLA (uncoated)	invalid	−0.8
HC	invalid	0.8
HA	invalid	0.7
	PLA-NATIVIA® NTSS 40 μm	
NATIVIA (uncoated)	1.1	−0.7
H-NATIVIA	1.8	0.8
HC-NATIVIA	2.2	1.2
HA-NATIVIA	0.8	0

Table 4. Results of the antibacterial tests performed according to ISO 16649-2/2007.

Sample	*Escherichia Coli*		*Listeria Monocytogenes*		*Salmonella Typhymurium*	
	24 h	**48 h**	**24 h**	**48 h**	**24 h**	**48 h**
			Inhibition (%)			
			PLA Hot-Pressed			
PLA -H	26	42	29	32	22	29
LA	43	70	49	54	37	49
LC	68	80	37	63	100	100
HA	50	77	60	**100**	51	78
HC	80	91	77	**100**	100	100
			PLA NATIVIA® NTSS 40 μm [1]			
H-NATIVIA	10	53	16	58	35	71
HA- NATIVIA	49	82	47	**100**	55	94
HC- NATIVIA	53	78	53	**100**	65	90

[1] As the antibacterial tests for the PLA hot-pressed samples evidenced the highest antimicrobial efficiency for the samples containing Chit-H, the commercial NATIVIA® foils were coated only with formulations containing Chit-H.

The results of the two antibacterial tests are presented as log reduction of the viable cells (first test—Table 3) and as inhibition percent (second test—Table 4). Regarding the efficiency of the coating procedure, it can be noted that there was higher antibacterial activity in the coated versus uncoated samples (Table 3), and also higher antibacterial activity on the chitosan–oil coatings versus chitosan coatings (Tables 3 and 4). Regarding the two types of chitosan and oils used for the coatings, the second antibacterial tests also showed higher antibacterial effect for Chit-H versus Chit-L, and for clove oil versus argan oil (Table 4).

More specifically, for both Chit-H–argan oil and Chit-H–clove oil combinations, the results obtained by the first test (Table 3) evidenced improved antibacterial activity of the coated PLA films compared with the uncoated films (for both NATIVIA and hot-pressed PLA films). Even the HA-NATIVIA sample (with 0 log reduction) had higher antibacterial activity against *S. aureus* (Table 3) compared with the uncoated NATIVIA film (with negative log reduction, i.e., cell proliferation). In the case of *E. coli* bacteria, antibacterial effect was observed even for the uncoated NATIVIA films because of the decreased pH by the free lactic acid, which was present in the film (Table 3).

Also, both tests showed that the combination Chit-H–oil had higher antibacterial effect than Chit-H coated alone (Table 3 and 4) and, therefore, indicates that improved antibacterial activity could be obtained by coating the PLA films with chitosan. In addition, the coaxial encapsulation of the clove and argan oils, electrospun together with the chitosan, led to enhanced antimicrobial activity for both NATIVIA and hot-pressed PLA films.

The second antibacterial test also showed higher antibacterial effect for Chit-H versus Chit-L, and for clove oil versus argan oil (Table 4).

The different antibacterial activities of the two types of oils can be explained by the different amounts of the compounds responsible for the antimicrobial activity in the two oils (primarily phenolic compounds [48–50]). Eugenol, which increases antimicrobial and antifungal activity, is the main component (~80%) of clove oil. On the contrary, the phenolic compounds in argan oil are found in much lower amount (<1%), which explains the lower antibacterial and antifungal activity of argan oil compared with clove oil [51,52].

The higher antibacterial activity of Chit-H compared with Chit-L cannot be explained by its higher MW or by its lower deacetylation degree (DD) because many authors reported an opposite result (i.e., lower antibacterial activity for higher MW and lower DD), as it is shown further on.

Indeed, it is reported in literature that chitosan had lower antibacterial activity for higher MWs [53,54], for both gram-positive bacteria (GP-b) [55] and gram-negative bacteria (GN-b) [56]. This effect is reported to take place for MWs higher than 100 kDa [57], and some authors even report low activity for MWs > 30 kDa [58]. As the MW of the chitosan types used in this work was higher than this threshold (50–200 kDa for Chit-L and 300–370 kDa for Chit-H) [57,58], we conclude that the higher antibacterial activity of Chit-H (for the same type of oil) cannot be attributed to its higher MW.

Moreover, the different DD for the two types of chitosan cannot explain the higher antibacterial activity for the Chit-H (for the same type of oil). There are many papers reporting higher antibacterial activity for higher DD of chitosan [59–64], due to the disruption of the cell wall which is caused by the higher number of free amino groups present in chitosan with higher DD [65]. Our results, showing lower antibacterial activity for the chitosan with higher DD (DD $\approx$ 87% for Chit-L compared to DD $\approx$ 77% for Chit-H) are opposed to the reports presented above, and also to another report which showed higher antibacterial activity for chitosan with higher DD and lower MW [66].

The higher antibacterial activity of Chit-H compared with Chit-L can be explained by the larger surface area of the rougher nanoscaled morphology of the coating layer for Chit-H samples compared with the Chit-L samples (according to SEM, TEM and AFM results).

In other papers it was reported that the chitosan nanoparticles (MW of 680 kDa) exhibit higher antibacterial activity against *E. coli* than the corresponding microparticles obtained from the same chitosan [56], because of the higher surface charge [67] and the larger surface area of the nanosized chitosan nanoparticles, which probably tightly adsorbed onto the bacteria cells surface, disrupting the normal functions of the membrane [56].

There are also studies showing that the bacterial attachment decreases when the exposed substrate surfaces (such as polydopamine [68]) have higher roughness [69,70], for both gram-positive bacteria (GP-b) and gram-negative bacteria (GN-b) [71]. Some authors have found that GP-b (which has a more rigid membrane than GN-b) are more sensitive to roughness than GN-b, due to the surface nano-irregularities, which limit the number of anchoring points for GP-b bacteria. In this way, the surface area in contact with the membrane is reduced, which in turn reduces the adhesion of GP-b to the rougher surface [72,73]. Additionally, it was hypothesized that air entrapped by the topographical features inhibited contact between *S. aureus* and the substrate [74]. Our results agree with these reports. Furthermore, among Chit-H samples, the GP-b *L. monocytogenes* (for the samples HA, HA-NATIVIA, and HC-NATIVIA after 48 h) had maximum inhibition. This confirms the above-mentioned results, which state that the GP-b are more sensitive to roughness, and make us conclude that the higher antibacterial activity of the Chit-H samples was attributed to the higher roughness of the coating.

3.5. Antioxidant Activity

It is known that it is advisable to perform the antioxidant tests with fresh DPPH solutions. However, if prolonged experiments are required (as are further presented in this section), the DPPH test can also be performed with solutions prepared tens of hours before testing without a substantial reduction in the activity of DPPH [75]. In order to test the stability of our DPPH solution, the absorbance of the DPPH (control) solution was measured over time. After 70 h (solution stored in the dark) the absorbance of the DPPH solution was about 80% of the initial value (with an almost linear decrease). This result is in line with other experimental results of other authors, which show that, after 24 h post preparation, the free radical activity of 0.1 mM DPPH methanolic solution was still very similar to that of the fresh preparation (about 20% activity was lost after about 120 h) [75].

It has been previously shown that it is possible to evaluate the antioxidant capacity of various compounds without preliminary solvent extraction. The reaction between the radical and the antioxidant occurs at the interface, when they come into contact, or by liquid diffusing into the interior of the reacting solid [76,77]. Using this method, it was possible to measure the antioxidant activity of solid matrices containing polyphenols [78], which are known to be found in argan [79] and clove oils [80]. In our study a similar approach was used: A small area of the coated films was cut into pieces and immersed into the methanolic DPPH radical solution. The antioxidant activity of the coated films was expressed by the relative decrease in DPPH absorbance at 517 nm:

$$\text{Relative Scavenging Activity (RSA)} = 1 - A_{SAMPLE}(t)/A_{DPPH}(t) \quad (1)$$

In order to eliminate the effect of decreasing DPPH absorption over time, the control was considered the absorbance of the control DPPH solution at the time of spectrophotometric analysis ($A_{DPPH}(t)$).

In the first test, a 1 cm^2 area, from the HC-NATIVIA and HA-NATIVIA freshly coated films, was cut into pieces and immersed for 24 h in 1.5 mL of DPPH solution in methanol (8 μg/mL). This first test evidenced the higher antioxidant activity of the samples containing clove oil. The RSA values were 0.29 for the HC-NATIVIA sample and 0.17 for the HA-NATIVIA sample. These results were in good agreement with the higher phenolic content in clove essential oil compared with argan oil. Due to high eugenol content, the concentration required to scavenge 50% of 2,2′-azino-bis 3-ethylbenzthiazoline-6-sulfonic acid (ABTS) free radicals (half inhibitory concentration) was much higher for argan oil (~5800 μg/mL) compared with clove oil (~8 μg/mL) [81]. At the same time, taking into consideration the low volatility and diffusivity of the components of argan oil (fatty acids), a very low antioxidant and antimicrobial activity of argan oil was expected.

In a second test, the sample with higher antioxidant activity (HC-NATIVIA) was chosen to measure the antioxidant activity over the immersion time. Before testing, the samples were stored for seven days at room temperature, in order to check the encapsulation and persistency of the volatile clove oil in the samples. The results were compared with the corresponding uncoated (NATIVIA) and chitosan coated film (H-NATIVIA). In this second test, a higher film area (4 cm^2) was cut and immersed into the same amount of radical solution with the same concentration. It can be observed that all the three samples had antioxidant activity (Figure 7) that was maintained over a fairly long time (tens of immersion hours) as a result of the clove oil encapsulation into the chitosan fibers. The concentration of biphenyl radicals continued to decrease (RSA increased) in the solution containing the immersed film relative to the control sample, even after 70 h of immersion. The antioxidant activity of the uncoated NATIVIA, which was lowest among the tested samples, was due to the residual lactic acid in the films.

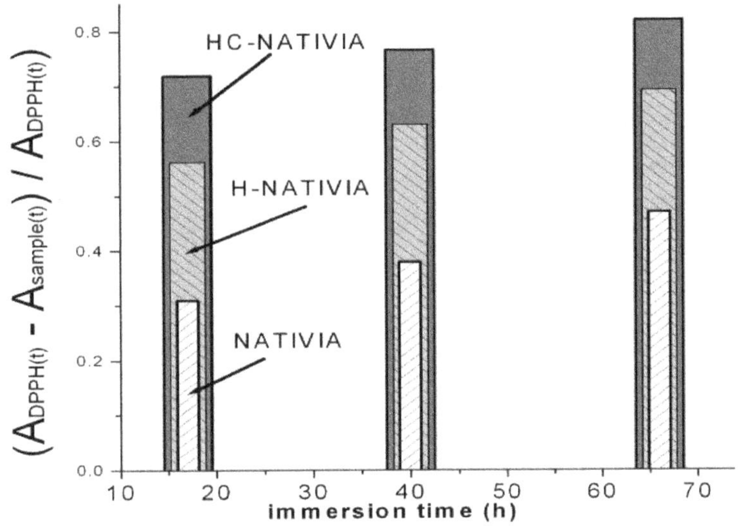

Figure 7. Antioxidant activity versus immersion time for the NATIVIA® films coated with Chit-H–clove oil (HC-NATIVIA), Chit-H (H-NATIVIA), and the uncoated film (NATIVIA).

Of these three samples, PLA-NATIVIA® films (HC-NATIVIA) coated with chitosan–clove oil had the highest antioxidant activity. In spite of the high volatility of the clove oil [81,82], the antioxidant activity of the HC-NATIVIA was maintained even after seven days of shelf storage. This shows that clove oil can be efficiently immobilized into the coating layer by encapsulation in chitosan. Therefore, the antioxidant tests confirmed the TEM results and the presence of the encapsulated clove oil into the chitosan fibers, which results in a combined effect obtained by simultaneous use of chitosan and clove oil.

4. Conclusions

The novelty of this study lies in providing a new and effective approach of nanocoating polymeric films with chitosan by coaxial electrospinning, with the advantage of simultaneous encapsulation of active vegetable and/or essential oils into the chitosan.

Polylactic acid (PLA) films were coated by coaxial electrospinning with formulations containing essential or vegetable oils (clove and argan oils, respectively) and encapsulated into chitosan in order to obtain biodegradable packaging materials with antimicrobial and antioxidant properties. When chitosan with high MW (Chit-H) was used, the coaxial electrospinning produced beaded chitosan nanofibers, with the oil distributed and encapsulated along the main chitosan fiber. Chitosan nanoparticles encapsulating the oil were obtained with low-MW chitosan (Chit-L). The roughness of the coating layer was higher for the samples prepared with Chit-H compared with samples prepared with Chit-L. The PLA films coated with chitosan–oil formulations had higher antibacterial activity than the films coated only with chitosan. The clove oil had higher antibacterial activity than the argan oil, for both types of chitosan, because of its higher phenolic content. Chit-H samples had higher antibacterial activity compared with Chit-L, for both types of oil, due to the higher specific surface area of the rougher nanofibrous morphology of the coating layer. As was expected, due to the phenolic content of clove oil, the chitosan–clove oil combination had higher antioxidant activity compared to the simple chitosan coating. Considering the volatility of the clove oil, its immobilization into the coating layer can be realized by encapsulation in chitosan. The TEM results and the antioxidant tests also confirm the encapsulation of the oil into the chitosan.

Author Contributions: In this study, the concepts and designs of the experiments were supervised by B.S.M. The manuscript text and the analysis of results were performed by C.V., B.S.M., and J.T.R. The experimental results were examined by B.S.M., L.S., A.-L.V., G.E.H., C.V., G.M.P., M.S., and J.T.R. All authors read and approved the final manuscript.

Acknowledgments: This research was funded by Romanian-EEA Research Program operated by MEN under the EEA Financial Mechanism 2009–2014 grant number (project contract) 1SEE/2014. The PLA NATIVIA® commercial films were kindly supplied by Taghleef Industries.

References

1. Ramos, M.; Jiménez, A.; Peltzer, M.; Garrigós, M.C. Characterization and antimicrobial activity studies of polypropylene films with carvacrol and thymol for active packaging. *J. Food Eng.* **2012**, *109*, 513–519. [CrossRef]
2. Rapa, M.; Mitelut, A.C.; Tanase, E.E.; Grosu, E.; Popescu, P.; Popa, M.E.; Rosnes, J.T.; Sivertsvik, M.; Darie-Nita, R.N.; Vasile, C. Influence of chitosan on mechanical, thermal, barrier and antimicrobial properties of PLA-biocomposites for food packaging. *Compos. Part B: Eng.* **2016**, *102*, 112–121. [CrossRef]
3. Munteanu, B.S.; Pâslaru, E.; Zemljic, L.F.; Sdrobis, A.; Pricope, G.M.; Vasile, C. Chitosan coatings applied to polyethylene surface to obtain food-packaging materials. *Cellul. Chem. Technol.* **2014**, *48*, 565–575.
4. Vasile, C.; Darie, R.N.; Sdrobis, A.; Pâslaru, E.; Pricope, G.; Baklavaridis, A.; Munteanu, S.B.; Zuburtikudis, I. Effectiveness of chitosan as antimicrobial agent in LDPE/CS composite films as minced poultry meat packaging materials. *Cellul. Chem. Technol.* **2014**, *48*, 325–336.
5. Ben Arfa, A.; Preziosi-Belloy, L.; Chalier, P.; Gontard, N. Antimicrobial paper based on a soy protein isolate or modified starch coating including carvacrol and cinnamaldehyde. *J. Agric. Food Chem.* **2007**, *55*, 2155–2162. [CrossRef] [PubMed]
6. Hanušová, K.; Dobiáš, J.; Klaudisová, K. Effect of Packaging Films releasing antimicrobial agents on stability of food products. *Czech J. Food Sci.* **2009**, *27*, S347–S349. [CrossRef]
7. Hauser, C.; Wunderlich, J. Antimicrobial packaging films with a sorbic acid based coating. *Procedia Food Sci.* **2011**, *1*, 197–202. [CrossRef]
8. Gamage, G.R.; Park, H.J.; Kim, K.M. Effectiveness of antimicrobial coated oriented polypropylene/polyethylene films in sprout packaging. *Food Res. Int.* **2009**, *42*, 832–839. [CrossRef]
9. Muriel-Galet, V.; Cerisuelo, J.P.; López-Carballo, G.; Aucejo, S.; Gavara, R.; Hernández-Muñoz, P. Evaluation of EVOH-coated PP films with oregano essential oil and citral to improve the shelf-life of packaged salad. *Food Control* **2013**, *30*, 137–143. [CrossRef]
10. Divya, K.; Vijayan, S.; George, T.K.; Jisha, M.S. Antimicrobial properties of chitosan nanoparticles: Mode of action and factors affecting activity. *Fibers Polym.* **2017**, *18*, 221–230. [CrossRef]
11. Gallstedt, M.; Brottman, A.; Hedenqvist, M.S. Packaging-related properties of protein- and chitosan-coated paper. *Packag. Technol. Sci.* **2005**, *18*, 161–170. [CrossRef]
12. Hong, S.I.; Lee, J.W.; Son, S.M. Properties of polysaccharide-coated polypropylene films as affected by biopolymer and plasticizer types. *Packag. Technol. Sci.* **2005**, *18*, 1–9. [CrossRef]
13. Ho-Lee, C.; An, D.S.; Jin-Park, H.; Lee, D.S. Wide-spectrum antimicrobial packaging materials incorporating nisin and chitosan in the coating. *Packag. Technol. Sci.* **2003**, *16*, 99–106. [CrossRef]
14. Tang, Z.X.; Qian, J.Q.; Shi, L.E. Preparation of chitosan nanoparticles as carrier for immobilized enzyme. *Appl. Biochem. Biotechnol.* **2007**, *136*, 77–96. [CrossRef] [PubMed]
15. Muriel-Galet, V.; Cerisuelo, J.P.; López-Carballo, G.; Lara, M.; Gavara, R.; Hernández-Muñoz, P. Development of antimicrobial films for microbiological control of packaged salad. *Int. J. Food Microbiol.* **2012**, *157*, 195–201. [CrossRef] [PubMed]
16. Bolumar, T.; Andersen, M.L.; Orlien, V. Antioxidant active packaging for chicken meat processed by high pressure treatment. *Food Chem.* **2011**, *129*, 1406–1412. [CrossRef]
17. Contini, C.; Katsikogianni, M.G.; O'Neill, F.T.; O'Sullivan, M.; Dowling, D.P.; Monahana, F.J. Development of active packaging containing natural antioxidants. *Procedia Food Sci.* **2011**, *1*, 224–228. [CrossRef]
18. Hoagland, P.D. Biodegradable Laminated Films Fabricated from Pectin and Chitosan. U.S. Patent 5,919,574, 6 July 1999.
19. Stoleru-Paslaru, E.; Tsekov, Y.; Kotsilkova, R.; Ivanov, E.; Vasile, C. Mechanical behavior at nanoscale of chitosan-coated PE surface. *J. Appl. Polym. Sci.* **2015**, *132*, 42344. [CrossRef]

20. Riccardi, C.; Zanini, S.; Tassetti, D. A Polymeric Film Coating Method on A Substrate by Depositing and Subsequently Polymerizing a Monomeric Composition by Plasma Treatment. U.S. Patent US20160122585A1, 5 May 2018.
21. Stoleru, E.; Dumitriu, R.P.; Munteanu, B.S.; Zaharescu, T.; Tanase, E.E.; Mitelut, A.; Ailiesei, G.L.; Vasile, C. Novel procedure to enhance PLA surface properties by chitosan irreversible immobilization. *Appl. Surf. Sci.* **2016**, *367*, 407–417. [CrossRef]
22. Munteanu, B.S. *Polymeric Nanomaterials in Nanotherapeutics*; Vasile, C., Ed.; Elsevier: Amsterdam, The Netherlands, 2018.
23. Cerisuelo, J.P.; Muriel-Galet, V.; Bermúdez, J.M.; Aucejo, S.; Catalá, R.; Gavara, R.; Hernández-Muñoz, P. Mathematical model to describe the release of an antimicrobial agent from an active package constituted by carvacrol in a hydrophilic EVOH coating on a PP film. *J. Food Eng.* **2012**, *110*, 26–37. [CrossRef]
24. Tihminlioglu, F.; Atik, İ.D.; Özen, B. Effect of corn-zein coating on the mechanical properties of polypropylene packaging films. *J. Appl. Polym. Sci.* **2011**, *119*, 235–241. [CrossRef]
25. Naraghi, M.; Arshad, S.N.; Chasiotis, I. Molecular orientation and mechanical property size effects in electrospun polyacrylonitrile nanofibers. *Polymer* **2011**, *52*, 1612–1618. [CrossRef]
26. Papkov, D.; Zou, Y.; Andalib, M.N.; Goponenko, A.; Cheng, S.Z.D.; Dzenis, Y.A. Simultaneously strong and tough ultrafine continuous nanofibers. *ACS Nano* **2013**, *7*, 3324–3331. [CrossRef] [PubMed]
27. Berrougui, H.; Cloutier, M.; Isabelle, M.; Khalil, A. Phenolic-extract from argan oil (*Argania spinosa* L.) inhibits human low-density lipoprotein (LDL) oxidation and enhances cholesterol efflux from human THP-1 macrophages. *Atherosclerosis* **2006**, *184*, 389–396. [CrossRef] [PubMed]
28. Kouidri, M.; Saadi, A.K.; Noui, A.; Medjahed, F. The chemical composition of argan oil. *Int. J. Adv. Stud. Comput. Sci. Eng.* **2015**, *4*, 24–28.
29. Chaieb, K.; Hajlaoui, H.; Zmantar, T.; Kahla-Nakbi, A.B.; Rouabhia, M.; Mahdouani, K.; Bakhrouf, A. The chemical composition and biological activity of clove essential oil, Eugenia caryophyllata (*Syzigium aromaticum* L. Myrtaceae): A short review. *Phytother. Res.* **2007**, *6*, 501–506. [CrossRef] [PubMed]
30. Seul, S.D.; Lim, J.M.; Ha, S.H.; Kim, Y.H. Adhesion Enhancement of polyurethane coated leather and polyurethane foam with plasma treatment. *Korean J. Chem. Eng.* **2005**, *22*, 745–749. [CrossRef]
31. Vasile, C.; Stoleru, E.; Munteanu, B.S.; Zaharescu, T.; Ioanid, G.E.; Pamfil, D. Radiation mediated bioactive compounds immobilization on polymers to obtain multifunctional food packaging materials. In Proceedings of the International Conference on Applications of Radiation Science and Technology, Vienna, Austria, 24–28 April 2017.
32. Mishra, K.; Ojha, H.; Chaudhury, N.K. Estimation of antiradical properties of antioxidants using DPPH assay: a critical review and results. *Food. Chem.* **2012**, *130*, 1036–1043. [CrossRef]
33. Shenoy, S.L.; Bates, W.D.; Frisch, H.L.; Wnek, G.E. Role of chain entanglements on fiber formation during electrospinning of polymer solutions: Good solvent, non-specific polymer–polymer interaction limit. *Polymer* **2005**, *46*, 3372–3384. [CrossRef]
34. Geng, X.; Kwon, O.H.; Jang, J. Electrospinning of chitosan dissolved in concentrated acetic acid solution Technical note. *Biomaterials* **2005**, *26*, 5427–5432. [CrossRef] [PubMed]
35. Elahi, M.F.; Lu, W.; Guoping, G.; Khan, F. Core-shell fibers for biomedical applications–A review. *J. Bioeng. Biomed. Sci.* **2013**, *3*, 1000121. [CrossRef]
36. Diaz, J.E.; Barrero, A.; Márquez, M.; Loscertales, I.G. Controlled Encapsulation of Hydrophobic Liquids in Hydrophilic Polymer Nanofibers by Co-electrospinning. *Adv. Funct. Mater.* **2006**, *16*, 2110–2116. [CrossRef]
37. Li, C.; Li, Q.; Ni, X.; Liu, G.; Cheng, W.; Han, G. Coaxial Electrospinning and characterization of core-shell structured cellulose nanocrystal reinforced PMMA/PAN composite fibers. *Materials* **2017**, *10*, 572–588.
38. He, C.L.; Huang, Z.M.; Han, X.J.; Liu, L.; Zhang, H.S.; Chen, L.S. Coaxial electrospun poly(L-Lactic Acid) ultrafine fibers for sustained drug delivery. *J. Macromol. Sci. Part B Phys.* **2006**, *45*, 515–524. [CrossRef]
39. Li, T.X.; Ding, X.; Sui, X.; Tian, L.L.; Zhang, Y.; Hu, J.Y.; Yang, X.D. Sustained Release of Protein Particle Encapsulated in Bead-on-String Electrospun Nanofibers. *J. Macromol. Sci. B* **2015**, *54*, 887–896. [CrossRef]
40. Gaharwar, A.K.; Mihaila, S.M.; Kulkarni, A.A.; Patel, A.; Di-Luca, A.; Reis, R.L.; Gomes, M.E.; van-Blitterswijk, C.; Moroni, L.; Khademhosseini, A. Amphiphilic beads as depots for sustained drug release integrated into fibrillar scaffolds. *J. Control. Release* **2014**, *187*, 66–73. [CrossRef] [PubMed]

41. Huan, S.; Liu, G.; Han, G.; Cheng, W.; Fu, Z.; Wu, Q.; Wang, Q. Effect of experimental parameters on morphological, mechanical and hydrophobic properties of electrospun polystyrene fibers. *Materials* **2015**, *8*, 2718–2734. [CrossRef]
42. Belgharza, M.; Hassanain, I.; Lakrari, K.; El-Moudane, M.; Satrallah, A.; Elhabib, E.; El-Azzouzi, M.; Elmakhoukhi, F.; Elazzouzi, H.; El-Imache, A.; et al. Kinematic Viscosity Versus Temperature for Vegetable Oils: Argan, Avocado and Olive. *Aust. J. Basic Appl. Sci.* **2014**, *8*, 342–345.
43. Siddiqui, N.; Ahmad, A. A study on viscosity, surface tension and volume flow rate of some edible and medicinal oils. *Int. J. Sci. Environ. Technol.* **2013**, *2*, 1318–1326.
44. Adams, T.; Grant, C.; Watson, H. A simple algorithm to relate measured surface roughness to equivalent sand-grain roughness. *Int. J. Mech. Eng. Mechatron.* **2012**, *1*, 66–71. [CrossRef]
45. Brassard, J.D.; Sarkar, D.K.; Perron, J. Synthesis of monodisperse fluorinated silica nanoparticles and their superhydrophobic thin films. *ACS Appl. Mater. Interfaces* **2011**, *3*, 3583–3588. [CrossRef] [PubMed]
46. Wan, Y.; Cao, X.; Zhang, S.; Wang, S.; Wu, Q. Fibrous poly(chitosan-g-dl-lactic acid) scaffolds prepared via electro-wet-spinning. *Acta Biomater.* **2008**, *4*, 876–886. [CrossRef] [PubMed]
47. Ahmad, A.; Shaheen, A.; Owais, M.; Gaurav, S.S. Antimicrobial Activity of Syzygium Aromaticum Oil and its Potential in the Treatment of Urogenital Infections. Available online: http://www.formatex.info/microbiology4/vol2/865-871.pdf (accessed on 12 October 2018).
48. Marchese, A.; Barbieri, R.; Coppo, E.; Orhan, I.E.; Daglia, M.; Nabavi, S.F.; Izadi, M.; Abdollahi, M.; Nabavi, S.M.; Ajami, M. Antimicrobial activity of eugenol and essential oils containing eugenol: A mechanistic viewpoint. *Crit. Rev. Microbiol.* **2017**, *43*, 668–689. [CrossRef] [PubMed]
49. Lotfi, N.; Chahboun, N.; El Hartiti, H.; Kabouche, Z.; El M'Rabet, M.; Berrabeh, M.; Touzani, R.; Ouhssine, M.; Oudda, H. Study of the antibacterial effect of Argan oil from Bechar region of Algeria on hospital resistant strains. *J. Mater. Environ. Sci.* **2015**, *6*, 2476–2482.
50. Dakiche, H.; Khali, M.; Abu-el-Haija, A.K.; Al-Maaytah, A.; Al-Balas, Q.A. Biological activities and phenolic contents of *Argania spinosa* L (Sapotaceae) leaf extract. *Trop. J. Pharm. Res.* **2016**, *15*, 2563–2570. [CrossRef]
51. Munteanu, B.S.; Stoleru, E.; Ioanid, E.G.; Vasile, C.; Mitelut, A.C.; Popa, M.E.; Tănase, E.E.; Mihai, A.L.; Drăghici, M.C.; Rosnes, J.T.; et al. The use of the electrospinnig technique to produce polymeric films/coatings with multifunctional properties. In Proceedings of the National Workshop "Ambalaje Alimentare Active", Bucuresti, Romania, 28 February 2017.
52. Miteluţ, A.C.; Popa, E.E.; Popescu, P.A.; Popa, M.E.; Munteanu, B.S.; Vasile, C.; Ştefănoiu, G. Research on chitosan and oil coated PLA as food packaging material. In Proceedings of the International Workshop "Progress in Antimicrobial Materials", Iasi, Romania, 30 March 2017.
53. Jing, Y.J.; Hao, Y.J.; Qu, H.; Shan, Y.; Li, D.S.; Du, R.Q. Studies on the antibacterial activities and mechanisms of chitosan obtained from cuticles of housefly larvae. *Acta Biol. Hung.* **2007**, *58*, 75–86. [CrossRef] [PubMed]
54. Liu, N.; Chen, X.G.; Park, H.J.; Liu, C.G.; Liu, C.S.; Meng, X.H. Effect of MW and concentration of chitosan on antibacterial activity of *Escherichia coli*. *Carbohydr. Polym.* **2006**, *64*, 60–65. [CrossRef]
55. Kaya, M.; Asan-Ozusaglam, M.; Erdogan, S. Comparison of antimicrobial activities of newly obtained low molecular weight scorpion chitosan and medium molecular weight commercial chitosan. *J. Biosci. Bioeng.* **2016**, *121*, 678–684. [CrossRef] [PubMed]
56. Mohammadi, A.; Hashemi, M.; Masoud-Hosseini, S. Effect of chitosan molecular weight as micro and nanoparticles on antibacterial activity against some soft rot pathogenic bacteria. *LWT-Food Sci. Technol.* **2016**, *71*, 347–355. [CrossRef]
57. Liu, X.F.; Guan, Y.L.; Yang, D.Z.; Li, Z.; Yao, F.L. Antibacterial action of chitosan and carboxymethylated chitosan. *J. Appl. Polym. Sci.* **2001**, *79*, 1324–1335.
58. Chang, S.H.; Lin, H.T.V.; Wu, G.J.; Tsai, G.J. pH Effects on solubility, zeta potential, and correlation between antibacterial activity and molecular weight of chitosan. *Carbohydr. Polym.* **2015**, *134*, 74–81. [CrossRef] [PubMed]
59. Tsai, G.J.; Su, W.H.; Chen, H.C.; Pan, C.L. Antimicrobial activity of shrimp chitin and chitosan from different treatments and applications of fish preservation. *Fish. Sci.* **2002**, *68*, 170–177. [CrossRef]
60. Hongpattarakere, T.; Riyaphan, O.S. Effect of deacetylation conditions on antimicrobial activity of chitosans prepared from carapace of black tiger shrimpPenaeus monodon. *J. Sci. Technol.* **2008**, *30*, 1–9.

61. Younes, I.; Sellimi, S.; Rinaudo, M.; Jellouli, K.; Nasri, M. Influence of acetylation degree and molecular weight of homogeneous chitosans on antibacterial and antifungal activities. *Int. J. Food Microbiol.* **2014**, *185*, 57–63. [CrossRef] [PubMed]
62. Takahashi, T.; Imai, M.; Suzuki, I.; Sawai, J. Growth inhibitory effect on bacteria of chitosan membranes regulated with deacetylation degree. *Biochem. Eng. J.* **2008**, *40*, 485–491. [CrossRef]
63. Omura, Y.; Shigemoto, M.; Akiyama, T.; Saimoto, H.; Shigemasa, Y.; Nakamura, I.; Tsuchido, T. Antimicrobial activity of chitosan with different degrees of acetylation and molecular weights. *Biocontrol Sci.* **2003**, *8*, 25–30. [CrossRef]
64. Mellegård, H.; Strand, S.P.; Christensen, B.E.; Granum, P.E.; Hardy, S.P. Antibacterial activity of chemically defined chitosans: Influence of molecular weight, degree of acetylation and test organism. *Int. J. Food Microbiol.* **2011**, *148*, 48–54. [CrossRef] [PubMed]
65. Andres, Y.; Giraud, L.; Gerente, C.; Le-Cloirec, P. Antibacterial effects of chitosan powder: Mechanisms of action. *Environ. Technol.* **2007**, *12*, 1357–1363. [CrossRef] [PubMed]
66. Jung, E.J.; Youn, D.K.; Lee, S.H.; No, H.K.; Ha, J.G.; Prinyawiwatkul, W. Antibacterial activity of chitosans with different degrees of deacetylation and viscosities. *Int. J. Food Sci. Technol.* **2010**, *45*, 676–682. [CrossRef]
67. Qi, L.; Xu, Z.; Jiang, X.; Hu, C.; Zou, X. Preparation and antibacterial activity of chitosan nanoparticles. *Carbohydr. Res.* **2004**, *339*, 2693–2700. [CrossRef] [PubMed]
68. Su, L.; Yu, Y.; Zhao, Y.; Liang, F.; Zhang, X. Strong antibacterial polydopamine coatings prepared by a shaking-assisted method. *Sci. Rep.* **2016**, *6*, 24420. [CrossRef] [PubMed]
69. Mitik-Dineva, N.; Wang, J.; Truong, V.K.; Stoddart, P.R.; Malherbe, F.; Crawford, R.J.; Ivanova, E.P. Differences in colonisation of five marine bacteria on two types of glass surfaces. *Biofouling* **2009**, *5*, 621–631. [CrossRef] [PubMed]
70. Colon, G.; Ward, B.C.; Webster, T.J. Increased osteoblast and decreased Staphylococcus epidermidis functions on nanophase ZnO and TiO_2. *J. Biomed. Mater. Res. Part A* **2006**, *78A*, 595–604. [CrossRef] [PubMed]
71. Puckett, S.D.; Taylor, E.; Raimondo, T.; Webster, T.J. The relationship between the nanostructure of titanium surfaces and bacterial attachment. *Biomaterials* **2010**, *31*, 706–713. [CrossRef] [PubMed]
72. Bagherifard, S.; Hickey, D.J.; de-Luca, A.C.; Malheiro, V.N.; Markaki, A.E.; Guagliano, M.; Webster, T.J. The influence of nanostructured features on bacterial adhesion and bone cell functions on severely shot peened 316L stainless steel. *Biomaterials* **2015**, *73*, 185–197. [CrossRef] [PubMed]
73. Izquierdo-Barba, I.; García-Martín, J.M.; Álvarez, R.; Palmero, A.; Esteban, J.; Pérez-Jorge, C.; Arcos, D.; Vallet-Regí, M. Nanocolumnar coatings with selective behavior towards osteoblast and Staphylococcus aureus proliferation. *Acta Biomater.* **2015**, *15*, 20–28. [CrossRef] [PubMed]
74. Truong, V.K.; Webb, H.K.; Fadeeva, E.; Chichkov, B.N.; Wu, A.H.F.; Lamb, R.; Wang, J.Y.; Crawford, R.J.; Ivanova, E.P. Air-directed attachment of coccoid bacteria to the surface of superhydrophobic lotus-like titanium. *Biofouling* **2012**, *28*, 539–550. [CrossRef] [PubMed]
75. Otohinoyi, D.A.; Ekpo, O.; Ibraheem, O. Effect of ambient temperature storage on 2,2-diphenyl-1-picrylhydrazyl; DPPH as a free radical for the evaluation of antioxidant activity. *Int. J. Biol. Chem. Sci.* **2014**, *8*, 1262–1268. [CrossRef]
76. Gokmen, V.; Serpen, A.; Fogliano, V. Direct measurement of the total antioxidant capacity of foods: the 'QUENCHER' approach. *Trends Food Sci. Technol.* **2009**, *20*, 278–288. [CrossRef]
77. Cömert, E.D.; Gökmen, V. Antioxidants bound to an insoluble food matrix: Their analysis, regeneration behavior, and physiological importance. *Compr. Rev. Food Sci. Food Safety* **2017**, *16*, 382–399. [CrossRef]
78. Tufan, A.N.; Celik, S.E.; Ozyurek, M.; Guclu, K.; Apak, R. Direct measurement of total antioxidant capacity of cereals: QUENCHER-CUPRAC method. *Talanta* **2013**, *108*, 136–142. [CrossRef] [PubMed]
79. Marfil, R.; Giménez, R.; Martínez, O.; Bouzas, P.R.; Rufián-Henares, J.A.; Mesías, M.; Cabrera-Vique, C. Determination of polyphenols, tocopherols, and antioxidant capacity in virgin argan oil *Argania spinosa*, Skeels. *Eur. J. Lipid Sci. Technol.* **2011**, *113*, 886–893. [CrossRef]
80. Cortés-Rojas, D.F.; de-Souza, C.R.F.; Oliveira, W.P. Syzygium aromaticum: a precious spice. *Asian Pac. J. Trop. Biomed.* **2014**, *4*, 90–96. [CrossRef]
81. Vasile, C.; Stoleru, E.; Irimia, A.; Zaharescu, T.; Dumitriu, R.P.; Ioanid, G.E.; Munteanu, B.S.; Oprica, L.; Pricope, G.M.; Hitruc, G.E. Ionizing Radiation And Plasma Discharge Mediating Covalent Linking

of Bioactive Compounds onto Polymeric Substrate to Obtain Stratified Composites for Food Packing. In Proceedings of the the 3rd RCM of the CRP on Application of Radiation Technology in the Development of Advanced Packaging Materials for Food Products, Vienna, Austria, 11–15 July 2016.

82. Vasile, C.; Pâslaru, E.; Sdrobis, A.; Pricope, G.; Ioanid, G.E.; Darie, R.N. Plasma assisted functionalization of synthetic and natural polymers to obtain new bioactive food packaging materials in Ionizing Radiation and Plasma discharge Mediating Covalent Linking of Stratified Composites Materials for Food Packaging. In Proceedings of the Part of Co-ordinated Project: 'Application of Radiation Technology in the Development of Advanced Packaging Materials for Food Products', Vienna, Austria, 22–26 April 2013.

6

Release of Graphene and Carbon Nanotubes from Biodegradable Poly(Lactic Acid) Films during Degradation and Combustion: Risk Associated with the End-of-Life of Nanocomposite Food Packaging Materials

Stanislav Kotsilkov [1,*], Evgeni Ivanov [1,2] and Nikolay Kolev Vitanov [1]

[1] Institute of Mechanics, Bulgarian Academy of Sciences, Acad. G. Bonchev, Block 4, 1113 Sofia, Bulgaria; ivanov_evgeni@yahoo.com (E.I.); vitanov@imbm.bas.bg (N.K.V.)

[2] Research and Development of Nanomaterials and Nanotechnologies (NanoTechLab Ltd.), Acad. G. Bonchev, Block 4, 1113 Sofia, Bulgaria

* Correspondence: kotsilkov@gmail.com

Abstract: Nanoparticles of graphene and carbon nanotubes are attractive materials for the improvement of mechanical and barrier properties and for the functionality of biodegradable polymers for packaging applications. However, the increase of the manufacture and consumption increases the probability of exposure of humans and the environment to such nanomaterials; this brings up questions about the risks of nanomaterials, since they can be toxic. For a risk assessment, it is crucial to know whether airborne nanoparticles of graphene and carbon nanotubes can be released from nanocomposites into the environment at their end-life, or whether they remain embedded in the matrix. In this work, the release of graphene and carbon nanotubes from the poly(lactic) acid nanocomposite films were studied for the scenarios of: (i) biodegradation of the matrix polymer at the disposal of wastes; and (ii) combustion and fire of nanocomposite wastes. Thermogravimetric analysis in air atmosphere, transmission electron microscopy (TEM), atomic force microscopy (AFM) and scanning electron microscope (SEM) were used to verify the release of nanoparticles from nanocomposite films. The three factors model was applied for the quantitative and qualitative risk assessment of the release of graphene and carbon nanotubes from nanocomposite wastes for these scenarios. Safety concern is discussed in respect to the existing regulations for nanowaste stream.

Keywords: graphene; carbon nanotubes; poly(lactic) acid; degradation; combustion; fire; risk analysis

---

1. Introduction

Manufactured nanomaterials are applied in various consumer goods in order to enhance their properties or to supplement novel functionalities. The industry has already utilized nanoclays, metal nanoparticles and carbon nanotubes and has managed to use them in products for a variety of applications, e.g., semiconductors, automotive, aerospace, electronics, energy, defense, sporting goods, and packaging [1]. Nanotechnology allows scientists to alter the structure of packaged materials on a molecular scale in order to give the materials the desired properties [2,3]. Nowadays, different types of carbon nanotubes and graphene in polymer nanocomposites are widely investigated for the development of smart, active and intelligent packaging that can improve the quality and safety of food, to solve the food storage problem and to inform the consumer about the quality of packaged food [2–4]. Graphene and multiwall carbon nanotubes (MWCNTs) in biodegradable polymers are the most continuous and potentially valuable nanoscale materials to have emerged in recent years, and are increasingly studied to enhance the thermal, mechanical, barrier properties and functionality of food packaging materials [1–3,5]. Among the many different biopolymers available, polylactic acid (PLA) is a promising alternative, especially for food packaging and biomedical packaging applications. However, a key challenge is to enhance the barrier properties of the PLA packaging. Graphene and its derivatives are identified as powerful candidates for gas-barrier materials because perfect graphene does not allow the diffusion of small gases through its plane [6–8]. Incorporation of graphene and carbon nanotubes into polymer matrices are also a promising nanotechnology approach to increase mechanical strength and improve thermal properties when properly dispersed in a polymer matrix [9,10].

At the same time, regulators at national, European and international level are still struggling to agree upon a unified and mutually accepted definition of a "nanomaterial". As the market of nanomaterial-based products is expected to triple by the year 2020 in more industrial sectors, the application of nanomaterials and nanoparticles is expected to grow proportionally [2,3,11]. As a result, environmental exposure to nanoparticles in air, water and soils is also expected to increase [11]. Therefore, more research efforts in nanoscience are needed to focus exclusively on the potential risk of nanomaterials graphene and carbon nanotubes with the increasing exposition of consumers and the environment to nanoparticle-containing packaging. The specific nanoscale size and shape of graphene and carbon nanotubes with a large aspect ratio and large surface area, airborne, non-soluble in water and absorptive in soil, will enhance the risk of their mobility in the environment [12,13].

Currently, very little is known about the release of the nanomaterials graphene and carbon nanotubes (CNT) incorporated into polymer nanocomposites. Although they are typically tightly bound in the matrix polymer, their release through the lifecycle of nanocomposites is possible [11]. Therefore, greater information is needed on the potential hazard associated with specific exposure scenarios. Few scenarios were identified and published in the literature [11,14,15] where CNTs might be released into the environment during the life cycle of polymer nanocomposites in the production, service life and disposal stages. Release during service the life of CNT-based composites is projected to be quite low and composed of polydisperse fragments with only a small fraction of free single nanotubes [15–17]. The general conclusion is that the CNTs form a network and are not easily detachable from the samples, but the CNT layer on the surface of the degraded composites could be a source of a high quantity of released free standing CNTs and thus, may pose a health risk [16,17]. Researchers [16,17] reviewed the potential release of CNTs into the environment during the service life where untrained humans are in contact with oil-based polymer nanocomposites. They considered three possible pathways for the release of CNTs: due to exploitation and use, degradation of the matrix due to weathering processes, and fire events. Duncan and Pillai [18] assumed the nanoparticle release paradigms to be: (i) via passive diffusion, desorption and dissolution into external liquid media; and (ii) by matrix degradation. However, currently there are no standard methods to measure what is released from use of products containing nanomaterials.

Studies on the release of graphene through the lifecycle of nanocomposites are very scare. Arvidsson, et al. [12] reported that the potential environmental and health risk of graphene at the production stage could be great, depending on the type of synthesis methods [12]. For most materials, degradation of the polymer matrix during the service life is associated with the greatest potential for release, with rates dependent on the specific characteristics of the polymer, carbon nanofiller, and environment [15,16]. In our previous studies [19–21], we have reported on the release of graphene nanoplatelets from food packaging materials into food simulants, due to the migration processes and the partial degradation of the biodegradable polymer matrix. Therefore, the released graphene nanoplatelets and their aggregates from the nanocomposite wastes during their degradation may pose a health risk [16,17]. The fate of graphene in polymer nanocomposite exposed to UV radiation was discussed and researchers concluded that graphene nanoparticles would be able to pass into the air or engage in the soil and groundwaters [22].

End-of-life aspects of nanomaterials have received far less attention than their preparation or application [23]. The two main strategies used for the end-of-life of thermoplastics products are recycling and burning to produce energy (thermal valorization). As the release of nanoparticles during grinding of nanocomposite wastes presents a potential risk, the incineration of nanocomposites has recently been accepted as a prospective waste management strategy, for which nanoparticle emission during burning must be addressed as a premise. So far, no detailed study has been published that investigates the release of graphene from nanocomposites due to accident fire or burning. Few publications have discussed the incineration and burning of CNT-based nanocomposites [24,25]. In contrast to incineration, where under high temperatures CNTs can be destroyed [24], a fire or burning of CNT nanocomposites in open air may not degrade all CNTs particles in composites, since the decomposition temperature of CNTs is much higher than that of the polymer matrix. A network of CNTs is formed in the char (residue ash), but measurements are still missing if CNTs can be released from the residue ash into the air. Bouillard, et al. [25] reported on the release of free CNTs and agglomerates of CNTs from acrylonitrile–butadiene–styrene (ABS) nanocomposites into the air during nanowastes combustion at quite low temperatures (about 400 °C). This information was found to be important to assess the environmental risks and the inhalation risks to people engaged in those practices.

In the present study, we investigate the release of graphene nanoplates and multi-walled carbon nanotubes from biodegradable poly(lactic) acid nanocomposite films at various temperatures, due to polymer degradation and burning. We analyze the risk from the release of graphene nanoplatelets and multi-walled carbon nanotubes at the end-of-life of the nanocomposite wastes, due to biodegradation at composting, combustion and accident fire. The thermogravimetric analysis (TGA), transmission electron microscopy (TEM), atomic force microscopy (AFM) and scanning electron microscope (SEM) were used to verify the release of nanoparticles from nanocomposite films. The three-factors model (C.E.L.) was applied for the quantitative and qualitative risk assessment of the release of graphene and carbon nanotubes from nanocomposite wastes for these scenarios' materials. Safety concerns are discussed in respect to the application of nanomaterials in food packaging, bio-medical packaging and others.

2. Materials and Methods

Nanocomposite of poly(lactic) acid polymer (PLA) filled with graphene nanoplatelets (GNP) and multiwall carbon nanotubes (MWCNTs) was supplied from Graphene 3D Lab, (Ronkonkoma, NY, USA) in the form of filament for 3D printing (FDM). Commercial neat PLA filament was also supplied. The matrix polylactic acid is CAS# 26100-51-6, Sigma-Aldrich, (St. Louis, MO, USA) with Mw~60,000 and 1.8% crystallinity. Disk samples having sandwich structure of 10 alternative layers of the nanocomposite and the PLA were 3D prepared with the dual extruder X400 Rep Rap printer (Feldkirchen, Germany). The printed samples were hot pressed to thin films with a thickness of 30 microns. The amount of nanocarbon filler in the final film samples is 3 wt. %.

The film samples were emerged in two aqueous-based solvents of 10 vol. % ethanol and 50 vol. % ethanol. Ultra-strong migration test was performed with heating at 90 °C for 4 h, followed by storage for 10 days at 40 °C and dynamic treatment for 1 min daily. The migration test regime was applied to simulate the conditions of high temperature treatment of nanocomposite packaging during their end-of-life as waste disposal.

Different visualization techniques were applied in order to identify migrants in the simulant media and to verify the film integrity after the migration test. TEM (EOL JEM 2100, Preabody, MA, USA) at accelerating voltage 200 kV and AFM (Bruker, Billerica, MA, USA) were used for the analysis of the dried colloids of migrated substances into the surrounding solvents. For preparation of the test samples, a micro-quantity of colloid after migration test was dropped on a copper TEM grid covered by a membrane from amorphous carbon (or on glass plate for AFM scan), and after that dried in a dust-free atmosphere at ambient conditions. The morphology of the film surface before and after the migration test was studied by scanning electron microscope (Philips 515, Eindhoven, Netherlands) with accelerating voltage 25 kV and 5 kV. Before the examination in the microscope, the samples were covered with metal coating for better conductivity of the surface and to avoid discharge effects.

Thermal stability and degradation of nanocomposite films were studied by TGA (Q50, TA Instruments, New Castle, DE, USA), in air atmosphere, at a heating rate of 10 °C/min in three different temperature ranges, from 30 °C to 500, 650 and 850 °C. The mass loss during heating and the amount of residue ash were analyzed. TEM was performed for analysis of the residue ash after burning at the three temperatures.

Risk assessment analysis was performed using the three factors method, 3F or C.E.L., i.e., grading the three risk analysis factors: Consequences (C), Exposure (E), and Probability/Likelihood (L).

3. Results and Discussions

3.1. Release of Graphene and Carbon Nanotubes Due to Degradation of PLA Polymer

The use of biodegradable packaging materials, such as PLA will contribute to sustainability and reduction of wastes via degradation [26,27]. Composting, for example, has the potential to transfer biodegradable waste, including biodegradable plastics, into useful soil amendment products by an accelerated degradation using a mixed microbial population in a moist, warm, aerobic environment under controlled conditions. Song, et al [26] found that biodegradable packaging materials are most suitable for single-use disposable applications where the post-consumer waste can be locally composted. However, special care should then be taken while handling local composting of biodegradable nanowastes to limit potential environmental risks due to the release of nanoparticles in the soil from the compost.

We discuss herewith if the GNP and MWCNTs release as single nanoparticles or large aggregates from the nanocomposite packaging films via degradation of the PLA matrix. The characterization of such release provides critical information for environmental nano-object exposure from biodegradable nanocomposites. In our previous studies [19–21], we have investigated the release of GNP and MWCNTs from the composite film GNP/MWCNT/PLA in alcoholic and acid food simulants, at high temperature migration conditions, such as: (i) strong static migration test (heating at 90 °C for 4 h); and (ii) ultra-strong dynamic migration test (heating at 90 °C for 4 h followed by subsequent storage for 10 days at 40 °C, including dynamic treatment for 1 min daily). The migration conditions were set accordingly with the prescription in EU Regulation 10/2011 (EU 2011) [28] and literature sources [29,30].

We have observed that large graphene nanoplatelets (GNP) of about 100–1000 nm in length and a few nanometers in thickness indeed migrate from the GNP/MWCNT/PLA film into the food simulants, due to the diffusion processes [19,20]. During the strong static migration test, the total amount of released substances (nanoparticles and organic matter) from the composite GNP/MWCNT/PLA films was estimated around 0.028–0.053 mg/cm^2, where the nanoparticle migrants are of 0.006–0.011 mg/cm^2, depending on the food simulants [20]. Therefore, the released substances after this test remain below the overall migration limit (OEL = 0.10 mg/cm^2) for food contact material accepted by the EU regulatory documents [28,31].

By contrast, during the ultra-strong dynamic migration test [20], the release of nanoscale size particles (100–1000 nm) from the GNP/MWCNT/PLA composite film is higher (0.5–0.7 number %) compared to (0.1–0.2 number %) nanoparticle migrants that were observed during the strong static migration test in the three food simulants, 3% acidic acid, and 10% and 50% ethanol. The larger size nanoparticle migrants (1–10 microns) were found also in higher amounts (3–5 number %) during the ultra-strong migration test, compared to 1–2 number % for the strong static test. Laser diffraction analysis was used in our previous studies [19,20] to detect the number and size distribution of the released nanoparticles from GNP/MWCNT/PLA films into the food simulants. In general, a threefold higher number of nanoparticles are observed to release during the ultra-strong dynamic migration test (heating at 90 °C for 4 h followed by storage for 10 days at 40 °C and dynamic treatment for 1 min daily) compared to the strong static test above. The detected nanoscale particles (100–1000 nm) in the food simulants are 0.5–0.7% while the micron scale agglomerates (1–10 μm) are 3–5%, compared to 0.1–0.2% nanoparticles and 1–2% agglomerates released after the strong static conditions [20]. This was associated with partial degradation of the PLA polymer matrix, which supports the diffusion of GNPs together with dissolved organic substances out of the film. However, the MWCNTs form the entangled network in the polymer film, which prevents their migration into food simulants if the polymer partially degrades.

Based on upper results, we discuss herein the scenario of nanoparticle release due to degradation of GNP/MWCNT/PLA composite films at the end-of-life stage of disposal or composting the wastes. Figure 1 presents the TEM micrographs (Figure 1a,b) and AFM scans (Figure 1c,d) of the dried migrants in 10% and 50% ethanol after an ultra-strong dynamic migration test. In Figure 2, the SEM micrographs of the film surfaces before and after such treatment are compared.

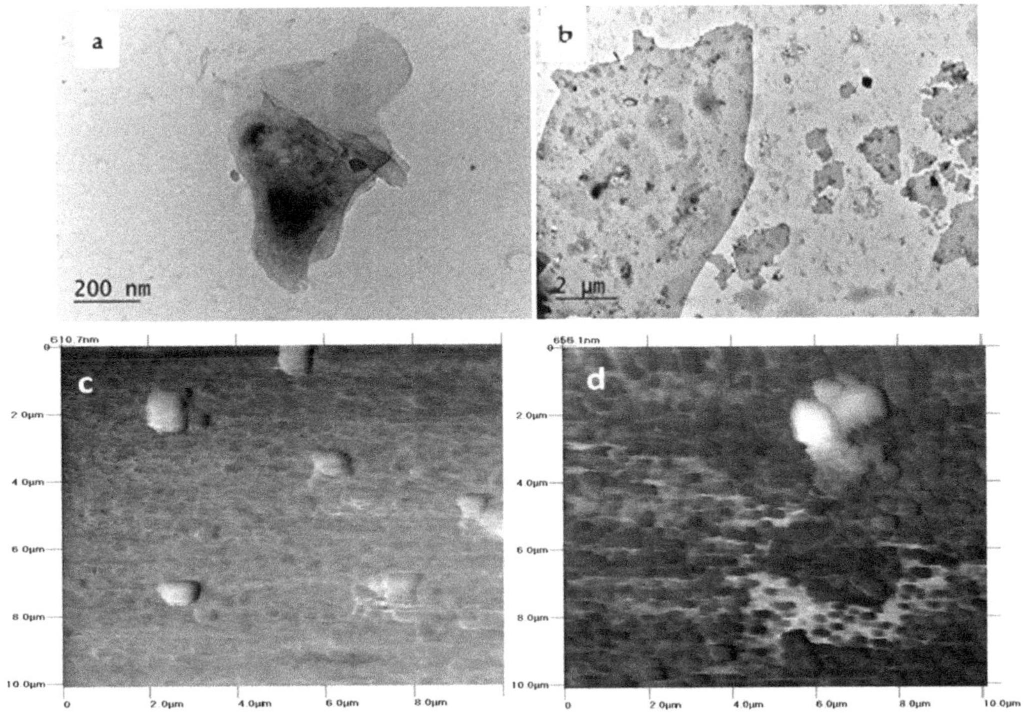

Figure 1. (**a**,**b**) transmission electron microscopy (TEM) micrographs and (**c**,**d**) atomic force microscopy (AFM) scan of dried migrants from GNP/MWCNT/PLA film after the ultra-strong dynamic test in 10% ethanol (first column) and 50% ethanol (second column).

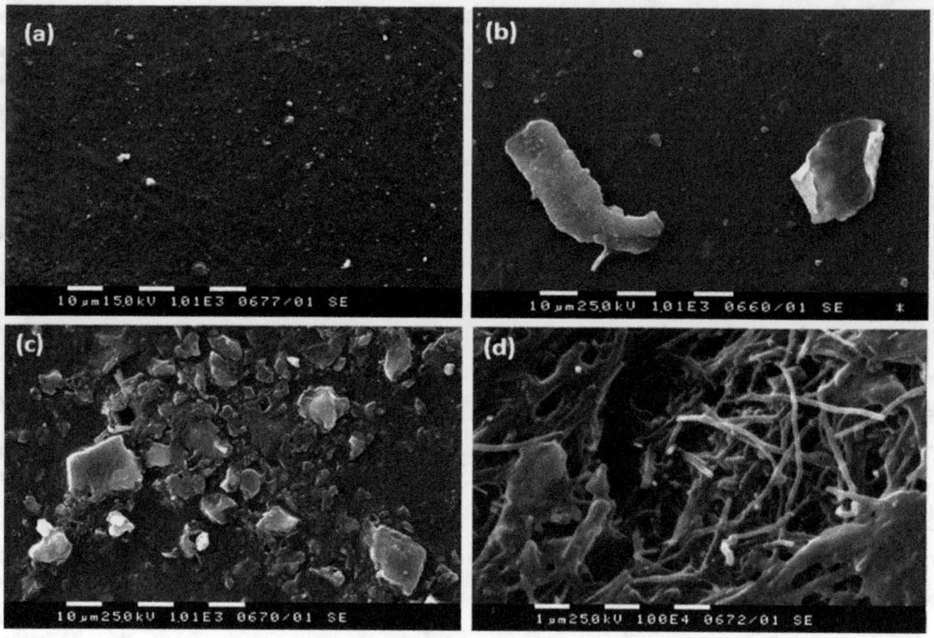

Figure 2. Scanning electron microscope (SEM) micrographs of the film surfaces: (**a**) reference GNP/MWCNT/PLA film before the test; (**b**) after the ultra-strong dynamic test in 10% ethanol, (**c**,**d**) after the ultra-strong dynamic test in 50% ethanol (in different places).

As seen from the TEM micrographs in Figure 1a,b, mainly small aggregates of GNPs below 500 nm are observed in a low amount in 10% ethanol (Figure 1a). While in a more aggressive media of 50% ethanol (Figure 1b), the amount of released GNPs increases apparently, and most of the platelets are in nanoscale size, 100–1000 nm; only a few aggregates of size 1–10 µm are visible. The AFM scans of dried colloids in 10% ethanol (Figure 1c) visualize the presence of a small amount of particles in a size of about 1 µm. However, in 50% ethanol (Figure 1d), the migrants are many small objects with a nanoscale size below 500 nm and a large aggregate of ~2 µm length and thickness of ~500 nm. Importantly, MWCNTs are not visible to release from the GNP/MWCNT/PLA nanocomposite films, as observed by TEM and AFM analysis. Following the release mechanisms of nanoparticles proposed by Duncan and Pillai [18], we assume that the physical changes of biodegradable PLA polymer due to polymer hydrolysis provokes the diffusion of the graphene nanoplatelets and the dissolved organic substances out of the film into the food simulant.

SEM analysis was performed in order to examine the GNP/MWCNT/PLA film integrity before and after the ultra-strong dynamic migration test in the two alcohol-based simulants. The film surface before the treatment (Figure 2a) is smooth and free from ingredients. For the film immersed in 10% ethanol (Figure 2b), a few graphene nanoplates of a size of above 10 µm are extracted on the film surface. By contrast, in 50% ethanol (Figure 2c,d), the integrity of the GNP/MWCNT/PLA composite film is destroyed due to the partial degradation of the PLA polymer and much extraction of substances on the film surface. It is visible that the fibrous MWCNTs formed an entangled network as the PLA polymer matrix dissolved, which prevented their release into the food simulant.

It may be concluded based on the above results, that if the biodegradable nanocomposite wastes containing GNPs and MWCNTs are disposed in landfills, nanoscale graphene platelets can certainly be released into the environment due to partial degradation and weathering. If such nanowastes are subjected to composting, the biodegradation process provides compost that is very likely to contain large amounts of nanoscale GNPs, as well as graphene aggregates and bundles of carbon nanotubes of micron size. Since the compost is intended to be used for soil improvement, those nanoparticles would penetrate into soil and groundwater, and there is a potential risk of falling into the food chain of different organisms.

3.2. Release of Graphene and Carbon Nanotubes Due to Burning of Nanowastes

Burning of nanocomposite wastes to produce energy (thermal valorization) has been recently discussed as a nanowaste management strategy, and thus the risks for nanoparticle emission during incineration of thermoplastic nanocomposites must be addressed and investigated [25]. Moreover, the treatment of such waste by accidental fire or burning in landfills (a common practice in underdeveloped regions) may pose questions associated with environmental and human risks due to the potential release of large amounts of nanoparticles into the environment. In principle, graphene and CNTs are combustible materials above 600 °C, and they can be easily transformed into CO/CO_2 during degradation [25]. The published results on this subject are very scarce, but a few papers [32–34] have reported that the combustion of polymer composites with CNTs could form residues (ashes) containing unburned CNTs. Moreover, the CNTs also may release in the combustion gas phase [25]. Therefore, in the present work, we classify burning of nanowastes as a scenario that may have a greater possibility to release airborne nanoparticles. The characterization of such a release may provide critical information for environmental and human accidental nanoparticle exposures.

TGA was performed to simulate the nanowaste combustion in three heating regimes: 30–500 °C, 30–650 °C and 30–850 °C, at a heating rate 10 °C/min in air atmosphere. The graphene nanoplatelets used in this study are one of the possible forms of graphene-related materials. In general, the obtained thermal stability and the release results will depend strongly on the particular graphene material used in the nanocomposite. The thermal decomposition of the neat PLA and the nanocomposite films (GNP/MWCNT/PLA) were analyzed by the thermal gravity analysis (TG) and Differential thermal gravity (DTG) curves, which present the weight loss (%) and its first derivative versus temperature (°C), as shown in Figure 3a,b, respectively. The initial decomposition temperature (T_{onset}), the decomposition peak temperature (T_p) and the residue ash (%) were evaluated from the TG/DTG curves and data are presented in Table 1. The T_{onset} of the nanocomposite is observed around 230 °C, while T_p appears at 361 °C, which are higher than 10 °C and 5 °C, respectively, compared to the neat PLA. This indicates that the thermal stability of PLA is improved by the addition of 3 wt. % mixed nanofillers, GNP and MWCNT. As might be expected, the weight loss increases with increasing the heating temperature (T_{max}) from 500 to 850 °C. The combustion of the neat PLA at 500 °C results in 0.3% residue from the initial weigh of the polymer sample and the ash consists of amorphous carbon (CB). In contrast, the combustion of GNP/MWCNT/PLA forms residue ash of 3.3% at 500 °C, 1.5% at 650 °C and 0.07% at 850 °C, containing mostly unburned nanoparticles, GNP and MWCNT. Therefore, further potential environmental problems may arise with handling such residue ash. Our study advances the observations in References [25,32–34] by showing that the amount of GNPs and MWCNTs in the residue ash decreases by increasing the burning temperature, this indicating increased decomposition of carbon nanoparticles by controlled incineration/combustion temperatures.

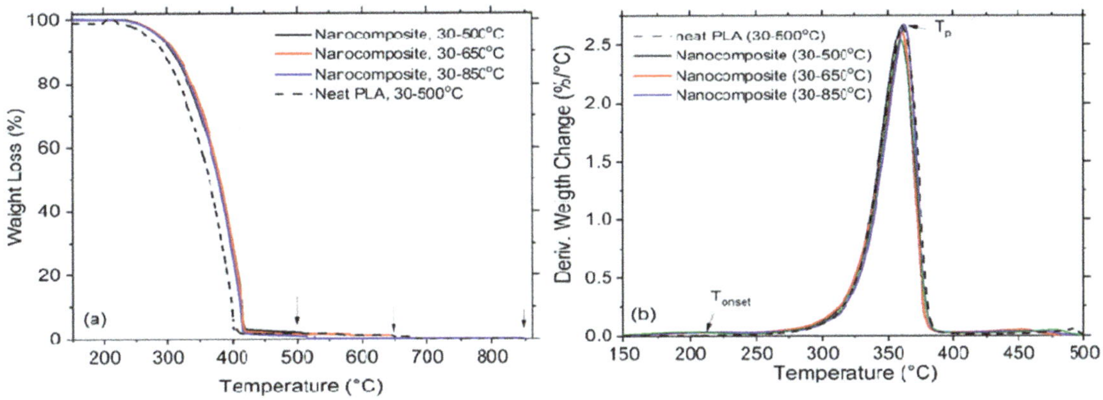

Figure 3. TGA (**a**) and DTG (**b**) curves of the neat PLA and GNP/MWCNT/PLA nanocomposite film compared to the neat PLA at three temperature regimes, 500, 650 and 850 °C, in an air atmosphere.

Table 1. Thermal characteristics of PLA and GNP/MWCNT/PLA nanocomposite by TG analysis.

Sample	Temperature of Burning T_{max}	T_{onset}, °C	T_p, °C	Weight Loss, % at T_{max}	Residue Ash, %
Neat PLA	500 °C	219.8	356.5	99.63	0.30
Nanocomposite	500 °C	230.7	360.2	96.70	3.30
Nanocomposite	650 °C	230.1	361.3	98.50	1.05
Nanocomposite	850 °C	230.3	362.8	99.98	0.07

TEM micrographs in Figure 4 visualize the content of the residue ash after combustion at the three temperatures in air atmosphere. As seen in the first column (Figure 4a), at 500 °C, the residue ash (3.3%) is completely composed of unburned single MWCNTs and GNPs, or their loose agglomerates. At 650 °C (Figure 4b), the residue decreases to 1.5% and consists mostly of single airborne particles, MWCNTs and GNP, and some soot CB nanoparticles of primary sizes of 10–30 nm. The nanotubes are about 30 nm in diameter and a few microns in length and are very similar to the original MWCNT size. Similar findings are observed for the GNP particles. While at 850 °C (Figure 4c), the amount of the residue ash strongly decreases to 0.07%, confirming that the carbonaceous fillers are mostly degraded. Indeed, the MWCNTs are missing in the residue ash, but unexpected content of GNP particles is observed and they are mainly displayed as fractal aggregates mixed with some soot nanoparticles.

The observations in Figure 4 reveal that large amounts of single isolated airborne MWCNTs (<50 nm diameter and >1 μm length) and GNPs (>100 nm), as well as their loose fractals (1–2 μm) can be released during burning in air atmosphere, addressing, therefore, a new kind of safety issue with regards to the combustion/incineration of nanowastes or accidental fires. The airborne particles of GNPs and MWCNTs may either stay in the char residues, or may be released in the gas phase during incineration or fire. Their fate depends on the local operating conditions of the burning process.

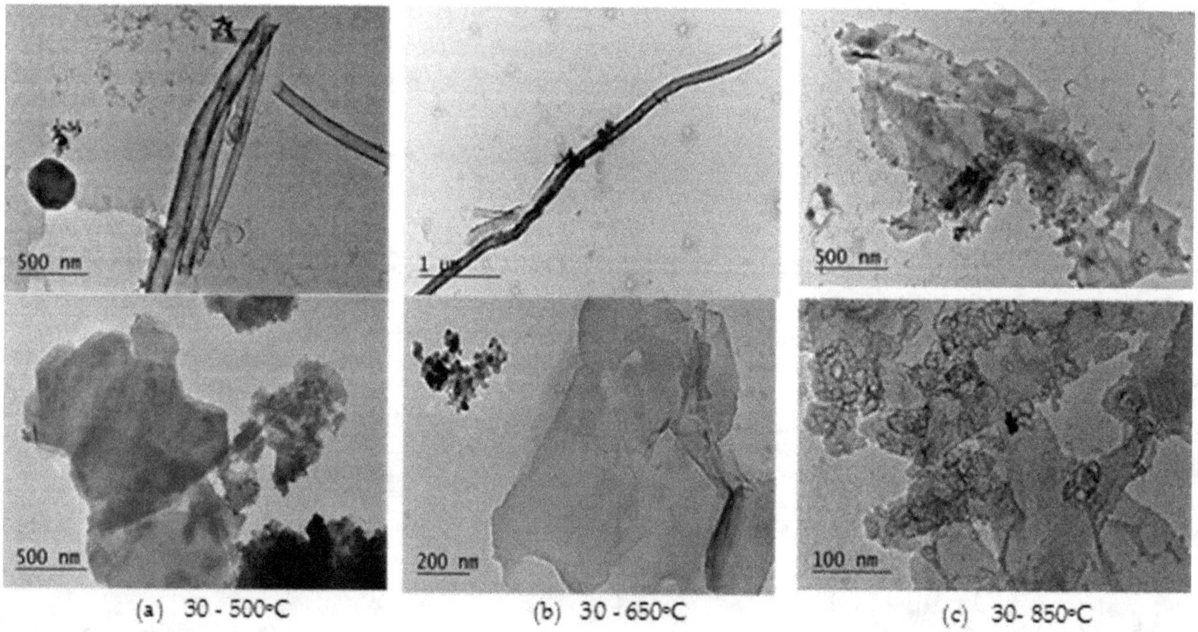

Figure 4. TEM micrographs of MWCNTs (first line) and GNPs (second line) observed in the residue ash at the three decomposition temperatures: (**a**) 500 °C, (**b**) 650 °C and (**c**) 850 °C in air atmosphere.

3.3. Risk Assessment Associated with End-of Life of Nanocomposite Food Packaging Materials

Risk could be defined as a combination of the probability of occurrence of an event and its consequences, establishing a negative outcome. The methodology we applied to analyze and quantify the risk is borrowed from the standards and guidelines presented in several regulatory documents, such as: Risk Management Standard ISO 31000:2009 [35], British Standard BSI 2007 on safe handling and disposal of manufactured nanomaterials [36], USEPA Guidelines for Ecological Risk Assessment [37], and the British CSIRO Safe Handling and Use of Carbon Nanotubes [38]. In this study, the risk is mainly defined according to standard ISO 31000:2009 [35] as a comprehensive process of analysis and categorization, where the risk could be assessed quantitatively or qualitatively, depending on the probability of occurrence of the possible consequences.

To quantify the risk (R) of the release of MWCNTs and GNP from 3 wt. % GNP/MWCNT/PLA film at the end-of-life, as food packaging wastes, we have adopted the three factors method (C.E.L.), i.e., grading the three risk analysis factors: Consequences (C), Exposure (E), and Probability/Likelihood (L). The C.E.L. model is a widely recognized method of analysis and quantitative risk assessment [39,40]. Therefore, we have applied it for risk assessment in the four most popular scenarios for treatment of the food packaging wastes: biodegradation, combustion, burning in open air and accidental fire. The risk analysis factors are defined according to the C.E.L. model [39].

Consequences (C) represent the undesired results of an event or series of events. In this work, consequences are determined from the amount of the released MWCNTs and GNPs, as graded according to the recommended exposure limit of Carbon Nanotubes and Nanofibers (μg) in air (1 m^3), REL = 1 $\mu g/m^3$, proposed by NIOSH [41]. The NIOSH REL is expected to reduce the risk for pulmonary inflammation and fibrosis. Thus, six grades from 1 to 100 [39] are used for the quantification of consequences (xREL = 1 to > 1,000,000), as shown in Table 2.

Exposure (E) shows how often a certain danger can occur and how much the system is often threatened by accidents. The exposure estimates are based on the E-classification method [39], with six steps in the range from 0.5 to 1 (Table 2).

Likelihood (L) shows how likely it is to have consequences. The following six steps of grades from 0.5–10 are used to quantify this factor, as shown in Table 2.

The risk (R) is defined as the quantity comprised of the product of the three parameters: consequences (C), exposure (E) and probability (L): $R = C \times E \times L$. The eligibility of risk to health and environment is classified in the following five risk areas presented in Table 2, namely: minimal, acceptable, high, very high and unacceptable (hazard), depending on the calculated values of risk (R) varying from <20 to >400 [39]. The end-results of the risk assessment determine the eligibility of the identified risk and the need to apply measures to prevent or limit it.

Table 2. The grades for Consequences (C), Exposure (E), Likelihood (L) and Risk (R) used in this study.

Consequences (C)	Exposure (E)	Likelihood (L)	Risk (R)
1 = minimal (≤100 REL)	0.5 = very rare (less than once a year)	0.2 = not imagine et al	<20 = minimal
3 = significant (100–1000 REL)	1 = rarely (once a year)	0.5 = almost impossible	20–70 = acceptable
7 = serious (1000–10,000 REL)	2 = sometimes (once a month)	1 = unbelievable, but long-term still possible	70–200 = high
15 = very serious (10,000–100,000 REL)	3 = happening (once a week)	3 = not be normal, but possible	200–400 = very high
40 = major damage (100,000–1,000,000 REL)	6 = regular (daily)	6 = completely possible	>400 unacceptable (hazard)
100 = crash (>1,000,000 REL)	10 = continuous	10 = almost certain	

Quantitative and Qualitative Risk Assessment for the Release of GNP and MWCNTs

For the quantitative risk assessment, we used the data obtained above for the release of GNP and MWCNTs from 3 wt. % GNP/MWCNT/PLA film, during the degradation in ultra-strong dynamic test, as well as the burning at three heating temperatures 500, 650 and 850 °C in air atmosphere. Importantly, for the biodegradation (i.e., composting, weathering), we assumed that the total amount of 3 wt. % GNPs/MWCNTs nanofiler may release from the GNP/MWCNT/PLA nanocomposite film in the form of agglomerates and single nanoparticles due to the full degradation (hydrolysis) of the PLA polymer.

While for the burning processes like: combustion, burning in open air and accidental fire, the released GNPs and MWCNTs will depend on the heating temperature. The results from thermogravimetric analysis in Table 1 are used to simulate the combustion at the three temperatures in air atmosphere. For analysis of risk to humans and the environment, the released GNPs and MWCNTs are estimated for 100 kg wastes. Moreover, we have assumed that the single airborne nanoparticles of GNPs and MWCNTs (≤100 nm) and CB nanoparticles, that may release in 1 m^3 air or soil are only 1% of the total amount of released nanoparticle agglomerates during the four scenarios studied.

The admissibility of the human and environmental risk is presented as a multiplication of the three factors (R = C × E × L). Table 3 presents the three risk analysis factors and the quantitative risk assessment for the four scenarios: biodegradation; combustion at 500, 650 and 850 °C; burning in open air (at 500 °C); and accidental fire (at 500 °C). As seen, the risk of GNPs and MWCNTs from the biodegradation is R = 270, while the risk from combustion varies within R = 100–1500 depending on the heating temperature (500–850 °C). However, the risk from the burning of wastes in open air (in landfills) is quite high (R = 180), while that from accident fire is low of R = 22.5, due to the very rare exposure.

Table 3. Risk assessment by CEL model for the release of GNPs, MWCNTs and CB from 3% GNP/MWCNT/PLA film during biodegradation, combustion, burning and accidental fire.

Scenario	Amount of Nano Wastes (kg)	Released Total Amount GNP/MWCNT/CB (kg)	Released GNP/MWCNT/CB Nanoparticles in 1 m^3 Air (μg/m^3)	Consequences (C) (× REL)	Exposure (E)	Likelihood (L)	Risk, (R) = C × E × L
Biodegradation	100	3	30,000	15	3	6	270 very high
Combustion 850–650–500 °C	100	0.07–1.05–3.3	70–10,500–33,000	1–15	10	10	100–1500 high-to-hazard
Burning of wastes at 500 °C	100	3.3	33,000	15	2	6	180 high
Accident fire 500 °C	100	3.3	33,000	15	0.5	3	22.5 acceptable

For qualitative risk assessment, we used the approach described by Aven [40], which represents the dependence of the "consequences" in a function on the "frequency" of occurrence, the last being a derivative of "exposure" and "likelihood". This dependence is shown in Figure 4, for the purpose of classification of the risk of MWCNTs and GNP release from 100 kg of 3 wt. % PLA/MWCNT/GNP nanocomposite wastes, under the scenarios of biodegradation, combustion (500–850 °C), burning in open air (at 500 °C) and accidental fire (at 500 °C).

The lines represent the "constant risk" at the following levels of risk: R = 20, R = 70, R = 200 and R = 400, and outline the five risk zones: minimum (R < 20); acceptable (20 < R < 70); high (70 < R < 200); very high (200 < R < 400) and hazard (R > 400). The experimental points represent the risk of exposure to MWCNTs and GNPs from the PLA nanocomposite film throughout the four scenarios of the waste treatment, listed in Table 3.

As shown in Figure 5, the scenario of total biodegradation of GNP/MWCNT/PLA nanowastes leads to a "very high" risk for release of GNPs and MWCNTs nanoparticles to the environment. This may pose questions associated with environmental and human risks due to local composting of post-consumer wastes, when those nanoparticles enter the soil. Combustion of nanowastes at heating temperatures of 500 °C may result in "unacceptable risk/hazard" from airborne GNPs and MWCNTs; however, by increasing the heating temperature to 850 °C, the risk decreases to "high". Particularly, burning of nanowastes in open air (e.g., in landfills), which is a regular practice in underdeveloped regions, results in "high" risk. Therefore, such practices must be addressed and limited by the regulators, as it may affect more people and cause significant damage to the environment in those regions. The scenario of accident fire lead to "acceptable" risk, but it will have a local negative effect; therefore, preventive measures for safety have to be taken into account.

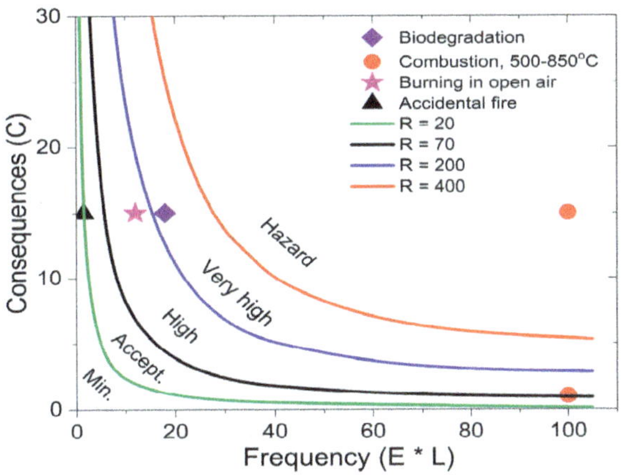

Figure 5. Qualitative risk assessment: consequences vs. frequency of occurrence of GNP and MWCNT exposure, related to four scenarios at the end-of life of 3 wt. % GNP/MWCNT/PLA packaging films: biodegradation, combustion, burning and accident fire. Risk is estimated for 100 kg nanocomposite wastes.

3.4. Safety Concerns

Bioplastic packaging is widely used these days for wrapping products from food to electronics to protect them from dust, bacteria and water vapor, and to maximize the lifetime of packaged products. Among the many different biopolymers used, poly(lactic acid) (PLA) possesses good mechanical property and cost-effectiveness necessary of biodegradable food packaging. However, PLA packaging suffers from poor water vapor and oxygen barrier properties compared to many petroleum-derived ones. A key challenge is, therefore, to simultaneously enhance both the water vapor and oxygen barrier properties by incorporation or encapsulation of graphene and carbon nanotubes into the PLA packaging [42,43]. Studies comparing the graphene nanoplates, GNPs and carbon nanotubes, shows CNTs are very rare, making it difficult to analyze their overall safety and risk [44,45]. In general, authors agree, that despite their common carbon structure, CNTs and GNPs are two very different nanomaterials, due to their different physical and chemical characteristics. Dimensions, surface chemistry and impurities are equally important for determining the aggregation, degradation and toxicological effects of CNTs and GNPs. Their shape (tubular vs. plane) and their dimensions (2D vs. 1D) are key structural differences. The CNTs tend to form entangled aggregates, and GNPs tend to stack in several layers. GNPs are characterized by a lower aspect ratio (length/width), greater surface area and better dispersion in most solvents, compared to CNTs. The colloidal dispersions of graphene can be obtained without metallic impurities, with high stability and less aggregation. All those characteristics could theoretically offer significant advantages of GNPs over CNTs, in terms of risk management and safety.

Our current study expands the upper safety concerns comparing GNPs and CNTs, by finding that large graphene nanosheets indeed release from PLA-based nanocomposite films at temperatures above the glass transition, during ultra-strong dynamic migration test. However, MWCNTs remain embedded in the polymer matrix if the PLA matrix is partly degraded. Moreover, GNPs and MWCNTs are found to remain unburned in the residue after combustion up to 650 °C, while at 850 °C only GNPs and carbon soot are found, but not MWCNTs.

In this research, we stress the safety concerns at the end-of-life of nanocomposite food packaging, related to different waste treatments, such as: biodegradation, combustion, burning in open air, and accident fire. Safety concerns may arise due to biodegradation and composting of nanowastes based on biodegradable polymers. Composting has the potential to transfer biodegradable plastics into useful soil amendment products by accelerated degradation. However, special care should be taken while handling local composting of biodegradable nanowastes to limit potential environmental risks due to the release of nanoparticles in the soil from the compost.

Combustion of nanocomposite wastes to produce energy has recently been discussed as a nanowastes management strategy, and thus the risks for nanoparticle emission during incineration of thermoplastic nanocomposites must be addressed and investigated. Our current study confirms that the combustion of PLA-based nanocomposites could form residue ashes containing unburned GNPs and MWCNTs; this is associated with "hazard" to "high risk", depending on the temperature. The amount of unburned nanoparticles may be controlled by increasing the heating temperature above 500–850 °C; however, single GNPs and MWCNTs may also release in the combustion gas phase. Such release can be a source of risk in accidental scenarios, like fire, uncontrolled incineration/combustion, or the absence of nano-filtration of the combustion gas phase. Safety concerns arise about common practice in some regions for the burning of nanowastes in open air, e.g., in landfills or single-use disposable systems, due to the gradual increase of nanowastes from food packaging. As shown in this study, such regular practice leads to "high" risk for humans and the environment from airborne nanoparticles, such as GNPs and MWCNTs.

Based on the above study, we may propose a few safety measures for the prevention of risk from the packaging containing GNPs and MWCNTs at the end-life waste treatment: (i) Material safety data sheet (MSDS should contain clear information if the nanomaterial is safe for composting; (ii) specific labeling for prevention from composting should be adopted and printed on the packaging; (iii) regulatory limitations imposing the control on the combustion processes and exhaust gases will contribute to safety and risk prevention; (iv) regulatory measures imposing the limitation of burning of nanowastes in open air in landfills are required.

4. Conclusions

The release of graphene nanoplatelets and multiwall carbon nanotubes from polylactic-based film at the end-of-life of wastes treatment was investigated during degradation and combustion/burning. The released airborne nanoparticles and the degradation of the nanocomposite film during an ultra-strong dynamic migration test were confirmed by different visualization methods (TEM, AFM, SEM). Thermogravimetric analysis in air atmosphere was used to simulate the combustion of nanocomposite wastes. It was found that single graphene nanoplatelets of nanosized thickness and length of 100–1000 nm, as well as their micron-size loose aggregates, indeed release in relatively large amounts from the PLA nanocomposite film at high temperature dynamic treatment due to partial degradation of the PLA polymer. However, the release of the entangled MWSNTs is possible only after full degradation (hydrolysis) of the PLA matrix polymer.

Combustion or burning at 500 and 650 °C result in residue ash, which contains mainly single airborne GNPs and MWCNTs, while at 850 °C, the small amount of residue ash (~0.07%) contains only GNPs and amorphous carbon soot. Therefore, the MWCNTs fully degrade at the heating temperature of 850 °C, while the GNPs still remain in the residue.

New concerns with the end-of-life nanostructured materials emerged by adopting the 3-factors, C.E.L. model for risk assessment. The consequences (C) were determined from the amount of the released MWCNTs and GNPs, and were graded according to the recommended exposure limit of carbon nanotubes, REL = 1 $\mu g/m^3$, proposed by NIOSH. The exposure (E) and likelihood (L) were estimated based on the E-classification method. Four scenarios are discussed: biodegradation by composting, combustion, burning in open air and accidental fire, which may lead to "very high", "hazard" and "high" risk, respectively. Such treatment of nanowastes may pose a potential release of GNPs and MWCNTs into the environment, with all their associated environmental and human riskspresently not accounted for. Safety measures are proposed for the end-of-life phase of nanowastes in order to avoid or prevent the risks.

Author Contributions: In this study, the concept, analysis of results and writing of the manuscript were performed by S.K.; the A.F.M. visualization was conducted by E.I.; the methodology and manuscript text were examined by N.V. and E.I. All authors read and approved the final manuscript.

Acknowledgments: E.I. acknowledged the support from the projects H2020-MSCA-RISE-734164 Graphene 3D and H2020-SGA-FET-GRAPHENE-2017-785219 Graphene Core 2. Authors thank I. Pertova for the TGA test.

References

1. Azeredo, H.M.C.D.; Mattoso, L.H.C.; McHugh, T.H. Nanocomposites in Food Packaging—A Review. In *Advances in Diverse Industrial Applications of Nanocomposites*; Reddy, B., Ed.; Intech Open Science: London, UK, 2011; pp. 57–78.
2. Silvestre, C.; Pezzuto, M.; Cimmino, S.; Duraccio, D. Polymer Nanomaterials for Food Packaging: Current Issues and Future Trends. In *Ecosustainable Polymer Nanomaterials for Food Packaging: Innovative Solutions, Characterization Needs, Safety and Environmental Issues*, 1st ed.; Silvestre, C., Cimmino, S., Eds.; CRC Press Taylor & Francis Group: Boca Raton, FL, USA, 2013; pp. 2–28.
3. Chandhry, Q.; Scotter, M.; Blackburn, J.; Ross, B.; Boxall, A.; Castle, I. Applications and implications of nanotechnologies for the food sector. *Food Addit. Contam.* **2008**, *25*, 241–258. [CrossRef] [PubMed]
4. Randviir, E.P.; Brownson, D.A.C.; Banks, C.E. A decade of graphene research: production, applications and outlook. *Mater. Today* **2014**, *17*, 426–432. [CrossRef]
5. Honarvar, Z.; Hadian, Z.; Mashayekh, M. Nanocomposites in food packaging applications and their risk assessment for health. *Electron. Phys.* **2016**, *8*, 2531–2538. [CrossRef] [PubMed]
6. Wu, L.L.; Wang, J.; He, X.; Zhang, T.; Sun, H. Using graphene oxide to enhance the barrier properties of poly(lactic acid) film. *Packag. Technol Sci.* **2014**, *27*, 693–700. [CrossRef]
7. Du, J.; Cheng, H.M. The fabrication, properties, and uses of graphene/polymer composites. *Macromol. Chem. Phys.* **2012**, *213*, 1060–1077. [CrossRef]
8. Cui, Y.; Kundalwal, S.I.; Kumara, S. Gas barrier performance of graphene/polymer nano-composites. *Carbon* **2016**, *98*, 313–333. [CrossRef]
9. Ivanov, E.; Kotsilkova, R. Reinforcement effects of carbon nanotubes in polypropylene: Rheology, structure, thermal stability, nano- micro- and macro mechanical properties. In *Handbook of Nanoceramic and Nanocomposite Coatings and Materials*; Makhlouf, A., Scharnweber, D., Eds.; Elsevier: Amsterdam, Netherlands, 2015; pp. 351–383.
10. Kotsilkova, R.; Ivanov, E.; Krusteva, E.; Silvestre, C.; Cimmino, S.; Duraccio, D. Evolution of rheology, structure and properties around the rheological flocculation and percolation thresholds in polymer nanocomposites. In *Ecosustainable Polymer Nanomaterials for Food Packaging: Innovative Solutions, Characterization Needs, Safety and Environmental Issues*, 1st ed.; Silvestre, C., Cimmino, S., Eds.; CRC Press Taylor & Francis Group: Boca Raton, FL, USA, 2013; pp. 55–86.
11. Novack, B.; Ranville, J.F.; Nowack, S.; David, R.M.; Fissan, H.; Morris, H.; Shatkin, J.A.; Stintz, M.; Zepp, R.; Brouwer, D. Potential release scenarios for carbon nanotubes used in composites. *Environ Int.* **2013**, *59*, 1–11.
12. Arvidsson, R.; Molander, S.; Sanden, B.A. Review of potential environmental and health risk of nanomaterial graphene. *Human Ecol. Risk Assess.* **2013**, *19*, 873–887. [CrossRef]
13. Lowry, G.V.; Gergory, K.B.; Apte, S.C.; Lead, J.R. Transformation of nanomaterials in the environment. *Environ. Sci. Tech.* **2012**, *46*, 6893–6899. [CrossRef] [PubMed]
14. Kingston, C.; Zepp, R.; Andrady, A.; Boverhof, D.; Fehir, R.; Hawkins, D.; Roberts, J.; Sayre, P.; Shelton, B.; Sultan, Y.; et al. Release Characteristics of Selected Carbon Nanotube Polymer Composites. *Carbon* **2014**, *68*, 33–57. [CrossRef]
15. Harper, S.; Wohlleben, W.; Doa, M.; Nowack, B.; Clancy, S.; Canady, R.; Maynard, A. Measuring Nanomaterial Release from Carbon Nanotube Composites: Review of the State of the Science. *J. Phys. Conf. Series* **2015**, *617*, 012026. [CrossRef]
16. Schlagenhauf, L.; Nuesch, F.; Wang, J. Release of Carbon Nanotubes from Polymer Nanocomposites. *Fibers* **2014**, *2*, 108–127. [CrossRef]
17. Petersen, E.J.; Zhang, L.; Mattison, N.T. Potential release pathways, environmental fate, and ecological risks of carbon nanotubes. *Environ. Sci. Technol.* **2011**, *45*, 9837–9856. [CrossRef] [PubMed]
18. Duncan, T.V.; Pillai, K. Release of engineered nanomaterials from polymer nanocomposites: Diffusion, dissolution, and desorption. *ACS Appl. Mater. Interfaces* **2015**, *7*, 2–19. [CrossRef] [PubMed]

19. Velichkova, H.; Petrova, I.; Kotsilkov, S.; Ivanov, E.; Vitanov, N.K.; Kotsilkova, R. Influence of Polymer Swelling and Dissolution in Food Simulants on the Release of Graphene Nanoplates and Carbon Nanotubes from Poly(lactic) acid and Polypropylene Composite Films. *J. Appl. Polym. Sci.* **2017**, *134*, 45469. [CrossRef]
20. Velichkova, H.; Kotsilkov, S.; Ivanov, E.; Kotsilkova, R.; Gyoshev, S.; Stoimenov, N.; Vitanov, N.K. Release of carbon nanoparticles of different size and shape from nanocomposite poly(lactic) acid film into food simulants. *Food Addit. Contam. A* **2017**, *34*, 1072–1085. [CrossRef] [PubMed]
21. Kotsilkov, S.; Ivanov, E.; Vitanov, N.K. Study on the release of graphene and carbon nanotubes at the end-of-life phase of polymer nanocomposites: Risk assessment and safety concerns. In Proceedings of the ISER 144th International Conference, New Delhi, India, 15–16 July 2018.
22. Bernard, C.; Nguyen, T.; Pellegrin, B.; Holbrook, R.D.; Zhao, M.; Chin, J. Fate of graphene in polymer nanocomposite exposed to UV radiation. *J. Phys. Conf. Series* **2011**, *304*, 012063. [CrossRef]
23. Musee, N. Nanowastes and the environment: Potential new waste management paradigm. *Environ Int.* **2011**, *37*, 112–128. [CrossRef] [PubMed]
24. Mueller, N.C.; Buha, J.; Wang, J.; Ulrich, A.; Nowack, B. Modeling the flows of engineered nanomaterials during waste handling. *Environ. Sci. Process. Impacts* **2013**, *15*, 251–259. [CrossRef] [PubMed]
25. Bouillard, J.; R'Mili, B.; Moranviller, D.; Vignes, A.; Le Bihan, O.; Ustache, A.; Bomfim, J.S.; Frejafon, E.; Fleury, D. Nanosafety by design: risks from nanocomposite/nanowaste combustion. *J. Nanopart. Res.* **2013**, *15*, 1–11. [CrossRef]
26. Song, J.H.; Murphy, R.J.; Narayan, R.; Davies, G.B.H. Biodegradable and compostable alternatives to conventional plastics. *Phil. Trans. R. Soc. B* **2009**, *364*, 2127–2139. [CrossRef] [PubMed]
27. Narayan, R. Rationale, drivers, standards, and technology for biobased materials. In *Renewable Resources and Renewable Energy*; Graziani, M., Fornasiero, P., Eds.; CRC Press Taylor & Francis Group: Boca Raton, FL, USA, 2006.
28. Commission Regulation, (EU). *No. 10/2011 of 14 January 2011 on Plastic Materials and Articles Intended to Come Into Contact with Food Text with EEA Relevance*; Official Journal of the European Union: Aberdeen, UK, 15 January 2011.
29. Mutsuga, M.; Kawamura, Y.; Tanamoto, K. Migration of lactic acid, lactide and oligomers from polylactide foodcontact materials. *Food Addit. Contam.* **2008**, *25*, 1283–1290. [CrossRef] [PubMed]
30. Xu, Q.; Yin, X.; Wang, M.; Wang, H.; Zhang, N.; Shen, Y.; Xu, S.; Zhang, L.; Gu, Z. Analysis of phthalate migration from plastic containers to packaged cooking oil and mineral water. *J. Agric. Food Chem.* **2010**, *58*, 11311–11317. [CrossRef] [PubMed]
31. Ebnesajjad, S. *Plastic Films in Food Packaging: Materials, Technology and Applications*, 1st ed.; Elsevier: Amsterdam, Netherlands, 2013; Volume 16, pp. 345–388.
32. Nyden, M.R.; Harris, R.H.; Kim, Y.S. Characterizing particle emissions from burning polymer nanocomposites. In Proceedings of the 21th BCC Conference on Flame Retardation; 2010; Volume 1, pp. 717–719.
33. Li, J.; Tong, L.; Fang, Z.; Gu, A.; Xu, Z. Thermal degradation behavior of multi-walled carbon nanotubes/polyamide 6 composites. *Polym. Degrad. Stabil.* **2006**, *91*, 2046–2052. [CrossRef]
34. Yang, S.; Castilleja, J.R.; Barrera, E.V.; Lozano, K. Thermal analysis of an acrylonitrile–butadiene–styrene/SWNT composite. *Polym. Degrad. Stabil.* **2004**, *83*, 383–388. [CrossRef]
35. ISO Guide. Risk Management. Geneva, Switzerland, 2009.
36. BSI Nanotechnologies. *Guide to Safe Handling and Disposal of Manufactured Nanomaterials*, Edinburg, UK, 2007.
37. *Guidelines for Ecological Risk Assessment*; US Environmental Protection Agency: Washington, DC, USA, 1998.
38. *Safe Handling and Use of Carbon Nanotubes*; Safe Work Australia: Canberra, Australia, 2012 March.
39. Vitanov, N.K.; Dimitrova, Z.I. Risk analysis. Technological, political and other risks. In *Qualitative Analysis of Risk*; Vanio Nedkov: Sofia, Bulgaria, 2015; Volume 1.
40. Aven, T.; Renn, O. *Risk Management and Governance. Concepts, Guidelines and Applications*; Springer: Berlin, Germany, 2010; Volume 16.
41. Howard, J. Occupational Exposure to Carbon Nanotubes and Nanofibers. In *Current Intelligence Bulletin*; National Institute for Occupational Safety and Health: Washington, DC, USA, 2013.
42. Seethamraju, S.; Kumar, S.; Bharadwaj, K.; Madras, G.; Raghavan, S.; Ramamurthy, P.C. Million-Fold Decrease in Polymer Moisture Permeability by a Graphene Monolayer. *ACS Nano* **2016**, *10*, 6501–6509. [CrossRef] [PubMed]

43. Norazlina, H.; Kamal, Y. Graphene modifications in polylactic acid nanocomposites: a review. *Polym. Bull.* **2015**, *72*, 931–961. [CrossRef]
44. Bussy, C.; Ali-Boucetta, H.; Kostarelos, K. Safety considerations for graphene: Lessons learnt from carbon nanotubes. *Accounts Chem. Res.* **2013**, *46*, 692–701. [CrossRef] [PubMed]
45. Shi, J.; Fang, Y. Biomedical Applications of Graphene. In *Graphene: Fabrication, Characterizations, Properties and Applications*, 1st ed.; Zhu, H., Xu, Z., Eds.; Academic Press: Cambridge, MA, USA, 2018; pp. 215–232.

7

Polymeric Nanocomposites and Nanocoatings for Food Packaging: A Review

Cornelia Vasile

Physical Chemistry of Polymers Department, Petru Poni Institute of Macromolecular Chemistry (PPIMC), Romanian Academy, 41A Gr. Ghica Alley, RO 700487 Iasi, Romania; cvasile@icmpp.ro

Abstract: Special properties of the polymeric nanomaterials (nanoscale size, large surface area to mass ratio and high reactivity individualize them in food packaging materials. They can be processed in precisely engineered materials with multifunctional and bioactive activity. This review offers a general view on polymeric nanocomposites and nanocoatings including classification, preparation methods, properties and short methodology of characterization, applications, selected types of them used in food packaging field and their antimicrobial, antioxidant, biological, biocatalyst and so forth, functions.

Keywords: polymer; nanocomposites; nanocoatings; food packaging; risks; smart nanomaterials

---

1. Introduction

Nanoscience and nanotechnology applications to the agriculture, food sector and food safety are relatively recent compared with their use in cosmetics and personal care products (60%), paints and coatings, catalysts and lubricants, security printing, drug delivery and pharmaceuticals, medical therapeutics and diagnostics, energy production, molecular computing and structural materials [1–4]. Over \$17.8 billion worldwide investment was made in 2010 for research and development in nanotechnology and this will grow to \$3 trillion by 2020. Food packaging sector counts for ~50% market value for all nanotechnology-enabled products. Annual growth rate is 11.65%. Nanotechnology is mainly applied in extremely high gas barriers packaging materials with antimicrobial properties and in nanoencapsulants for the delivery of nutrients, flavours, or aromas [5,6].

According to regulation (EC) = No 1935/2004 a good packaging should have the following functions: "to protecting the food from dirt or dust, oxygen, light, pathogenic microorganisms, moisture and a variety of other destructive or harmful substances. Packaging must also be safe under its intended conditions of use, inert, cheap to be produced, lightweight, easy to dispose or to reuse, able to withstand extreme conditions during processing or filling, impervious to a host of environmental storage and transport conditions and resistant to physical abuse" [7].

Nanomaterials are applied in packaging and food safety are in various forms such as: polymer nanocomposites with high barrier properties, intelligent packaging, nanocoatings, surface biocides, active packaging, silver nanoparticles as antimicrobial agents, nutrition and nutraceuticals, nanosensors and assays for the detection of food relevant analytes (gases, small organic molecules and food-borne pathogens) and bioplastics.

Main recent development areas in packaging field are:

(I) *Improved packaging* which offers improved mechanical properties as flexibility, enhanced barrier properties against water, gases, taint, durability, temperature/moisture stability, and so forth;

(II) *Active/bioactive food packaging* offers antimicrobial, antioxidant or biocatalytic functions. It can be obtained by the incorporation of active/bioactive compounds into matrices used in existing packaging materials, or by the application of coatings with the mentioned functionality through physical or chemical surface modification. Active packaging is applied in food packaging, pharmaceuticals and consumer goods in order to improve shelf life, safety, or quality of packaged foods. Coating option is advantageous because the bulk properties of the packaging materials are preserved almost intact by using a minimum amount of active agent required to impart efficacy and therefore also cost is reduced.

(III) *Smart/intelligent packaging* as a promising area for active packaging coating is developed by manufacture of nano(bio)sensors which can indicate quality of foodstuffs, of nano(bio)switch to release preservatives and nano-coatings as antimicrobial, antifungal, antioxidant, barrier coatings, external stimuli responsive materials and self-cleaning food contact surfaces. Intelligent inks such as nanoparticles and reactive nanolayers allow analyte recognition at nanoscale. Printed labels are applied to indicate: temperature, time, pathogen, freshness, humidity, integrity [8]. Smart packaging may monitor various parameters such as: temperature, oxygen, pH, moisture and so forth [9–11] of packaged products.

2. Polymeric Nanocomposites

Polymeric nanocomposites are mainly obtained by dispersing of nanoscale filler into a polymeric matrix.

Polymers commonly used in food packaging as either matrices or substrates for bioactive coatings are synthetic undegradable as low density and high density polyethylene (LDPE and HDPE), polypropylene (PP), polyethylene terephthalate (PET), ethylene vinyl alcohol (EVOH) copolymer, polyamide (PA), polystyrene (PS) and degradable ones as polyhydroxyalkanoates (polyhydroxybutyrate (PHB), poly(hydroxybutyrate-*co*-hydroxyvalerate) (PHBV)), poly(lactic acid) (PLA), polycaprolactone (PCL), polyvinyl alcohol (PVOH) and biopolymers as polysaccharides and proteins with critical issue on water resistance, migration and permeability and so forth These properties may be improved by adding reinforcing compounds (nanofillers), forming bionanocomposites [12]. Because of the different characteristics, their application in food packaging is specific. For example, HDPE is used as milk bottles and bags, LDPE for trays and general-purpose containers, PP has a high melting point, being ideal for hot-fill liquids, films and microwavable containers, while PET is clear, tough and has good gas and moisture barrier properties, so it is used for soft drink bottles, beverages and mineral waters, carbonated drinks and so forth [13]. While synthetic polymers generate huge waste quantity and have an environmental negative impact, by using degradable polymers these problems are avoid, because they are biocompatible and safe to use. However, most degradable polymers are in the research stage or their cost is high, some food packaging based on PLA (the cheapest bio-based material) and poly(butylene adipate-*co*-terephthalate) (PBAT) based packaging is used in Europe, Japan and North America [14,15]

Moreover, recently, a series of biodegradable nanocomposite films based on PBAT and reinforced with an organophilic layered double hydroxide OLDH (0.5−4 wt %) nanosheets were scale-up fabricated. Xie et al. reported such materials with "outstanding thermal, optical, mechanical and water vapour barrier properties a 37% reduction in haze and a 41.9% increase of tensile strain at break better than the pure PBAT film and commercial polyethylene packing materials" [16].

The active agents used in active packaging and coatings include antimicrobials, antioxidants and enzymes which control microbial growth, inhibit oxidative degradation reactions and offer a targeted biocatalysis that maintains food safety and quality and controls spoilage.

The most used nanofillers and nanoreinforcements are: clay montmorillonite (MMT) and kaolinite and silicate nanoplatelets (two-dimensional layers, which are 1 nm thick and several microns long), silica nanoparticles, carbon nanotubes, graphene nanosheets, silver, zinc oxide, titanium dioxide, copper and copper oxides, starch nanocrystals, cellulose nanofibres and nanowhiskers, chitosan and chitin whiskers and others. In respect with their geometry isodimensional nanoparticles have three nanometric external dimensions while nanotubes or whiskers are elongated structures with two external dimensions on nanometre scale and the third is larger (Figure 1) [17–19]. Other classification is zero-dimensional, one-dimensional, two-dimensional and three-dimensional nanostructured materials [20]. Nanofibres may be also known as nanowire (electrically conducting nanofibre), nanotube (hallow nanofibre) and nanorods (solid nanofibre) [12].

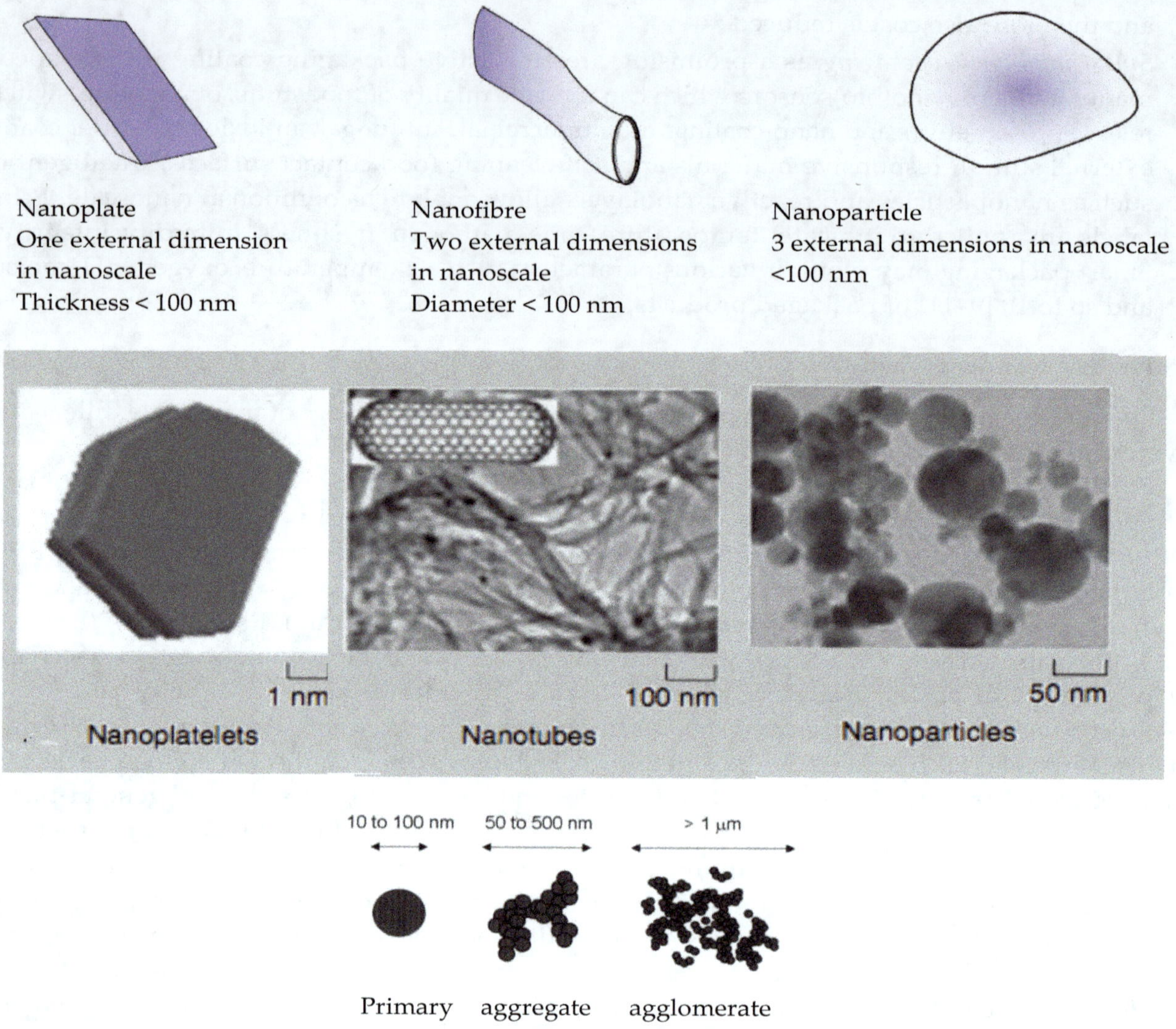

Figure 1. Representative images for nanofillers for nanocomposites. adapted from [17–19].

When only one dimension is in the nanometre range, the composites are known as polymer-layered silicate nanocomposites [21,22]. The role of nanofillers and nanoreinforcements in composites consists in the enhancement of the mechanical, thermal (glass transition, melting and degradation temperature) and barrier properties by increasing path length for gas diffusion, changes in surface wettability and hydrophobicity and so forth For food packaging they may offer antimicrobial activity, oxygen scavengers, enzyme immobilization, biosensing and so forth Nanocomposites usually represent significant changes in properties at low loads <5 wt % of nanofillers and <2% volume. Also, nanocomposites improve the stability of sensory properties, such as flavour, better maintenance of colour and texture, increased product stability through the food chain and less spoilage. It is well-known that the interfacial region in nanocomposites is extremely large, therefore the interaction of the polymer with the nanoparticles is strong and this will change the polymer mobility and relaxation dynamics, decreases chain mobility due to nanomodification and that means an increase in Tg [23].

Nanocomposites may improve both thermal and environmental dimensional stability. Two factors are considered responsible for improving the thermal stability in thermoplastic nanocomposites with respect to the neat matrix: (i) the chemical composition and morphology of the nanocomposites is different from that of the pristine polymer; (ii) the thermal motion of the polymer molecules is restricted by the filler nanoparticles. Incorporation of nanofillers in thermoplastics usually led to a marked increase in the melt viscosity, especially in the range of low frequencies.

Several types of polymer nanocomposites are known depending on the nanofiller used as mentioned above. Examples include UV absorbers to prevent UV degradation of plastic polymers (e.g., nano-titanium dioxide, iron oxides, silica, alumina), titanium nitride (TiN) used to improve the strength of materials, nano-calcium carbonate-polymer composites, biodegradable starch and/or polylactic acid with various nanoclay composites, gas barrier coatings and so forth [24–28].

2.1. Preparation Methods

The preparation methods for polymer nanocomposites should assure a good dispersion of nanofiller into matrix to obtain intercalated or exfoliated structures [29–33]. The exfoliated nanocomposites are also classified as ordered and disordered exfoliations. Partial exfoliation is an intermediate morphology between intercalation and exfoliation. Significant variations in physical and mechanical properties of polymer nanocomposites are directly correlated with the differences in morphology [34].

There are two main strategies to obtain nanostructures and consequently polymer nanocomposites and nanocoatings—Table 1—so called "top down" or ex situ synthesis (attachment of nanoparticles to the polymer matrices prepared into different step) and "bottom up" or in situ synthesis methods (e.g., in situ polymerization, spin coating, casting) [35]. In the "top-down" approach the nanoparticles are produced in a separate step and then they are dispersed in polymer (direct mixing of the filler into a polymeric matrix, ball milling and application of severe plastic deformation.), melt, or monomer solution which is then polymerized or nanostructures are generated by mechanical disintegration (milling) of the previous prepared material. In the "bottom-up" approach, the nanostructures are built up in a chemical process consisting in generation of nanoparticles inside polymer matrix or monomer polymerizing solution by chemical, thermal or photolytic decomposition of some precursors. Bottom-up approaches may be in gas-phase or in liquid phase such as: sol-gel processing, chemical syntheses, spraying, plasma or flame sputtering, spinning to make thin polymer fibres, chemical vapour deposition (CVD), laser pyrolysis, atomic or molecular condensation, electrodeposition, supercritical fluid synthesis, and so forth [36], while among the top-down techniques it is possible to include the energetic mechanical milling, sonication and so forth.

The preparation method determines both the concentration and distribution of the nanofillers and nanoreinforcements into the polymer [37].

The in-situ polymerization technique was widely used to obtain different kinds of nanocomposites such as those based on poly (vinyl acetate), nylon 6, in situ synthesis of poly(urethanes) PU in in dimethylformamide (DMF), emulsion polymerization of styrene in water and many others [38,39].

Solution mixing or *solvent casting* involves the vigorous stirring or ultrasonication of the nanoparticles in a polymer solution before casting in a mould and then evaporating the solvent. Both water and organic solvents can be used to prepare nanocomposites with either thermoplastics or thermosets. The removal of organic solvent after casting has environmental implications. The polarity of the solvent is a critical characteristic which influences the intercalation of the polymer into the space between the clay platelets. Nanocomposites are prepared by *emulsion polymerization* of acrylates in silica sols. By using this technique, it is possible to achieve a homogeneous distribution of silica nanoparticles into polymer at silica content up to 50% in the nanocomposites.

Table 1. Methods for preparation of polymer nanocomposites.

Method	Method to Obtain Nanocomposites	Results
In situ polymerization: emulsion and miniemulsion polymerization	dispersing fine nanofiller and nanoreinforcement in a monomer	monomers interact with the nanofiller surface and form a uniform suspension
Solution casting and latex method or solvent processing	1. filler dispersal in the polymer solution 2. solvent evaporation 3. freeze-drying and hot-pressing 4. freeze-drying, extruding and hot-pressing the mixture. 5. surfactant addition 6. grafting of long chains onto nanofiller surface	sandwiched multilayer structures forming of filler-rich layers polymer polarity-based results The dispersion of nanoparticles in the nanocomposite film strongly depends on the processing technique and conditions
Direct addition/extruder Blending, melt processing Shear mixing	1.addition of filler directly into the melted polymer 2. blending either by (mechanical mixer and extruder)	target: to obtain uniform distribution of the nanofillers in a polymer matrix
Deposition or layer (LBL) assembly	Layer-by-layer deposition	sequential substrate dipping in clay and polycation solutions were adopted to make the coating Multilayer films
Dispersion and chemical reaction	UV-curing in presence of photo-initiator; casting, evaporating the solvent	cross-linked nanocomposites
Electrospinning	development of electrospun nanofibers	nanofibres with different morphologies

At industrial scale of the *extrusion* or *melt processing*, the compounding of the components of a nanocomposite material is performed in a single or twin-screw extruder where the mixture of polymer and the nanofiller is a melt state. The shear and elongational stress applied with mixer during process help to break the filler agglomerates and uniformly dispersing them into the polymer matrix. As end products, films and other items can be formed from blend by profile injection moulding, extrusion, blow moulding. The intrinsic lack of thermal stability of many active/bioactive compounds which can be lost through degradation and evaporation during the heat transfer and high shear that also could degrade both the polymer and the nanofillers (as carbon nanotubes) limit the applicability of this preparation method. The homogeneous distribution of active agents into matrix is difficult. Processing at high shear or sonication techniques are used to deaggregate or exfoliate the clusters and to increase the surface area exposed to the polymer. Polymers able to interact with nanofillers give a good dispersion.

Shear mixing: Low-shear mixing or high-shear mixing can be used for incorporating solid nanoparticles into a liquid polymer. Under these conditions and if the nanoparticles are compatible with the selected polymer, the mixing will disrupt the nanoparticle aggregates and disperse the polymer matrix into the nanoparticle layers or onto the nanoparticle surfaces. The dispersion degree of the nanofiller and the filler/matrix interaction can be generally improved by (i) using surface-treated filler to reduce aggregation phenomena and (ii) incorporating a compatibilizer or a tensioactive agent within the polymer matrix (such as maleic anhydride grafted polypropylene).

LBL and electrospinning techniques are described in the nanocoatings section.

2.2. *Types of Polymer Nanocomposites Used in Food Packaging*

Some types of nanocomposites used in food packaging are given in Table 2 [12,33].

2.2.1. Montmorilonites (MMT) Containing Nanocomposites

The polymer–clay morphologies are classified in (1) tactoid, (2) intercalated and (3) exfoliated. In the *tactoid structures* due to the poor affinity of nanoclay with the polymer, the interlayer space of the clay gallery will not expand, the components do not mix each other and nanocomposites are not formed. In *intercalated structures* a moderate affinity between polymer and clay exists hence they are characterized by moderate expansion of the clay interlayer but the shape remains unchanged, polymer chains are able to penetrate the basal spacing of clay, mixing of components being homogeneous.

A high affinity between polymer and nanoclay promotes separation of the clay clusters into single sheets within the continuous polymer matrix resulting a homogeneous dispersion as the *exfoliated structures* which determine important changes in properties of the nanocomposites in respect with those of neat polymer.

Generally, MMT behaves as an effective reinforcement/filler, offering a high surface area and large aspect ratio. However, in many cases the compatibility with the matrix should be improved both by organophilization of the filler, the use of surfactants or compatibilizers. By these ways better exfoliated structure and better mechanical properties are achieved for nanocomposites. Clay layers constitute a barrier to gases and water, forcing them to follow a tortuous path, nanocomposites showing improved oxygen and water vapour barrier properties and some of them as Cloisire 93A, Cloisite 30B and Dellite High Pressure Sodium HPS offer also antimicrobial activity against both Gram-positive and Gram-negative bacteria, excepting fungus *Candida albicans* [40]. In many cases the mechanical properties are improved, glass transition and thermal degradation temperatures increase but some nanoclays decrease transparency of the films [12].

Table 2. Some examples of nanofillers intended to be used in nanocomposites for food packaging (for many examples see also recent reviews on this subject [41–44].

Type of Nanofiller	Matrix	Preparation	Properties/Applications	Ref
Organoclay	LDPE and HDPE	melt mixing using PE grafted with maleic anhydride as compatibilizer-exfoliation	oxygen permeability of PE decreases gradually with the clay concentration, reaching a maximum reduction of ~30% for 15 wt % MMT; dynamic moduli increase showing pseudo solid-like behaviour at clay concentrations higher than 8 wt %.	[45]
Nanolayers of Nanoter™ from NanoBioMatters LTD Spain	PE	melt processing	very good barrier properties	[46]
4% MMT	EPDM	melt processing	decreased N_2 permeability by 30%	[46]
Bentonite	PLA	solution casting	improve strength and modulus; decreased elongation at break	[47]
5% MMT	PVOH	casting	90% reduction in water permeability retaining optical clarity	
MMT	proteins and polysaccharides	casting	60% reduction in water permeability	[46]
1.1%–4%–10% Various unmodified and organically modified MMT, Cloisite 25A, Nanoter™	PLA, PCL, PHA, PHBV Strach	monolayer packaging	reduction of oxygen and water permeability	
MMT	chitosan films	solvent casting	exfoliated and intercalated structure depending on MMT amount, Tensile strength of a chitosan film was enhanced and elongation-at-break decreased	[48]
MMT	poly (ε-caprolactone) (PCL)	electrospinning	improved mechanical properties even elongation at break	[49]
Anionic sodium MMT exfoliated	cationic polyacrylamide on a PET substrate	Layer-by-layer (LbL) self-assembly multilayer film	oxygen transmission rate (OTR) decreased as a function of number of bilayers deposited, until a negligible value–below 0.005 $cm^3/(m^2$ day atm)—for a 30-bilayer film, microwaveable and with a good optical transparency (higher than 90%), it was presented as a good candidate for aluminium foil replacement in food packaging.	[50]
5% Clay ZnO stabilized with sodium carboxymethylcellulose	thermoplastic starch; gelatinized starch film glycerol plasticized-PEA starch	melt extruded	improve the mechanical strength of biopolymers, decreased water vapour permeability by using only 5% (*w/w*) of clays; the highest exfoliation and best improvement in mechanical properties, exfoliated clay.	[51,52]

Table 2. *Cont.*

Type of Nanofiller	Matrix	Preparation	Properties/Applications	Ref
1 wt % and 5 wt % of modified (surfactant-modified) and un-modified cellulose nanocrystals,	PLA	solvent casting method in the presence of surfactant	reductions of 34% in water permeability for the cast films with 1 wt % of surfacant modified-CNC; good oxygen barrier properties; the migration level of the studied nano-biocomposites was below the overall migration limits required by the current normative for food packaging materials in both non-polar and polar simulants.	[53]
Up to 3% cellulose nanocrystals CNCs	PLA	extrusion, twin-screw extruder	water vapour permeability decreased gradually with increasing addition of CNCs up to 3%; good oxygen barrier properties; enhanced barrier and mechanical properties	[54]
Cellulosic nanoparticles in chloroform and layered silicates; whiskers	PLA	dispersion in non-aqueous medium, casting; addition of PVOH	An improvement in storage modulus over the entire temperature range for both nanoreinforcements together with shifts in the tanδ peaks for both nanoreinforcements to higher temperatures; reduction in the oxygen permeability for the bentonite nanocomposite but not for the MCC nanocomposite. The amount of light being transmitted through the nanocomposites was reduced compared to pure PLA indicating that both nanoreinforcements were not fully exfoliated	[47,55]
cellulose whiskers extracted from PEA hull fibres with different hydrolysis times, which resulted in different aspect ratios.	PEA starch matrix		The composite produced by using the whiskers with the highest aspect ratio exhibited the highest transparency and best tensile properties; enhances thermomechanical properties, reduces the water sensitivity and keeps biodegradability; Tg increases; moisture resistance improved	[56,57]
cellulose whiskers	PVOH	solution casting, water solvent	the modulus increased by orientation of reinforcement under magnetic field	[58]
cellulose nanowhiskers	k/l carrageenan	casting	high crystallinity, enhanced water barrier of carrageenan	[59]
	PLA	casting	improved barrier properties to gases and vapours, fully renewable biocomposites for biopackaging	
Bacterial cellulose nanowhiskers	EVOH or PLA	electrospinning	increased thermal stability	
Bacterial cellulose nanowhiskers	EVOH	melt compounding	enhanced barrier and mechanical properties of EVOH	

Table 2. *Cont.*

Type of Nanofiller	Matrix	Preparation	Properties/Applications	Ref
30 wt % of Straw cellulose whiskers	poly(styrene-*co*-butyl acrylate) latex film	freeze-drying and moulding a mixture of aqueous suspensions	modulus more than a thousand times higher than that of the bulk matrix	[60]
Aqueous suspensions of polysaccharide (cellulose, chitin or starch) nanocrystals tunicin (the cellulose extracted from a tunicate–a sea animal) whiskers, wheat straw or sugar beet cellulose nanocrystals, potato starch nanocrystal and squid pen and Riftia tubes chitin whiskers	hydrophobic polymers as: styrene and butyl acrylate [poly(S-*co*-BuA)] poly(β-hydroxyoctanoate) (PHO) polyvinylchloride (PVC), waterborne epoxy, natural rubber (NR) and polyvinyl acetate (PVAc), poly(styrene-*co*-hexyl-acrylate)	dispersion of these nanocrystals in non-aqueous media is possible using surfactants or chemical grafting long chain surface chemical modification; mixing and casting the two aqueous suspensions, freeze-drying and hot-pressing or freeze-drying, extruding and hot pressing; mixture extrusion methods, miniemulsion polymerization	films; larger latex particle size results in higher mechanical properties	[30]
aqueous suspension of polysaccharide nanocrystals	hydrosoluble or hydrodispersible polymers as: reinforced starch, silk fibroin, poly(oxyethylene), polyvinyl alcohol, hydroxypropyl cellulose, carboxymethyl cellulose or soy protein isolate	mixing and casting the aqueous solutions, freeze-drying and hot-pressing		
starch nanocrystals	waterborne polyurethane	solution casting; chemical grafting of starch nanocrystals	enhanced strength, elongation and Young's modulus. The chemical grafting of the starch nanocrystals StNs did not affect positively the strength and elongation, because such a treatment inhibited the formation of physical interaction and increasing network density in nanocomposites	[61]
Coating of cotton and tunicin whiskers by a surfactant phosphoric ester of polyoxyethylene (9)-nonyl phenyl ether leads to stable suspensions in toluene and cyclohexane or chloroform	atactic polypropylene, isotactic polypropylene, or (EVA)	dispersion in non-aqueous medium, casting	decreased mechanical properties	[62,63]
Chitin whiskers	protein isolate thermoplastics	solution-casting technique	improved not only the tensile properties (tensile strength and elastic modulus) of the matrix but also its water resistance	[64]
chitin whiskers	chitosan films	solution-casting technique	improved chitosan films tensile strength until a whisker content of 2.96%, while higher increases of whiskers contents resulted in decreasing strength. Improved water resistance of the films.	[65]
chitosan–tripolyphosphate (CS–TPP) nanoparticles	hydroxypropyl methylcellulose (HPMC) films	solution-casting technique	improved mechanical and barrier properties of the films	[66]
carbon-based graphene, 20 to 60 nm in thickness and 0.5 to 25 μm in diameter, at 1 to 5 wt % loading	poly(methyl methacrylate) (PMMA)	dispersion at 30 °C by high speed shearing methods	heat resistant, high barrier nanocomposites promising in food packaging; increase the glass transition temperature of PMMA	[67]

MMT containing nanocomposites show many advantages in respect of neat corresponding polymeric matrices; they are lighter, stronger, more heat-resistant, offer improved barrier against gases, moisture and volatiles materials, antimicrobial properties and so forth. They are produced by Nanocor Inc. (Arlington Heights, IL, USA) and Southern Clay Products, Inc. (Gonzales, TX, USA), Mitsubishi Gas Chemical (New York, NY, USA) and other units using as matrices Nylon-6, PE, PET and EVOH. Films, PET bottles and multilayer films from nanocomposites are obtained to prolong the shelf life of a variety processed meats, cheese, confectionery, cereals and boil-in-bag foods and so forth [8,68,69].

2.2.2. Bionanofibrils

Biopolymer nanofibrils with dimensions of angstrom to hundreds of nanometre scales arise from renewable resources, are widely available at low-cost, offering exceptional biocompatibility, biodegradability, flexibility and the availability of multiple reactive sites for introducing novel functionalities. Because of the natural origin and well-defined supramolecular assemblies and geometries they exhibit exceptional mechanical properties mainly a unique combination of strength and toughness. Their biological functions allow the interactions with the surrounding environment performing a variety of specific functions in living systems and also serving as building blocks stabilized by non-covalent interactions in material science. The nanofibrillar biopolymers are potential candidates for high-performance and functional bionanocomposites as strong, sustainable, stimuli responsive and biocompatible materials for a wide range of applications such as environmental, energy, food packaging, optical and biomedical [70,71]. The combination of biological and synthetic components frequently shows synergistic effects because of strong interfacial interactions that significantly enhance the structural performance and facilitate added functionalities of nanocomposites. In some bionanocomposites, novel synthetic nanoparticles (graphene, carbon nanotubes, mineral nanoparticles, metallic nanoparticles and so forth) were efficiently combined with biological components to achieve superior electrical and thermal conductivity, controlled gas barrier properties, complex actuation and unique optical properties [72]. The biopolymer nanofibrils as cellulose nanofibrils (CNF), chitin nanofibrils (ChNF), silk nanofibrils (SNF) and collagen nanofibrils (CoNF) are most abundant nanofibrils in nature. The nanocellulose and silk nanofibrils, are representatives of classes of polysaccharides and polypeptides [73].

Both top-down and bottom-up strategies have been developed to exfoliate and regenerate bionanofibrils.

Nanocellulose is a term referring to nano-structured cellulose. This may be either cellulose nanocrystal (CNC), cellulose nanofibers (CNF) also called microfibrillated cellulose, or bacterial nanocellulose, which refers to nano-structured cellulose produced by bacteria (BC). They are renewable abundant raw materials derived from wood and plants. As biodegradable fibre-reinforcement for polymeric bionanocomposites, they also offer excellent tensile properties, non-toxicity, high thermal conductivity and optical transparency as indispensable requirements for advanced food packaging.

Cellulose-based nanoreinforcements (as cellulose microfibrils or nanofibers and microcrystalline cellulose). Cellulose microcrystals may contain also amorphous areas. Nanofibres have nanosized diameters (dimension range of 2–20 nm, depending on the bionanofibril origin) and lengths in the micrometre range. Whiskers are crystalline parts and are known as nanocrystals, nanorods, or rod like cellulose microcrystals with lengths ranging from 500 nm up to 1–2 μm and about 8–20 nm or less in diameter and high aspect ratios. Microcrystalline cellulose has 200–400 nm in length and an aspect ratio of about 10 [12]. Their dimensions affect the properties of nanocomposites. Nanocellulose can be extracted via enzymatic pre-treatments, tempo oxidation and chemical extraction, while CNC are extracted generally via acid hydrolysis with sulfuric acid [74–76].

Polysaccharide nanoparticles as aqueous suspensions hydrosoluble (or at least hydrodispersible) or latex-form polymers are obtained. The dispersion of nanocrystals into non-aqueous media is also possible using surfactants or chemical grafting. These are other possibilities for processing of the nanocomposites. The surface of the polysaccharide nanocrystals is covered with reactive hydroxyl groups, which provide the possibility of their extensive chemical modification using grafting agents bearing a reactive end group and a long compatibilizing tail [30]. The surface chemical modification of the cellulose biofibres leads to improved interfacial adhesion between the fibres and the matrix, resulting in enhanced mechanical properties and thermal stability. CNC coatings are also transparent, nontoxic and sustainable. CNCs are highly crystalline and easily dispersed in water, therefore the structure can be controlled to eliminate free volume and to improve barrier properties of the materials. They offer to coatings a higher density and packing that reduces diffusion pathways and drastically improves oxygen, carbon dioxide and water vapour permeability properties which are similar with those of EVOH commercial food packaging material. Their use is also associated with biodegradability and sustainability [77]. *Bacterial cellulose (BC) nanofibrils* randomly-oriented as web-like form show full potential as reinforcement. To orient the nanofibrils a controlled stretching of BC hydrogel was performed by Rahman and Netravali [78,79]. They obtained a BC-reinforced soy protein soy protein isolate (SPI) by vacuum-assisted SPI resin impregnation into BC hydrogel and then stretching the resin impregnated BC hydrogel. Aligned BC nanofibrils were produced by cultivating them inside polydimethylsiloxane (PDMS) tubes followed by wet-stretching. Due to higher BC nanofibrillar orientation, the stretched BC-SPI green composites showed significant improvement in their tensile properties. Cellulose nanofibrils were produced from commercial bleached cellulose pulps after 30 passages in a SuperMasscolloider Grinder (MASUKO SANGYO CO., Kawaguchi, Saitama, Japan), or using mechanical processes [80]. Films were prepared by a casting method. BC-reinforced PLA showed enhanced barrier and mechanical properties [81].

Hybrid nanocomposites, where two different nanoparticles CNC/metallic nanoparticles CNF/clay and CNC/clay) are incorporated into the polymer have been also obtained. [82] PLA-based nanocomposites with CNF, 1–5% nanocrystalline cellulose (CNC) and 1–5% nanoclay (Cloisite™ 30B) for food packaging use led to up to a 90% reduction in the oxygen transmission rate (OTR) and 76% reduction of water vapour transmission rate (WVTR), significant increase in thermomechanical resistance and increased crystallisation kinetics. [83,84] The incorporation of nanoparticles will increase the tortuosity within the material and that this increases barrier properties.

The corona treatment leads to the quality improvement of fabrics and composites. It was applied with good for surface modification and physical and mechanical properties of nanofibril containing films useful for printing and packaging materials such as eucalyptus and pinus nanofibril films [85]. After the corona discharge and subsequent exposure to oxygen from air hydroxyl, carbonyl and other functional groups are formed by breaking C–C bonds and interaction with active species from plasma (see below). After such treatment, the Eucalyptus nanofibril films showed increased tensile strength due to their higher crystalline index and nanofibril dimensions.

Chitin nanofibrils were directly extracted from crab shells through a green deprotonation-assisted liquid exfoliation procedure. By aqueous re-dispersibility after freezing/air/thermal drying, hybridization with other two-dimensional nanomaterials, (e.g., graphene and transitional metal dichalcogenides) was performed. Plasticized PLA with 1 and 5 wt % chitin nanocrystals (ChNC) nanocomposite after orientation, showed a 'shish-kebab' morphology in the drawn tapes. Singh et al. reported improved mechanical and thermal properties. The tensile strength increased and the elongation at break increased from 5% to 60% for the nanocomposite with 5 wt % ChNC and a draw ratio of 3 because of the synergistic effects of the ChNC in the nanocomposite and their alignment of the ChNC together with the polymer chains induced by the solid-state drawing [86].

Silk fibres. A variety of artificial spinning methods have been applied to produce regenerated SNFs. Ling et al. developed a bioinspired spinning method [87]. They prepared nematic silk microfibril solution, highly viscous and stable by partially dissolving silk fibres into micro-fibrils. The hierarchical structures in natural silks was maintained in solution which was spun into regenerated silk fibres by direct extrusion in the air. In this way, polymorphic and hierarchical regenerated silk fibres with physical properties beyond natural fibre construction were generated. By further functionalization with a conductive silk/carbon nanotube coating, responsive materials to changes in humidity and temperature can be produced.

The following surface modification were applied: Poly (ethylene glycol) PEG-grafted rod-like cellulose microcrystals [88], alkenyl succinic anhydride acylating the surface of cellulose nanocrystals, crab shell chitin whiskers [89], waxy maize starch nanocrystals [90] and tunicin whiskers used then to reinforce some plastics [91,92] their performance depending on type of matrix.

Starch nanocrystals show platelet morphology with thicknesses of 6–8 nm. Their incorporation lead to improve some properties of PVOH [93] and pullulan [94].

2.2.3. Other Types of Nanofillers

Carbon nanotubes (*CNT*). Tensile strength/modulus of PVOH [95], PP [96], PA [97] are improved by addition of CNTs.

Silica nanoparticles ($nSiO_2$) improve mechanical and/or barrier properties of PP [98], mainly when maleic anhydride grafted polypropylene (PP-g-MA) is used as a compatibilizer [99], of PVOH [100] or of starch matrix [101].

Haloysite (HAL) represents a class of natural, non-toxic, non-swelling silicates, stable even at very high temperatures and show a good encapsulation and release capacity for bioactive compounds (anticorrosion, antimicrobial, drugs, flame retardant, microcrack self-healing). They have a wide variety of applications including also as fillers and reinforcements for polymers. The HAL particles can adopt a variety of morphologies, as elongated tubules known as haloysite nanotubes (HNT), short tubular, spheroidal and platy particle shapes, the first one being the most common [102]. HAL can be incorporated into different polymers to obtain non-degradable and degradable nanocomposites with versatile properties some of them being attractive for potential uses in food packaging [103]. The HNT-incorporated polymer nanocomposites can be prepared by various processing routes such as melt blending, melt spinning, water-assisted extrusion process, methods involving hydrogen bonding and charge transfer mechanism, solution casting, co-coagulation or co-curing processes, in situ or emulsion polymerization techniques for nondegradable nanocomposites and solution casting, coagulation, electrospinning for HNT-containing degradable nanobiocomposites. HNTs containing nanocomposites exhibit remarkable mechanical performance and their properties can be tailored by HAL surface modification [104], by nanocomposites composition and by the preparation methods optimization. HAL nanofillers act as ethylene gas absorbers. This gas causes softening and aging of fruits and vegetables; limits the migration of spoilage-inducing gas molecules within the polymer matrix. Active food packaging materials as HNT/PE nanocomposite films showed higher ethylene scavenging capacity and OTR and WVTR than neat PE films. Due to their ethylene scavenging activity and good water vapour and oxygen barrier properties, the nanocomposite films slow down the ripening process of bananas and retain the firmness of tomatoes and also slowed down the weight loss of strawberries and aerobic bacterial growth on chicken surfaces. HNT/PE nanocomposite films contribute to food safety improving the quality and shelf life of fresh food products [105]. Starch/halloysite/nisin nanocomposite films inhibited activity of *Listeria monocytogenes*, *Clostridium perfringens* and *Staphylococcus aureus* in skimmed milk agar being active and useful barrier to control food contamination [106]. HNTs loaded with lysozyme, as nano-hybrid antimicrobial was incorporated into a poly (ε-caprolactone) (PCL) matrix at 10 wt % and then the films were submitted to a cold drawn process. The films exhibit a controlled release of lysozyme being for specific active packaging requirements [107].

Graphene as atomically thin carbon sheets improve physical properties of host polymers at extremely small loading. Nanocomposites containing graphene are obtained by top-down strategies starting from graphite oxide, solvent- and melt-based strategies [108]. Such nanocomposites offer extremely high barrier performance at even very low graphene loadings, which is interesting for food packaging [109].

Metallic nanoparticle incorporated into food contact polymeric matrices enhance mechanical and barrier properties, prevent the photodegradation of plastics and are effective antimicrobials (as salts, oxides and colloids, complexes) such as AgNP [110]. Copper, zinc and titanium NP are also useful in food safety and technology. Copper is an efficient sensor for humidity and titanium oxide has resistance to abrasion and UV-blocking performance [111].

Lignin nanoparticles were tested for their antioxidant, antimicrobial activities and potential biological being effectively used as active agents towards different pathogen strains when is combined with PVA, chitosan CH, natural polymers and so forth [112,113].

Multifunctional nanocomposites contain usually a mixture of micro- and nanostructured materials such as essential oils, natural extracts and metal nanoparticles with antimicrobial/antioxidant agents/biological functions and they are also interesting for food contact materials.

2.2.4. Bioplastics–Biopackaging

Increased use of the synthetic packaging materials (as films, bottles, trays and so forth) has led to serious ecological problems due to their non-biodegradability and waste accumulation. Biopolymers derived from natural resources as plants (starch, cellulose and their derivatives, proteins), animal (proteins, polysaccharides) and microbial products, such as polyhydroxybutyrate have been extensively studied in the bio-based nanocomposites field [114] because of their good biocompatibility, degradability [115] and recyclability [116]. The starch-clay biodegradable nanocomposites were investigated for several applications including food packaging. Some restrictions are encountered on the use of biodegradable or natural polymers in food packaging since they have poor barriers and mechanical properties, high permeability to gases, such as oxygen and water vapour because of their hydrophilic nature with a few exceptions, such as caseinates, which show excellent barrier performance, even better than PET [117]. Nanoreinforcements can improve barrier properties and have a positive impact on the oxidation stability, thermal and mechanical characteristics and eventually bionanocomposites show the good biodegradability comparing with conventional polymeric matrices. Polymer cross-linking and graft copolymerization of natural polymers with synthetic monomers/polymers are other alternatives of promising values in biodegradable packaging. The complete replacement of synthetic polymers with degradable ones is almost impossible to achieve and even unnecessary, at least for specific applications the use of biodegradable polymers is desirable. No doubt, In the future biopackaging will be increasingly developed [46,118].

Protein-based nanocomposites. In commercial applications, the following animal derived proteins used are: casein, whey protein, collagen, egg white and fish myofibrillar protein. The plant-based proteins also under large scale application consideration include soybean protein, zein (corn protein) and wheat gluten. Compared with non-ionic polysaccharide films, protein films, due to their more polar nature and more linear (non-ring) structure and lower free volume offer better oxygen barrier properties and lower water vapour permeability [119]. The transparent films based on whey protein show good oxygen barrier and TiO_2 or ZnO ware added to form nanocomposites with improved antimicrobial properties [120–122] used to fabricate food-grade, biodegradable packaging materials. Chen et al. and Yu et al. showed that the addition of small amounts (<1 wt %) of TiO_2 nanoparticles significantly increased the tensile properties of whey protein film (from 1.69 to 2.38 MPa). Soy protein nanocomposite films exhibited reduced water vapour permeability, improved elastic modulus and tensile strength compared with neat polymer [123,124]. Zein as biodegradable polymer is used in the food industry as a coating agent [125]. It is less water sensitive than other biopolymers but shows high water vapour permeability and low tensile strength when compared with commodity polymers. To ameliorate its inherent brittleness, plasticizers may be used.

2.2.5. Degradability and Recyclability of Nanocomposites

Bionanocomposites (BNC) can be recycled/valorised or treated together with other organic wastes in composting facilities and produce compost, as valuable soil conditioners and fertilizers and so forth [126]. They are high-performance biodegradable materials, based on plant, animal and other natural materials, therefore these are safely decomposed into CO_2 consumed during plant photosynthesis, water and humus through the activity of microorganisms [127]. This behaviour refers to with BNC with degradable polymeric matrices like poly(lactic acid) (PLA), poly(butylene adipate-*co*-terephthalate) and thermoplastic starch which undergo degradation after various time period. Different biodegradation conditions can be considered: hydrolytic, composting, enzymatic, according to the final applications and the post-use of the new developed materials [128].

Combining the individual advantages of starch/synthetic or natural polymers and nanoparticles (cellulose nanocrystals, clay, TiO_2, layered silicate and so forth) starch-based completely biodegradable materials with potential applications in food industry as edible films were prepared. [129].

The effect of the nanofillers on degradability and compostability of bioplastics depends on their type. PLA/Laponite biocomposites showed the greatest microbial attachment on the surface (biofilm is formed), while PLA/ organo-montmorillonite (OMMT) had the lowest biofilm formation because of the inhibitory effect of this NP during biodegradation test. Nanoclays influence the polymer bacterial degradation due to the affinity of the bacterium towards clay. The addition of nanoclays during composting process increases the PLA degradation rate due to the presence of hydroxyl groups from the silicate layers of these clays. As an example, due to the hydrophilic nature of nanocellulose, the cellulose nanocrystals (CNCs) increased the disintegrability rate of PLA.

PLA bio-nanocomposite films containing nanoclay with organo-modified (with methyl, tallow, bis-2-hydroxyethyl, quaternary ammonium montmorillonite), Halloysite nanotubes and Laponite® RD, MMT showed a significantly higher mineralization of the films in comparison to the pristine PLA, mainly attributed to the reduction in the PLA lag time. PLGA/CNT films were degraded by hydrolytic degradation [130]. CNTs insignificantly modifies the kinetics and the mechanism of the hydrolytic erosion with respect to the neat PLGA, while CNTs functionalized with carboxylic groups accelerated the hydrolytic degradation and the weight loss of the PLGA matrix of the nanocomposites.

MWCNTs reduce the biodegradation rate of PCL in the presence of *P. aeruginosa*, systematically as the CNT loading increased from 0.1 to 10% w/w [131].

3. Polymer Nano-Coatings in Food Packaging

A coating is defined as a coherent layer formed from a single or multiple applications of a coating material to a substrate. Nanocoatings are ultra thin layers on the nanoscale <1–100 nm thick built-up onto surfaces [132,133]. Nanocoatings do not modify the surface topography, they do not fill in defects or make a smooth surface like a paint does, and they also do not stand up to abrasion and wear. They are used to impart a particular chemical or physical function(s) to a surface as gas-barrier coatings, hydrophilic/hydrophobic or oleophobic properties, improve corrosion resistance and enhance insulating or conductive properties [134].

Nanocoatings can be built up with thickness of one molecular or multiple molecular layers. Nanocoatings are applied to various substrates such as metals, glass, ceramics, polymers and so forth (Figure 2) [135,136]. Some nanocoatings are polymers, either polymerized in-situ or prior to application. "*Smart coatings*" are coatings with multifunctional additional functions some also assuring thermal insulation, controlled release of active ingredients or self-healing functions.

Additives are added to a coating material in very small quantities. They can modify a large variety of properties, for instance, flow behaviour, surface tension, gloss, structure, UV and weather resistance. Depending on the desired function, nanotechnology-based functional coatings uses some nanomaterials as in the case of antimicrobial coatings containing silver, titanium dioxide, zinc oxide, some organic bioactive agents as polysaccharides, proteins, spice and herb extracts as essential and vegetable oils, bacteriocins (ex: Nisaplin®, nisin, pediocins), organic acids for food packaging or hygienic surfaces. Nanomaterial-containing coatings offer high performance materials and better processing properties than conventional coatings (e.g., increased indentation resistance, high elasticity, fast drying, no expansion after contact with water, high water vapour permeability).

3.1. Types of Nanocoatings

Several types of nanocoatings are known in food packaging applications as: nanocoating inside package, outside package, sandwiched as a layer in laminated multilayer packaging films, polymer with nanocoating with high barrier properties, edible coatings and films on a wide variety of foods which serve as moisture, lipid and as gas barriers [137], vacuum deposited aluminium coatings on plastic films, coating of the surfaces of glass food and beverage containers (bottles, jars) with organosilanes, nanosurface biocides and so forth

Nanosilica-coated films onto PET, Nylon and so forth, replace polyvinylidene fluoride (PVDF) coated films and oxide evaporated films, for processed meat products (beef meta, sausage, ham and so forth, fresh food like rare fish, sushi, dried fish and so forth, bakery, sandwiches, snack, candy, nut products with high fat) and so forth [28], offering oxygen and moisture barrier and aroma preservation increased self-life and decreased production cost, good printability and laminating machinability, are eco-friendly, no dioxines are formed during incineration, excellent mechanical and optical properties (retaining bulk properties of base film).

Nano surface biocides (using as nanoparticles: nanosilver as metallic Ag, $AgNO_3$, ZnO, TiO_2, MgO) are obtained by incorporation such nanomaterials with antimicrobial properties onto packaging surface. Such food contact nano surface containing biocides maintain the hygienic conditions when applied to reusable food containers such as boxes and crates and inside liners of refrigerators and freezers.

The most known coating technologies are immobilization using covalent (non-migratory) and non-covalent (migratory), layer-by-layer deposition, photografting chemistries and embedding for controlled release, schematically represented in Figure 2 [135–140].

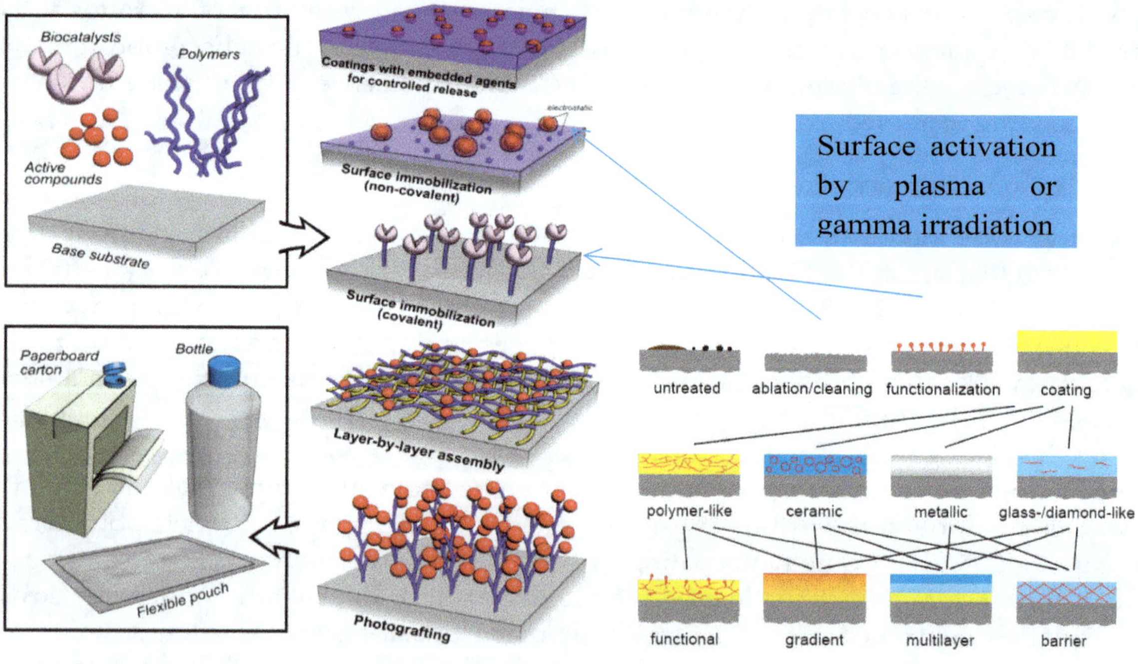

Figure 2. Summary of coating technologies (adapted from [135–140]).

In coating technology which uses the embedded agents for controlled release, the active agent is intended to migrate to the packaged good (embedding, non-covalent immobilization, some layer-by-layer deposition techniques) while in non-migratory technologies, the active agent is intended to remain stable in the packaging matrix (in covalent immobilization, some layer-by-layer deposition techniques, photografting). Some techniques to prepare nanocoating are similar with those for nanocomposites obtaining, some are adapted and other are specific for this purpose. In some cases, the nanocoatings application required specialized equipment and complex procedures, such as, chemical vapour deposition or plasma spray techniques.

The active agents are expected to migrate and exert their specific antimicrobial, antioxidant, biocatalytic, or nutraceutical functions within a packaged food. Active compounds are incorporated into polymeric materials by extrusion/blending and pip-coating techniques or solution casting.

3.2. Coatings Procedures

3.2.1. Solution Casting

Solution casting consists of dissolving the polymer in a suitable solvent and simultaneously incorporating the active compound of interest, followed by pouring this mix solution onto an inert surface. By the solvent evaporation, the films with desired functionality (antimicrobial, antioxidant, pharmaceutical, biocatalytic and so forth) result. It is mainly used at laboratory scale and can be translated at industrial scale by coupling with a wet-coating station and roll-to-roll system [141]. However, it has some limitations both for practical and commercial application because the most of the interesting polymers used in food packaging dissolve only at high temperatures and in certain organic solvents [142], which will affect both the stability and effectiveness of many active/bioactive compounds of interest. Therefore, the incorporation of active compounds through solution casting has mainly applied when biodegradable polymers dissolved at milder temperatures, such as poly(lactic acid), poly(butylene adipate-*co*-terephthalate) and cellulose derivatives, PVOH and so forth [143]. Butnaru et al. obtained films by casting and dried in a vacuum oven with controlled temperature using solutions of highly viscous chitosan from crab shells where food grade vegetable oils (clove, thyme and rosehip seeds oils) were incorporated were homogenized by ultrasonication [144].

3.2.2. Extrusion

In commercial extrusion, the active compounds with polymeric materials in melted state, in a single or twin screw extruder by heat transfer where are incorporated a blend is obtained from which films can be formed. It also has serious limitations due to the intrinsic lack of thermal stability of many active compounds as it was already mentioned above, which can be lost either through degradation and evaporation. During the heat transfer involved in these unit operations the homogeneous distribution of active agent into matrix is difficult. The active compound release is controlled by packaging material characteristics such as the degree of affinity between it and the matrix, morphology and porosity and also by medium temperature. Multilaminated system in which the layer that harbours the active compound is covered by an adjacent layer could act as a barrier slowing release rate. In such case the bulk properties as tensile and thermal are changed therefore the coating technique is preferred.

3.2.3. Sol-Gel Procedure

Sol-gel procedure is extensively used in obtaining of nano-based coatings. A viscous colloidal-sol is applied to a surface by dip, spray or spin coating processes onto surface. The thickness of the obtained layers varies from a few nanometres thick to 0.5 and 3 μm. The process occurs at low temperature, in a short time, is low energy consuming and more important without disruption of the structure or functionality of biocomplexes or organic aggregates, the nanocoating as a gel may exhibit a porous 3D network [145].

3.2.4. Spraying Solution with Compressor Gun

There several types of sparaying guns as manual and automatic spray guns with air compressors, etc. Manual spray guns are available in various configurations to suit a wide range of applications and processes. HVLP (high volume low pressure) gun offers the following advatages: the spraying is controllable, high transfer efficiency, can be used for a wide variety of materials, it is inexpensive to maintain because needs a reduced preparation and clean-up, is portable, the material consumption is reduced by 40%.

Their applications are: painting, food industry, pharmaceutical, dyestuff coating, oil, lubricating, rubber solution spray, textile spray, water based materials/paints, etc.

They are already applied on a large scale both by individuals and industrially.

3.2.5. Surface Immobilization

Surface immobilization of the bioactive compounds was performed for a wide range of inert hydrophobic packaging polymeric materials. Surface modification of polymers can be performed either by physical, chemical (wet) or biological methods. Physical methods are preferred over chemical techniques because they offer greater precision, ease of process control and environment friendliness. Classical physical methods for modifying polymer surfaces include flame and corona treatment, ultraviolet light, gamma-ray, ion-beam techniques, low-pressure plasma and laser treatment. Physical methods either modify the surface layer or an extraneous layer is deposited on the top of the existing material—*coating*. The main advantage of plasma treatments is that the modification is restricted to the uppermost layers of the substrate, thus not affecting the overall desirable bulk properties. To achieve approximately similar surface modification, gamma irradiation should apply only in mild controlled conditions [146,147].

The physical methods are widely used due to their easier industrial scalability, no liquid reagents of any kind are needed in their application, thus is avoided the accumulation and handling of harmful waste [148]. The main effect of the surface activation methods is the formation of reactive species on the polymer surface like oxygenated groups as carbonyl, hydroxyl and carboxylic acid groups or nitrogen containing groups depending on discharge gas used in plasma device or after air exposure. On activated polymer surface bioactive compound can be either covalent or non-covalent immobilized Figure 2. Non-covalent way is based on electrostatic interactions when substrate and bioactive compound show opposite net charges, while ligand–receptor interactions, like in the case of biotin–avidin [149] are specific for affinity immobilization.

Covalent attachment/conjugation of bioactive compounds to functionalized polymer surfaces is realized by: hydrophilic, bifunctional, and/or branched spacer molecules. The main advantage of covalent immobilization is high stability of the formed layer and the active compound does not migrate into the packaged food product to affect its quality and safety. Active agents can be covalently linked directly to the polymer surface or by use of a spacer or crosslinker, which either share a permanent covalent bond between the polymeric substrate and the bioactive compound, or promotes covalent bond formation between the activated substrate and the bioactive compound without forming part of that link ("zero-length" crosslinkers). It can also perform optimization of the surface functionalization by using a polyfunctional agent, increased number of functional groups, tethering a bioactive compound to a solid substrate via a spacer molecule, covalently attach a bioactive compound to the functionalized polymer surface via an intermediary. The immobilization technologies are potential versatile as once functional groups are introduced to the polymer surface, a range of bioactive agents (e.g., enzymes, peptides) can be immobilized through standard chemical or bioconjugation techniques [150].

Plasma and gamma rays assisted nanocoating and immobilization. Plasma treatment and gamma irradiation are energetic processing methods that create reactive spots to which either oxygen- or nitrogen containing groups can be implemented on surface and functional groups and radicals are created depending on the type of discharge gas/atmosphere used inside reactors. The energy levels of entities generated in cold plasma are comparable to the bond energies of organic compounds and thus can facilitate surface functionalization reactions. The extent of functionalization can be controlled by selecting the discharge parameters [139]. On polymer surfaces three main effects can be obtained depending on the treatment conditions: a cleaning effect, an increase of microroughness surface and functionalization. According to the literature data acidic surfaces are obtained by O_2 and air plasma activation of inert surface of synthetic polymers [151] and amphoteric surfaces are obtained by N_2 plasma activation [152,153], in all cases after exposure to air for the same period of time of about 1 min. Nitrogen containing functionalities originate from plasma treatment itself and also, very likely, from the post-reaction with the venting gas. The oxygen atoms are both in the form of alcohol (C–OH) or/and carbonyl (C=O) groups, epoxy groups and radicals, while the nitrogen atoms appear as amino ($C–NH_2$) groups. In order to develop new multifunctional active polymeric surfaces with special properties for food packaging applications based on inert polymer such as *polyolefins or polylactic acid and polyalkanoates*, the following solutions have been applied:

a. Activation of polymeric substrates by non-solvent, environment friendly methods by using of gamma-ionizing radiation or cold plasma gas discharge.
b. Stable layers have been deposited onto activated polymeric substrates using different coupling agents for covalent linking of active/bioactive formulations.
c. Selected bioactive compounds were: chitosan/chitin, lactoferrin, vitamin E, natural vegetable oils with high content of antioxidant compounds as phenols or flavonoids mixtures.

Covalent grafting of bioactive compounds onto inert polymeric surfaces occurs by a two-step procedure:

Step I. *Polymeric surface activation.*

Surface activation techniques are applied in order to introduce the desired type and quantity of reactive functional groups which are able then to covalently attach a bioactive compound. The polymeric films/papers were exposed both to corona and to high-frequency plasma (air or nitrogen were used as discharge gas). The most important feature of irradiated polymers is the possibility of grafting for certain structures, because the radical sites are available for coupling. γ-exposure of polymeric films/papers was carried out in an irradiation machine at various irradiation doses from 2–30 kGy absorbed in air, at room temperature. Surface functionalization because of implementation of polar groups and radicals onto substrate surface provides desired changes in physical properties of the substrate surface (e.g., wettability, improved adhesion and biocompatibility, interaction with surrounding media) [154]. Although physical adsorption may be useful in some applications, covalent immobilization provides the most stable controllable bonds, between the bioactive compound and the functionalized polymer surface [155]. Optima conditions of activation by each method or compound to be linked were established by varying the exposure parameters and determination of the dependence of the surface wettability and morphology on these [146,147,156–162].

Step II. *Covalent bonding/grafting of bioactive compound onto activated polymeric surfaces.*

After the corona, plasma discharge or γ-irradiation pre-treatment, the polymer surface was enriched with oxygen-containing groups, such as carboxyl, carbonyl, hydroxyl, ester groups and/or nitrogen containing groups. The covalent attachment between a polymeric surface and an active compound is achieved by formation of amide, ether, ester and thioester bonds, created between the hydroxyl, amine, imine, carboxylic acid and thiol groups of the active compounds of interest which may possess intrinsically (or are incorporated in their structure) and the functional groups created on the substrate [156,157,163,164]. After cold plasma/γ-ray exposure the polymeric films/fibres/papers were removed from the treatment chamber and then immersed into the bioactive compound solution (chitosan (CHT), grape seed oil, clove oil, etc.). The dipping/immersion, spreading or electrospinning/electrospraying of such solution onto activated surface lead to coat of the activated surface. For covalent bonding of CHT or other bioactive compounds onto activated surfaces, the Ethyl-3-[3-dimethylaminopropyl] carbodiimide hydrochloride) (EDC), -*N*-hydroxysuccinimide (NHS) or 1′-Carbonyldiimidazole (CDI) coupling agents were used.

Comparing various methods of chitosan deposition onto cold plasma and γ-irradiation activated PE surface it was established that for the same concentration of the chitosan solution the most efficient method was the immersion in respect with homogeneity of surface and thickness of deposited layer but the electrospraying appeared to be more versatile method, because it allows a more precise control of the chitosan content deposited onto the surface by varying the deposition time.

Attenuated Total Reflectance Fourier TransformIinfrared spectroscopy (ATR-FTIR), X-ray photoelectron spectroscopy (XPS) and potentiometric titration proved CHT presence on the PE surface and also new bonds are evidenced mainly amide groups or ester groups because of interaction of $-NH_2$ or –OH groups of CHT with carboxyl group implemented on PE or PLA surface by radiation activation of surface. Other reactions were also possible because of the high reactivity of the active species created by plasma exposure/gamma irradiation and the functional groups of chitosan [157,165] Chitosan coating improved the oxygen barrier properties of PE (Oxygen Transmission Rate (OTR) of PE decreased 3 times being of 1066 mL/(m^2·day) while OTR of the uncoated PE is 3833.36 mL/(m^2·day). The potentiometric and polyelectrolyte titrations, zeta potential determination showed that some amount of chitosan desorbed faster from the surface until equilibrium was reached and also that the grafted chitosan layer was more stable than the physically adsorbed one. In the case of grafting, a thin chitosan layer was irreversibly immobilized on the PE surface. The obtained materials combine the antibacterial, antifungal, bioadhesivity, biocompatibility, biodegradability of chitosan [146,150,156,157,163–167]. with the biological functions and antioxidative activity of vitamin E. Vitamin E is also known for its activity to prevent cell membrane damage by inhibiting peroxidation of membrane phospholipids and disrupting free radical chain reactions induced by formation of lipid peroxides [168].

The pH responsiveness was evidenced was evidenced by contact angle measurements using buffered solutions with pH varying in the 2–11 range and it appears by sudden switching from hydrophilic to hydrophobic surface at critical pH $\approx$ 6. The contact angle of the polyethylene surface remains constant over the entire studied pH range (Figure 3).

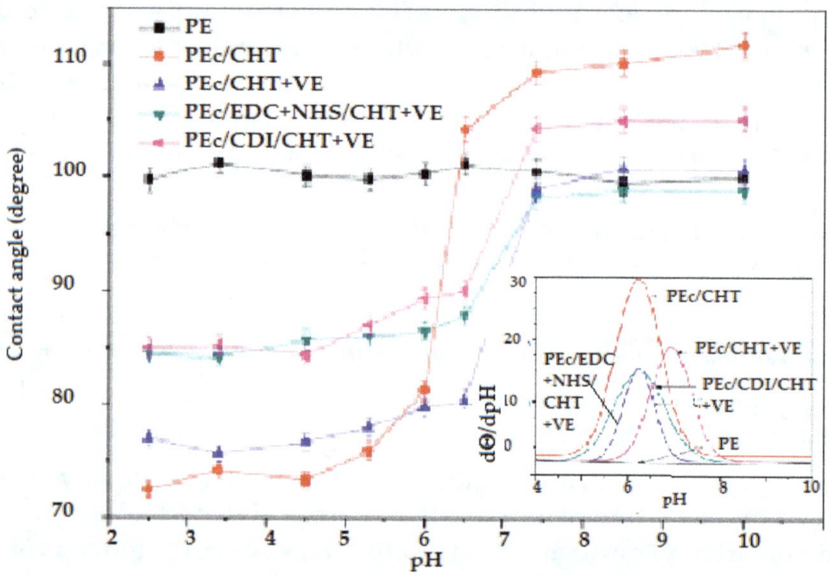

Figure 3. Contact angle (θ) versus pH titration curves and derivative curves of dθ/dpH (insert) for chitosan/vitamin E-coated PE films adapted from [157].

Also, vegetable oils (clove oil, argan oil, thyme oil, tee tree oil and so forth) and combinations chitosan/vegetable oils have been tested to combine antimicrobial activity of CHT with those of vegetable oils and their antioxidant, biological function and nutrition [169]. They were covalently immobilized both on the activated substrate of PE and PLA and the results indicated their synergism in preventing microbial growth on fresh crude cheese, beef meat, poultry minced meat and fresh apple juice [136,140,146]. Both plasma activation and γ-irradiation have great influence on the antimicrobial activity being known that they are relatively simple and quite safe microbial sterilization techniques that are utilized in a variety of applications for their low operating costs and non-polluting capabilities. In the case of lactoferrin immobilization it was established that gamma irradiated samples showed higher activities than those nitrogen plasma activated [150].

Lignocellulosic materials usually display a very low microbial resistance and microbial contaminations might be an additional issue to be taken into account. Conferring antibacterial activity to lignocellulose-based products may represent a main functional property which is useful not only for advanced food packaging [113] but also for textile applications. This can be realized by incorporation of aldehydes, epoxy, carboxylic acids and so forth [160,170] followed by attachment of some phenolic structures which can react at different extents with plasma/γ-irradiated activated surfaces and therefore they will be grafted on the lignocellulosic material surface to develop covalently bound antimicrobial products. The cellulose/chitin mixture and kraft paper were modified using different types of plasma: air, oxygen, nitrogen and argon activation followed by reaction with various phenolic compounds such as p a-hydroxybenzoic cid (HBA), galic acid (GA) eugenol (Eu) and grape seed oil (GO) and rosehip seed oil (RO) and it was found efficient for inhibiting growth of microorganisms onto fresh crude cheese and beef meat. [160,171]. The antimicrobial activity was increased up to 100%. The rosehip seeds oil imparted the best antimicrobial properties to cellulose/chitin mix substrate. As functionalizable substrate is a biodegradable polymer and surface treatment was realized by physical methods of activation (cold plasma and gamma irradiation in mild conditions), the obtaining of the materials with desired surface properties avoided the harmful environmental concerns.

3.2.6. Wet Methods

Wet Methods involve both corrosive liquids to which the polymer substrates are directly exposed, like piranha solution (dissolved hydrogen peroxide and sulfuric acid), combined sodium hydroxide and sulfuric acid, chromic acid, potassium permanganate and nitric acid [172] and graft copolymerization. The first method has not a wide applicability but graft copolymerization received a wide scientific and applicability interest.

3.2.7. Photografting

Photografting method is a free radicals photografting procedure which occurs onto a polymer surface exposed to UV light in the 315–400 nm range in the presence of photoinitiators (e.g., benzophenone, anthraquinone, thioxantone, phenyl azide, polymers, curcumin) and monomers which usually bear ketone groups [173]. The active agent can be directly incorporated during photografting, or by its subsequent immobilization after grafting of a polymer chain with reactive functional groups (e.g., acrylic acid).

3.2.8. Biological Methods

Biological methods consist in: pre-adsorption of proteins, drug, enzyme immobilization, cell seeding and pre-clotting [174].

3.2.9. Chemical Vapour Deposition (CVD)

In typical *CVD*, the substrate is exposed to one or more volatile precursors, which react and/or decompose on the substrate surface to produce the desired deposit [175]. Diamond-like carbon (DLC) has been investigated as coating materials for industrial applications due to its outstanding characteristics such as high hardness, wear resistance, chemical inertness and biological compatibility. The demand for polymeric materials has also been increasing and highly functionalized polymers are much desired for wider applications to the production materials ranging from biomaterials to eco-friendly materials. In order to modify the polymers for highly-functionalized and highly biocompatible materials, DLC can be effectively nanocoated on polymeric materials, especially on polymeric surfaces through CVD methods. By nanocoating such DLC, a new type of polymer/DLC nanocomposite could be established for the use in biomedical devices and food packaging.

3.2.10. Atomic Layer Deposition (ALD) Method

VTT Technical Research Centre of Finland developed a fully recyclable nanocoating for food and pharmaceuticals [176] which is thin, light and air-tight, conformal and pinhole-free coating which follows the contours of the coated material, with excellent gas permeation resistance for packaging materials by ALD method. These thin bio-based packaging materials have gas permeability properties similar to those of existing dry food packages and pharmaceutical blister packs protecting the products from humidity, drying or oxidation. ALD coating allows integrating different functions into the packaging material, such as prevent water, oxygen, humidity, fats and aromas from permeating the packaging and protect the surface from stains and bacterial growth. The ALD technology improves the humidity tolerance and performance of bio-polymers, reducing the need for oil-based plastics.

3.2.11. Layer-by-Layer Assembly

By layer-by-layer assembly deposition can prepare active packaging coatings by the incorporation of active agents either between layers or within the structure of an individual polyelectrolyte [177]. Onto a polymeric activated substrate can be deposited by mutual attraction of polyelectrolytes with opposite charges or even by covalently attached of a polyelectrolyte such as proteins, polysaccharides, synthetic polymers of successive alternate layers [135,178]. Deposition can be accomplished either by submersion of the substrate into polyelectrolyte solutions or by spraying of solutions onto the substrate. The depositions are the presence optimized by adjusting the pH of their solutions for a full protonation or deprotonation, by maximizing of charge [179]. Between layers can be either only electrostatic interactions [180], and/or covalent bonds through the use of crosslinkers. The number of deposited layers is limited by a saturation state characterized by a thickness and stability of the system [181,182]. Jokar et al. produced coatings by the layer-by-layer deposition method. LDPE films were ultrasonically washed with acetone, functionalised in dilute aqueous solution of polyethylenimine for 10 min and sequentially dipped into anionic silver colloid dispersions containing PEG-capped silver NPs or cationic chitosan for 10 min. Silver particles were 18–32 nm and 22–30 nm in size in melt-blended and coated composites, respectively and their migration in simulated media as water, 10% ethanol, 3% acetic acid and apple juice depends on silver concentration and temperature [183]. Covalently bonded multi-layered films were assembled also using the LBL techniques by using polymers bearing functional groups capable of reacting with one another or with a bi-functional agent (diamines, diimides or dialdehydes). Covalent cross-linking increased the modulus and stability of the films.

3.2.12. Ultrasonic Nozzle Systems for Nanotechnology Coating Applications

Sono-Tek ultrasonic nozzles are suitable to depositing precision nanotechnology coatings for spray applications in research as well as production volume spray processes [184]. Uniform nanolayers of solutions onto any size or width substrate (including moving webs of material) are deposited and nano-suspensions are homogenously dispersed the onto a substrate, including suspensions that tend to agglomerate easily which are continuously break due to the mechanical vibration of the nozzle which is much efficient than air spraying. Sono-Tek systems employ the ultrasonic nozzle technology which enables to be easily scaled up from R&D to high volume production.

3.2.13. Plasma Nano-Coating of Beverage Cans

Interior and possibly even exterior coating of the aluminium cans and tubes is required in the food industry because the direct contact between the product and the packaging can lead to aluminium corrosion and the food could spoil. For example, the pasteurization increases corrosive impact on packaging. PlasmaPlus® coating technique uses the plasma jets deposit micro-fine glass-like nano-coats which form a highly effective protective film on the packaging with minimum material usage. Plasma coats provide a perfect quality imprinting [185]. Nobile et al. [186] produced the coating in polyethylene oxide on the surface of PE film by plasma-based vapour deposition. Transmission electron microscopy (TEM) imaging revealed AgNPs about 90 nm in size deposited on the films. The migration of silver ions from nanosilver coatings onto PE films studied by immersion in distilled water, acidified malt extract broth and apple juice at 44 °C for 5 days. It was found that the silver content in the solutions was the highest in the distilled water simulant (1–1.9 mg/kg), with lower concentrations in the malt extract broth and apple juice (0.2–0.38 mg/kg).

3.2.14. Electrospinning/Electrospraying

Electrospinning/electrospraying has recently gained much attention because of its versatility in processing a wide range of polymer and biopolymer materials. It has the ability to produce fibre/particle diameters within the submicron and nano range using electric field which cannot be achieved by using conventional-generating technologies. Nanofibers have wide applications in industry such as food packaging coatings, composite materials, carbon nano tubes, inorganic fibres and tissue scaffolds, filtration, medical, membranes and so forth [187]. Electrospinning has been used to create a well-dispersed "concentrate" or master batch that will act as a carrier of well dispersed nanoparticles. It is an electro-hydrodynamic process, comprising electrospraying and electrospinning techniques which produces nanostructured fibre-based and particle-based materials both in laboratory and industrial scale. The devices used for the mono-axial and co-axial electrospinning/electrospraying method is presented in Figure 4.

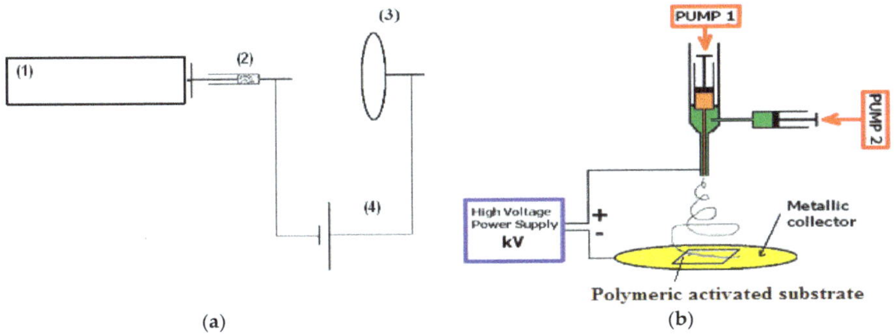

Figure 4. Representative experimental set-up of the uniaxial (**a**): (1) pump; (2) syringe; (3) collector; (4) high direct voltage (0 to 30 kV) power supply; or coaxial (**b**) electrospinning device used for immobilization of vegetable oil into chitosan and bioactive layer deposition onto polymeric activated substrate adapted from [146]; Vasile et al. (2016).

In monoaxial electrospinning the device consists of a direct current high voltage supply, a metal plate collector sustaining the polymeric substrate, a pump and a syringe oriented perpendicularly to the metal plate with the low diameter needle (Figure 4a). A high direct voltage (0 to 30 kV) between the metal plate and the syringe needle is applied. The polymer solution is extruded through the needle tip by the syringe pump at a constant flow rate. At the point of ejection (the needle tip), is created a polymer jet because of the electric charge repulsion outgoing the solution surface tension. Based on electrostatic field between nozzle (needle tip) and collector, the dry polymer or solution are collected on the surface of collector screen. On the collector a very thin polymer layer is deposited, whose morphology as nanofibres or nanoparticles, depends on rheological properties of the solution [156] and composition of electrospinnable solutions [157,188]. This device has some disadvantages such as drying which can be controlled by the distance between nozzle and collector.

In co-axial electrospinning two syringes are used one of them being provided with a co-axial needle oriented with the needle perpendicular to the metal plate. It offers the possibility of depositing a bioactive layer/coating containing two bioactive compounds onto activated surface achieving the encapsulation of less stable or liquid compound (as vegetable oils) into a polymeric one so core-shell morphology is obtain which can increase the coating functionality and protection of the labile bioactive component and also the use of nanomaterials in other technology. Encapsulation of the vegetable oils in chitosan was achieved by coaxial electrospinning method and immediately a coating of the nanoparticles/nanofibres was deposited onto polymeric substrate [146,189]. This technique may also be used with one of the two syringes with the polymeric solution and the other one with the same solvent able to dissolve the polymeric matrix. This approach is to improve the processability, being particularly interesting to scale up the electrospinning technique from lab scale to the industrial sector.

Electrospun fibres have a very high specific surface and porosity and they are able to encapsulate active substances. A wide range of applications are known for such fibres as improvement of physico-chemical and functional properties of biopolymer based materials by means of a controlled release of active compounds or by enhancing the dispersion of nano-additives into the biopolymer matrices. The nano encapsulation/ entrapment can reduce the amount of active ingredients needed to obtain certain functionality and also their controlled release [158,188,190]. Nanofiber-based systems for antimicrobial food packaging and food contact surface applications contain antimicrobial fibre-base mats as coating of the packaging material, improve their activity in maintaining an optimal effect during the food storage period. Furthermore, the high surface to volume ratio of electrospun mats offer an efficient and prolonged delivery of loaded bioactive compounds.

An antimicrobial active multilayer system was obtained from a commercial polyhydroxyalkanoate substrate (PHA) and an electrospun PHA coating with in situ-stabilized silver nanoparticles (AgNPs) with satisfactory thermal, mechanical, barrier and antibacterial properties. It reduced the bacterial population of *Salmonella enterica* below the detection limits at very low silver loading of 0.002 ± 0.0005 wt %. Also, polylactic acid (PLA)/Silver-NP/Vitamin E bionanocomposite has been prepared by electrospinning and they showed a good antimicrobial activity against both gram positive and gram-negative bacteria [190]. As a result, this method provides an innovative route to generate fully renewable and biodegradable antimicrobial materials for food packages and food contact surfaces [191]. The choice of the manufacturing process depends on specific application requirements of the coating. As an example, the percent reduction against the fungus *A. Níger* after 14 days, for the three multilayer samples coated with AgNP/polyethylene nanocomposites with the silver nanocomposite layer of $AgNO_3$ suspension (black) all at 1.0 wt % of silver content depends on the coating method, the most efficient being spraying one which gave a reduction of more than 70% while laminated and casting method reduction were of 20% and 45%, respectively [192].

4. Applications

Nanocomposites and nanocoatings applications in food packaging depend both on the properties of matrices/substrates and activity of ingredients. Nanoclay composites (usually montmorillonites) are used as flexible and rigid food packaging because of their excellent barrier characteristics in packaging of processed meats, cheese, confectionery, cereals and boil-in-the bag foods, items obtained by extrusion-coating process together with the paperboard are useful for fruit juice and dairy products and the beer and carbonated drinks bottles are manufactured by co-extrusion processes. For example, the PET bottles contain nanocore and nylon beer bottles with nanoclays show improved barrier characteristics. The active agents functions include antimicrobial, antifungal, antioxidant, biocatalyst, external stimuli responsiveness, additives that absorb undesirable component from food such as ethylene, moisture and odour and so forth They are incorporated into matrix or coating to enhance the quality and shelf life of the food and are released in a controllable manner, the procedure being much efficient then direct addition into the food because they may react with other food components and loss their activity or affect food sensorial characteristics.

There are many formulations including such antimicrobials, antioxidants and biocatalysts, some of them being summarized in Tables 3 and 4.

Table 3. Formulations including antimicrobials as bioactive compounds intended to use in food packaging.

Active Agent	Substrate or Matrix	Technique	Characteristics/Observations	Ref.
		Antimicrobials		
Essential oils; adhesion promotors (e.g., acrylic or vinyl resins or nitrocellulose) and fixatives	common packaging materials	different coating or spraying	controlled release	[193]
Bacteriocins (ex: Nisaplin® nisin, pediocins); spice and herb extracts; Organic acids, microencapsulated nisin	oolycaprolactone, alginate, crosslinked chitosan/cellulose nanocrystal	crosslinking reaction under γ-irradiation	nanocellulose/PCL and alginate/cellulose nanocrystal based edible films and ready-to-eat meat	[194]
Silver nanoparticles AgNP	poly(3-hydroxybutyrate-*co*-3-hydroxyvalerate	extrusion/blending		[195]
Silver	poly(L-lactide)	melt-compounding	sustained release of antimicrobial silver ions in food applications	[196]
AgNP/peptide/cellulose nanocrystals	PLA	solvent casting	significant inhibition of microbial growth; migration rates below values reported by international imposed limits	[197]
AgNP/kaolinite	PLA and poly(butylene adipate-*co*-terephthalate)	blow films	the composite showed its biodegradation extent of 69.94% (after 90 days), offering good biodegradability for use as a material for the production of degradable plastic bags. The ageing, hydrolytic degradation and biodegradation of PLA-based films could be tailored by Ag kaolinite incorporation	[198]
AgNPs	multilayer films of PHBV3 and electrospun fibres	compression-moulding, 180 °C and 1.8 MPa for 5 min; coated with PHBVs and PHBVs/AgNsP ultrathin fibre mats produced by electrospinning followed by an annealing step; electrospun coating (~0.71 ± 0.01 mg/cm^2)	multilayer materials for food packaging and food contact surface applications; efficient antimicrobial materials; bactericidal effect against *Salmonella enterica*	[191]
AgNP	PVOH	mixing a colloidal solution	improved thermal properties, enhancing stability and increasing Tg.	[199]
2 wt % Ag-NPs	polyamide 6	thermal reduction of silver ions during the melt processing of a PA6/silver acetate mixture	effective against E. coli	[200,201]
TiO_2/Ag grafted with γ-aminopropyltriethoxysilane	PVC	mixing	good antibacterial properties by photocatalytic bacterial inactivation	[202]
nano-Ag 35%, nano-TiO_2 40%, kaolin 25%) (30%)	PE	high-speed mixer; extruded by a twin-screw extruder and then film was obtained	Ag-NPs retarded the senescence of jujube, a Chinese fruit.	[203]
ZnO:Cu/Ag	PLA	melt processing	Antimicrobial materials	[189]

Table 3. *Cont.*

Active Agent	Substrate or Matrix	Technique	Characteristics/Observations	Ref.
		Antimicrobials		
cellulose nanocrystal/silver nanohybrids bifunctional nanofillers—10 wt % CNC-Ag	poly(3-hydroxybutyrate-*co*-3-hydroxyvalerate)	solution casting	high performance nanocomposites with improved thermal, mechanical and antibacterial properties against both *Gram-negative E. coli* and *Gram-positive S. Aureus* of PHBV. Reduced water uptake and water vapour permeability; lower migration level in both non-polar and polar simulants because of increased crystallinity and improved interfacial adhesion. Great potential applications in the fields of food, beverage packaging and disposable overwrap films.	[204]
30 μm thick coating containing 2–6% cinnamon essential oil	PP	spreading	controlled release, total inhibition *against Aspergillus* flavus and niger and *Penicillium roqueforti* and *Penicillium expansum*	[205]
4% thyme and oregano essential oils	corona treated LDPE	ionizing treatment and directly extrusion	controlled release, inhibit *Escherichia coli* 0157:H7, *Salmonella* Typhimurium and *Listeria monocytogenes*, changes in barrier properties	[206]
Organic-inorganic hybrid coatings, polyvinyl alcohol with improved water resistance	poly(ethylene terephthalate) and oriented polypropylene	sol-gel technique condensation of hydroxylated monomers and polymers into a network	multilayer materials for packaging applications	[207]
Antifungal agent natamycin was embedded in a tetraethyl orthosilicate/EVOH gel	plasma treated PLA films	sol-gel	antimicrobial coatings; controlled release; inhibit mould growth on cheese stored for 30 days	[208]
Nanosilver or chitosan	LDPE	melt-blended and layered deposited silver	silver ion migration from the nanocomposites into the food simulants and apple juice was less than the cytotoxicity-level concentration (10 mg kg^{-1}) in all cases over 30 days	[209]
Antimicrobial photocatalysts (e.g., TiO_2)	PLA multi-layered hybrid coatings	sol-gel technique	Controlled release, antimicrobial	[210]
Polydiacetylene liposomes containing cinnamaldehyde as liposome-encapsulated cinnamaldehyde	amine-functionalized silane monolayer on piranha treated glass or amine-functionalized PLA films	nanoencapsulation and immobilization of cinnamaldehyde	Controlled release, efficient against *Bacillus cereus*	[211]
Sorbic acid and/or lauric arginate ester and chitosan and nisin	PLA or corona treated PLA	directly coated with the solutions, or treated with solution-coated polylactic acid (PLA) films	2–3 logarithmic reductions of *Listeria innocua* (2–3 logarithmic reductions), *Listeria monocytogenes* and *Salmonella* Tiphymurium/negative effect on CO_2 gas barrier properties. There was no significant difference in the effectiveness of antimicrobial films versus the coatings. Antimicrobial packaging may be used alone, or in combination with flash pasteurization, in preventing foodborne illness due to post processing contamination of ready-to-eat meat products.	[212,213]

Table 3. *Cont.*

Active Agent	Substrate or Matrix	Technique	Characteristics/Observations	Ref.
		Antimicrobials		
Peptide nisin entrapped in polyethylene oxide brushes grown on silane modified silicon wafers which protected nisin.	polyethylene oxide	entrapment	inhibition of Gram positive bacterium *Pediococcus pentosaceous* over a period of seven days	[214]
Pullalan powder was rendered cationic by reaction with an amine terminated silane as 3-aminopropyltrimethoxysilane	antimicrobial pullulan	immobilization	inhibits *Staphylococcus aureus* and *Escherichia coli*	[215]
Enzyme lysozyme	covalently attached onto UV-ozone treated EVOH films	immobilization via carbodiimide chemistry	inhibits *Listeria monocytogenes*	[216]
(3-Bromopropyl)triphenylphosphonium	poly(butylene adipate-*co*-terephthalate) functionalized with a quaternary phosphonium compound, (3- bromopropyl) triphenylphosphonium	immobilization by azide-alkyne "click" reaction	inhibits *Escherichia coli*	[217]
SO_2	multi-layered film made of PA and PE was subjected to atmospheric plasma treatment (Ar, Na_2O and SO_2) on the PA side of the films	immobilization	inhibits *Escherichia coli* (82%), *Staphylococcus aureus* (86%), *Listeria monocytogenes* (63%), *Bacillus subtilis* (79%) and *Candida albicans* (75%)	[218]
Chitosan (polycation) and κ-carrageenan (polyanion)	aminated PET	layer-by-layer assembly	improved gas barrier properties	[219,220]
Lysozyme: κ-carrageenan alternated with two layers of the antimicrobial enzyme lysozyme	amine-functionalized PET films	layer-by-layer assembly	improved oxygen and water vapour permeability	[221]

Table 4. Formulations including antioxidants and biocatalysts as bioactive compounds intended to use in food packaging.

		Antioxidants		
Citrus oil	plasma treated PET trays	spray deposition of citrus oil in methanol	controlled release; antioxidant activity with cooked turkey meat and retained activity after six months of storage	[222,223]
Rosemary extract	LDPE plastic wrap or a polymeric carrier	applied direct onto LDPE plastic wrap or with a polymeric carrier	controlled release; 0.45 mg rosemary cm^{-2}	[224]
α-Tocopherol	paperboard using a vinyl acetate-ethylene copolymer as a carrier for controlled release	solvent casting coating at a concentration of 3%	antimicrobial and antioxidant coating; controlled release	[225]
gallic acid	chitosan	immobilization by carbodiimide assisted reaction	reduced oxidation of peanuts	[226]
Metal oxide coatings Aluminium oxide or silicon oxide	biaxially oriented polypropylene and polyethylene terephthalate film substrates	reactive evaporation using an industrial high-speed vacuum deposition technique 'boat-type' roll-to-roll metallizer	transparent barrier coatings based on aluminium oxide or silicon oxide fulfil requirements such as product visibility, microwaveability or retortability reduce oxygen diffusion	[227,228]
Tannic acid and poly(allylamine hydrochloride)	glass slides	layer-by-layer assembly	the number of bilayers increases in overall scavenging activity	[229]
Caffeic acid	polypropylene packaging materials coating	photografting	prevent oxidative degradation of ascorbic acid in orange juice	[230]
Acrylic acid	metal chelating active packaging coatings PP films, PP-g-PAA	photografting	prevent lipid oxidation in food emulsions	[231–233]
Hydroxamic acid photografted polyhydroxamate chelators Plant-derived phenolic compounds, metal chelating	PP	*in situ* polymerization of a mixture of catechol and catechin and oxidative polymerization with laccase and in alkaline saline and photografting	surface adhesion properties upon polymerization; non-migratory iron chelating active packaging material biomimetic iron chelating active packaging material, inhibit oxidation of food	[234–236]
Lignosulfonate	Alginate	Solution casting	Antioxidant and UV protective films	[113]
		Biocatalysts		
Lactase blended into polyethylene oxide nanofibers	oxygen scavenging	electrospinning-enzymes or other active agents are incorporated into polymer nanofibers	controlled release, retained up to 93% of free enzyme activity	[237,238]
Lactase	attachment of lactase to polyethylene films;	immobilization β-galactosidase bound to amine-functionalized PE films by a dialdehyde tether; polyethylene glycol tether size influences the attachment	reduce milk lactose in package; retained activity of immobilized lactase	[239,240]
Lactase conjugated to nanomaterials	lactase immobilization onto nanostructures	lactase was attached to carboxylic acid functionalized magnetic nanoparticles 18 nm, 50 nm and 200 nm in diameter using carbodiimide chemistry.	retained activity of immobilized lactase; reducing the particle size of magnetic nanoparticles can increase the activity retention of conjugated lactase	[241]
Lactase and polyethylenamine, glutaraldehyde	lactase covalently bound to low-density polyethylene	layer-by-layer assembly	more enzyme is immobilised but diffusion is difficult	[242]

Table 4. *Cont.*

Glucose oxidase	electrospun polyvinyl acetate/chitosan/tea extract fibres	electrospinning	reduce oxygen in packaged foods	[243]
Glucose oxidase	chitosan	LbL films	biosensors	[244]
Oxalate oxidase, oxygen-reducing enzymes in coatings and films for active packaging	extrusion-coated liner of polypropylene on top of the coating.	entrapment in a latex polymer matrix	protective packaging gas carbon dioxide; oxygen scavenging in active packaging; retained catalytic activity through entrapment in a latex polymer matrix	[245]
Laccase-catalysed reduction of oxygen	it was possible to use lignin derivatives as substrates for the enzymatic reaction.	laccase-catalysed reaction created a polymeric network by cross-linking of lignin-based entities,	resulted in increased stiffness and water-resistance of biopolymer films	[238]
Catalase	layered haemoglobin, PS	Immobilization	create a physical and chemical protective barrier	[229,246]
Fungal naringinase	cross-linking naringinase to polyvinyl alcohol and alginate	immobilization	bitterness reduction in grapefruit juice	[247,248]
Protease, trypsin and endoproteinases	PP, PVOH and PS	photografting, carboiimide chemistry -covalently couple enzymes (via amine groups) to carboxylic acid groups poly(ethylene) glycol methacrylate and 4-vinyl-2,2-dimethylazolactone	antibody analysis in enzyme reactors	[249]
Hydrolase urease	photografted polytetrafluoroethylene	photografting	remove urea from beverages and foods	[250]
Glucose oxidase catalase for oxygen scavenging activity	low-density polyethylene and paper board multilaminate or combinations of LDPE, PP and PLA	industial lamination	scaled-up production in tetra pack pilot plan, scavenge oxygen to improve food shell life	[251]
Polysiloxane-based healing agents	acrylate matrix	polymer coatings coaxial electrospinning	self-healing polymer coating systems	[252]

4.1. Antimicrobial/Antibacterial

Nanomaterials are more and more used to target bacteria in textile industry, marine transport, medicine and food packaging as antibacterial coatings and other materials [253]. Antimicrobial packaging role is to control the growth of pathogenic and/or spoilage microorganisms in packaged products. The diminishing and control bacterial colonization by the modification of the surfaces can be achieved [254]. As antimicrobial agents, essential oils, organic acids, peptides, enzymes and biopolymers are applied and nanoparticles like AgNP, MgO, CuO, Cu, ZnO, Cd selenide/telluride, chitosan, carbon nanotubes are used. Antimicrobial plant extracts from many plants and fruits such as thyme, clove and tea tree, rosemary oil or powder [169], sea buckthorn (*Hippophaë rhamnoides* L.) leaves and inner bark of pine trees (*Pinus silvestris*) [255] used in food packaging provide a healthy alternative. They contain aromatic and phenolic compounds that are responsible for their antibacterial properties and antioxidant activities. They can be applied as components in chitosan films or nanofibres obtained by electrospinning deposited on PLA [256]. Such essential or cold press oils exhibited selective antibacterial and/or antifungal effect both as a solvent extract and as a coating against three food spoilage fungi—*Fusarium graminearum*, *Penicillium corylophilum* and *Aspergillus brasiliensis*—and three potential pathogenic food bacteria as *Staphylococcus aureus*, *Escherichia coli* and *Listeria monocytogenes* and also against *Pseudomonas aeruginosa.*

TiO_2 and ZnO are non-toxic and approved by FDA as GRAS. TiO_2 nanoparticles exhibiting bactericidal and fungicidal effect against *Salmonella choleraesius*, *Vibrio parahaemoliticus*, *Staphylococus aureus*, *Diaporthe actinide* and *Penicilinum expansum.* It was ultrasonically dispersed ethylene-vinyl alcohol copolymer films to obtain active packaging. It showed excellent bactericidal property under UV light [257]. Because of their very small dimensions, the nanomaterials are able to attach many biological molecules, therefore showing a greater efficiency [258]. ZnO is efficient both against Gram-positive and Gram-negative bacteria. NanoZnO coated films exhibit antimicrobial effects against *L. Monocytes* and *S. Enteritis.*

Have been proposed several mechanisms for the antimicrobial activity of nanoparticles as: directly interactions with microbial cells and interruption of the trans-membrane electron transfer, disrupting/penetrating the cell envelope, oxidizing cell components or producing secondary products as reactive oxygen species (ROS) or dissolved heavy metal ions. Secondary processes also can occur. For example, the antibacterial activity of silver ion is reduced by the protein rich food because it can bind the cysteine, methionine, lysine and arginine [259]. Nanosilver antimicrobial activity is explained by adhesion to the cell surface, degrading lipopolysaccharides and damaging the membranes, largely increasing permeability, by penetration inside bacterial cell, damaging DNA [260] and releasing antimicrobial Ag^+ ions by AgNPs dissolution which then bind to electron donor groups in biological molecules containing sulphur, oxygen or nitrogen [261]. Silver ions can be released to destroy food spoilage. They show also a high temperature stability and low volatility. Nanosilver is useful as coatings component where the antimicrobial action occurs at the surface as on containers, mugs, dishes, cutlery, fridges, chopping boards and so forth [262]. It was found effective against numerous species of bacteria including *E. Coli*, *Enterococcus faecalis*, *Staphylococcus aureus* and *Epidermidis*, *Vibrio cholera*, *Pseudomonas aeruginosa* and *putida* and *fluorescens* and *oleovorans*, *Shigella flexneri*, *Bacillus anthracis* and *subtilisi* and *Cereus*, *Proteus mirabilis*, *Salmonella enterica* typhmurium, *Micrococcus luteus*, *Listeria monocytogenes* and *Klebsiella pneumoniae* [1,263] some of them presenting resistance to other potent chemical antimicrobials. Migration studies for nanosilver containing PE composites or coatings [264] in different simulants as malt broth, apple juice, distilled water and 10% ethanol showed good results.

The NanoBioMatters highly efficient antimicrobial product and BactiBlock® technology are based on silver-functionalized nanoclay. MgO and ZnO nanoparticles could provide cheap, safe alternative to expensive nano-sized silver as a heavy metal not suitable for human contact. The combinations of $\alpha-$tocopherol with nisin or chitosan as coating or biocomposites confer both antimicrobial and antioxidative properties [136,140,225].

Antimicrobial activity of the chitosan and engineered nanoparticles is explained by several mechanisms including interactions between positively charged chitosan and negatively charged cell membranes, increased membrane permeability and rupture and leakage of intracellular material. They are ineffective at pH values above 6 because of the absence of protonated amino groups [265]. Another two antimicrobial mechanisms as chelation of trace metals by chitosan, inhibiting enzyme activities and, in fungal cells, penetration through the cell wall and membranes to bind DNA and inhibit RNA synthesis were also proposed [167].

Carbon nanotubes also have antibacterial properties. They are efficient against *E. coli*, possibly because puncture microbial cells [266] but is possible to be cytotoxic to human cells, to skin and lungs in processing stages rather than for consumers, therefore their migration to food must be carefully controlled.

4.2. Antioxidant

Oxidative degradation of food products happens during transport and storage producing lipid rancidity, colour loss (e.g., oxidation of carotenoids, chlorophyll, anthocyanins) and vitamin degradation. The antioxidants (e.g., free radical scavengers, metal chelators, singlet oxygen quenchers, oxygen scavengers) are commonly incorporated in food packaging by blending with polymers during processing (mixing, extrusion) [267,268]. The antioxidant active packaging technologies include oxygen scavengers, manufactured as sachets or labels and by the development of migratory and non-migratory antioxidant coatings. In the first case is possible a controlled release of an antioxidant (migration is 20 times less than the legal limits for the European Union) and sometimes does not requiring direct contact with the food [269]. Non-migratory antioxidant coatings may also be applied by covalent immobilization by means of the functional groups on the surface of packaging materials and shows advantages that do not alter sensorial properties of the packaged food product [270].

4.3. Biocatalytic

Enzymes onto solid support materials provide biocatalytic coatings for active packaging by catalysing some reactions. Biocatalysts are used in ingredients production and breakdown of undesirable components which are harmful or may decrease product quality. Biocatalysts are very sensitive, therefore the immobilization method onto and into solid supports must preserve their thermostability, optima pH and solvent stability and usually so-called "in-package processing" is applied to avoid changes during processing, transport and storage. Antimicrobial enzymes (e.g., lysozyme) are incorporated into active packaging coatings via blending, non-covalent binding for controlled release and covalent immobilization [150,271,272]. Their compatibility with matrix can be achieved by surface functionalization and cross-linking techniques [273]. The thermostability and pH stability of the bound enzyme determines the success of the coating method.

4.4. Barrier Applications of Polymer Nanocomposites

A critical issue in food packaging is that of migration and permeability [170,274] to atmospheric gasses, water vapour, or substances contained within both in the food being packaged or even the packaging material itself. As examples, in packages for fresh fruits and vegetables high barriers to migration or gas diffusion are undesirable because their shelf-life depends on oxygen for sustained cellular respiration while the plastics for carbonated beverage containers must have high oxygen and carbon dioxide barriers to prevent oxidation and decarbonation of the beverage contents. Though polymers possess numerous advantages their major drawback is an inherent permeability to gasses and other small molecules. PET provides a good barrier to oxygen (6–8 nmol/ms GPa) while this is very low for HDPE (200–400 nmol/ms GPa). The situation is opposite in respect with barrier against water vapour [170]. Oxygen and carbon dioxide barriers are necessary for plastics used for carbonated beverage containers. EVOH exhibits excellent oxygen transmission rate (OTR) values under dry conditions but under very humid conditions (relative humidity > 75%) it can possess OTR values more than an order of magnitude higher due its swelling and plasticization in the presence of diffused water molecules [275]. Bio-derived polysaccharide (starch, chitosan, pullulan) based packaging show even larger dependence of their OTR on humidity level, which has severely limited their usefulness. Thermoplastic biopolymers like PLA or polycarprolactones (PCLs) have good tolerance to moisture [276]. Because of the tortuosity effect created by the presence of highly crystalline CNCs (1% CNC) in the PLA-based nanocomposites, which increased degree of crystallinity all barrier properties are significantly improved (WVP (lower ~ 40%; OP ~75% than neat PLA films). PLA/CNC with 1% nanofiber showed also improved of elongation at break [277–279]. Because of this specific dependence, complex multilayer films (by co-extrusion) [280] or polymer blends and composites are often preferred [281]. A high oxygen barrier, water sensitive material like EVOH was sandwiched between two layers composed relatively hydrophobic polyethylene [172]. Multilayer films and polymer blending require the use of additional additives and adhesives that complicate their regulation by federal agencies and arise recycling difficulties. Polymer nanocomposites (PNCs) should solve most of the above-mentioned problems.

4.5. *Stimuli Responsive Nanocomposites/Nanocoatings*

Stimuli responsive nanocomposites/nanocoatings prepared using "stimuli-responsive, "smart," "intelligent" or "environmentally sensitive" polymers [282–285]. These polymers respond to environmental changes by changing their conformation, solubility, hydrophilic/hydrophobic balance, reaction rate, swelling and release behaviour (for hydrogels) and molecular recognition. Their response to the external changes are reversible, "switching" may occur repeatedly and occurs in a narrow interval of stimuli variation. Responsive polymers have found applications as free chains dissolved in aqueous solutions, as covalently or noncovalently crosslinked hydrogels and immobilized adsorbed or surface–grafted onto solid surfaces [286]. Surfaces modified with stimuli-responsive polymers dynamically modify their physico-chemical properties in response to changes in their environmental conditions. Stimuli-responsive surfaces have been created by the immobilization of reversible thermally responsive elastin-like polypeptide, on a glass surface [287]. The distance-dependent colorimetric properties of gold nanoparticles, enable to determine the effect of different variables on the lower critical solution temperature (LCST) at the solid–water interface. A poly(ethyleneimine)–poly(dimethylsiloxane) mixed brush switched spontaneously from the hydrophilic state in water to the hydrophobic one in air, therefore materials show a poor adhesion in a changeable environment. The surface coatings fabricated from mixed poly (ethylene oxide) (PEO)–poly(dimethylsiloxane) brushes or from fluorinated nanoparticles were adaptive to liquid and vapour environments so that the surfaces were spontaneously transformed to non-sticky states in air and in water. Smart and self-healing coatings have been also prepared. Colloidal particles obtained by the emulsion copolymerization of acrylate and fluorinated acrylate monomers form stratified film morphologies in superhydrophobic surfaces [288]. Thermoresponsive poly(*N*-isopropylacrylamide)-based microgels and assemblies have a diverse range of applications for example, biosensing, smart coatings and so forth [289].

Sensory Packaging is a category of devices used in intelligent packaging as nanosensors to monitor and report on the condition of food, communicate the degradation of product or microbial contamination and give information on history of storage, avoiding inaccurate expiration date increasing security and food safety. It helps food industry and retailers to now if the food package has been opened or tampered. This could be achieved by using a nanocrystalline indicator in form of an oxygen intelligence ink printable on most surfaces. This ink contains an UV light activated nanocrystalline particles of a semiconductor (usually TiO_2) or nanophosphourus particles. These particles are white in daylight but will fluoresce when exposed to light of certain wavelengths. The colour pigments can be also used as micro-colour codes in packaging and labelling applications. The development of sensory packaging which monitor the conditions of pharmaceuticals and foods are those affected by changes in temperature, humidity and shock. Nanobarcodes developed by Nanoplex are written into fluorescent microspheres by photobleaching. Coating on PET bottles will expand the packaging of sensitive juice or beer beverages

5. Possible Risks

Consumers are hesitant to buy nanotechnology foods or food with nanotechnology packaging. However, some results suggest that nanotechnology packaging is more beneficial than nanotechnology foods. The main concern with the application of the nanotechnology in food packaging is related to the fact that the very small sizes of the nanoparticles have different chemical and physical properties in respect of macroscale materials and it is possible they could cause health problems. New packaging materials must have good barrier properties to oxygen, carbon dioxide, water vapour and flavour compounds. This controlled release packaging is another example of a nanomaterial application in active packaging. Nanoclays can be used as carriers for the active agents with high efficacy because it is highly dispersed in the polymeric matrix and, hence, exposed more efficiently to the substance on which it is required to act. Potential implications for consumer safety are migration of nanoparticles. They can enter into body through ingestion, inhalation or dermal contact leading to health effects of exposure to some insoluble, persistent nanoparticles. Such health effects are currently not known. Nanoparticles also migrate to foodstuffs with possible adverse effects on food quality. Other concern is with degradability of bio-polymers and formation of degradation products with possible adverse effects. Also, may appear potential environmental impacts of nano-polymer composites and some problems with end-of-life treatments as recycling, re-use and disposal. Detecting the migration of nanomaterials requires more sensitive analytical techniques due to the complexity of the nanomaterials and because they represent only a very small portion of the bulk food. Migration into food can cause both undesirable organoleptic changes (migration of TiO_2 into lipid matrix results in rancidity) and in some cases nanoparticles of bioactive compounds are intended to be released deliberately.

A considerable number of migration studies were found for nanosilver containing polymer composites or coatings. Overall the results from these studies suggest the production method of nanocomposites (e.g., incorporation or coating, surfactant modification), starting silver concentration, temperature, time and choice of contact medium are all factors which may have an effect on the extent of silver ion migration into food simulants. In general, the rate of migration increases when nanosilver is coated onto the food packaging material or surfactants are added, when the storage temperature and length of storage increases and the acidity of the contact medium increases. There appears to be a specific time of storage, after which a steady state release of silver is achieved. This is supported also by a repeat contact migration experiments, which found silver migration decreased considerably (by an order of magnitude) after first contact. Nevertheless, there is some evidence to suggest that if silver nanoparticles do migrate into food/food simulants, they would most likely dissolve quickly into ionic silver. The majority of the migration studies found for nanosilver food packaging composites have shown levels of migration of ionic silver into foods and food simulants below the European specific migration limit (SML) of 0.05 mg Ag/kg food, suggesting low consumer exposure and subsequently low risk of adverse effects. However, there are also several studies, in which migration exceeded this limit. This indicates that for new food packaging products containing nanosilver, the migration experiments should be conducted in each particular case. Migration of intact nanoparticles into food simulants is negligible, implying consumer exposure to these materials is likely low. This suggests there is low potential for safety issues related to the nanosize level of the materials incorporated into food packaging. If they migrate in nanoparticulate form, it would be anticipated at the resulting low concentration in food that many of the metal oxide nanoparticulates would likely dissolve into their ionic forms upon contact with acid foods or stomach acid. Theoretically, potential consumer exposure to nanomaterials incorporated into food packaging will appear if: they migrate into foodstuffs or drinks from the packaging, or if the nanocomposite polymers degrade and 'dissolve' into food or drinks. Migration can theoretically occur if nanoparticles desorb from the surface of the packaging material due to weak bonding at the surface (only really relevant for coatings), diffuse into foods as a result of a concentration gradient, or dissolve resulting in ions released into food [209,290]. However, currently, there are no internationally protocols or standards concerning nanomaterials characterization or to assess their implications on the consumers health [291–293]

6. Commercial Level

The companies producing nanoreinforced food packaging materials are Color Matrix Corporation manufacturing Imperm a high barrier nylon multilayer bottles films; Mitsubishi Gas Chemical Comp. Nanocore producing NANO-N-MXD6 for PET bottles; Lanxess for Durethan KU2-2601; Honeywell polymer with Aegis NC and OX as coating PA6 on paperboard and coinjection PET bottles [33], NanoSealTM Barrier Coating and Bairicade XTTM Barrier Coating (from NanoPack Inc, Wayne, PA, USA) described as a water based coating comprised of a master batch and a liquid dispersion of clay platelets ('nano' or particle size is not mentioned). The coating is applied to traditional packaging films to enhance gas barrier properties and is stated to be approved for indirect food contact (i.e., used with dry and moderately dry food applications).

Nanocor Inc. Deveoped Imperm® as multilayer PET bottle and sheets is used to improve barrier properties and Nylon MXD6 Nylon MXD6-Ultra Barrier System [294], Duretham® KU 2-2601 (LanXess GmbH, Cologne, Germany) nylon nanocomposites for films and paper coating designed for medium barrier applications, requiring excellent clarity; Aegis® (Honeywell, NJ, USA) a polymerized nanocomposite film incorporating an active oxygen scavengers and passive nanocomposite clay particles.

"Micro-pore-plugging" film as single nanolayer on PET, manufactured by Tera-Barrier, provides transparency and stretchability for applications in food, medical and other packaging markets. Coating of 400 nm is obtained by slot die and has WVTR (40 °C and 100% RH) of 5×10^{-2} g/(m^2 day), transparency 86% (PET 87%) at 550 nm, stretchability > 5%, average roughness 10.3 nm. Sidel trade name used for less sensitive foods are a coating of about 200 nm thick obtained inside PET bottle wall by acetylene plasma.

Barrier films are manufactured by Clairiant and Nanocore from PP and respectively Nylon with organoclay for packaging or PET beer bottle.

Purdue University developed a large-scale manufacturing process that may change the way some grocery store foods are packaged. This new manufacturing process uses cellulose nanocrystals as advanced barrier coatings for food packaging. The Purdue manufacturing technique is a roll-to-roll manufacturing process using waterborne polymer systems being scalable [77]. By their studies it appreciated that food packaging is a growing billion-dollar market and overall predicted growth is expected to reach 6 percent by 2024. Advanced barrier coatings, which help to protect grocery items such as foods and beverages, are growing by as much as 45 percent each year.

7. Conclusion and Future Trends

The global trade of active food packaging was estimated to be around US $12,000,000,000.00 in 2017 [295]. The active food packaging market is dominated by oxygen scavenging and moisture absorption applications [296,297]. Consideration must also be given to requirements of different regulatory agencies, under the jurisdiction of the Food and Drug Administration (in the United States), the European Food Safety Authority (in the European Union), or for each country. Non-migratory technologies produced by either covalent immobilization, cross-linked layer-by-layer deposition, or some photografted coatings offer a potential regulatory benefit. Migration testing using standardized simulants (water, 3% acetic acid, 15% ethanol, olive oil, iso-octane and 95% ethanol) must be performed to quantify levels of migrants in packaged product systems [298]. A benefit to coatings over bulk material modification is that bulk material properties should remain intact. However, the influence of the coating on processability, thermomechanical properties, barrier properties and seal strength must be characterized [299]. Rigorous application tests must also be performed to ensure that neither material conversion steps nor end use result in delamination of active coatings. Many of the coatings technologies have potential for scalability to roll-to-roll, high throughput coating operations. Finally, while incorporation of active agents and specialized packaging processes will indeed increase material cost, the opportunities for new products, enhanced safety and reduced waste of packaged goods highlight the potential for increasing product value through smart integration of active biodegradable packaging coatings.

Nanotechnology has the potential to generate new food products and new food packaging. Analyses of individual data showed that the importance of naturalness in food products and trust were significant factors influencing the perceived risk and the perceived benefit of nanotechnology foods and nanotechnology food packaging [300–303].

Nanotechnology has been found to be a promising technology for the food packaging industry. It has proven capabilities that are valuable in packaging foods, including improved barriers; mechanical, thermal and biodegradable properties; and applications in active and intelligent food packaging including anti-microbial agents and nanosensors, respectively. However, the use of nanocomposites in food packaging might be challenging due to the reduced particle size of nanomaterials and the fact that the chemical and physical characteristics of such tiny materials may be quite different from those of their macro-scale counterparts. Migration studies must be conducted to determine the amounts of nanomaterials released into the food matrices [292]. Nanoscale dimensions can increase significantly the physical interactions, physico-chemical and chemical interfaces in materials [304]. The morphologies obtained for the nanocomposites and the ability to modify the interfaces are essential to maximize the properties. Surface treatment and intensive in optima conditions mixing are key solutions determining the nanomaterials performance. The combinations between nanofillers/matrix and bioactive compounds allow wide possibilities of mechanical, thermal, optical, electrical, barrier properties and multifunctionality to create good food packaging materials. Coating technology has undergone a wide variety of changes in the last few years to become a flexible coating process for innovative surface functions. Today, barriers and multifunctional surfaces can meet complex requirements for a wide range of applications in a variety of sectors. New technological solutions for upgrading smart products as other emerging trends are developing.

Author Contributions: C.V. wrote the paper.

References

1. Duncan, T.V. Applications of nanotechnology in food packaging and food safety: Barrier materials, antimicrobials and sensors. *J. Colloid Interface Sci.* **2011**, *363*, 1–24. [CrossRef] [PubMed]
2. Vasile, C. (Ed.) *Polymeric Nanomaterials for Nanotherapeutics*; Elsevier: Amsterdam, The Netherlands, 2019; ISBN 978-0-12-813932-5.
3. Grumazescu, A.M. (Ed.) *Food Preservation, Nanitechnology in the Agri-Food Industry*; Elsevier: Amsterdam, The Netherlands, 2017; Volume 5, ISBN Hardcover 9780128043035.
4. Chaudhry, Q. Nanotechnology Applications for Food Packaging. 2017. Available online: http://ilsi.eu/wp-content/uploads/sites/3/2016/06/S2.3-WSPM12-Chaudhry.pdf (accessed on 21 June 2017).
5. Huang, Q.; Given, P.; Qian, M. *Micro/Nano-Encapsulation of Active Food Ingredients*; American Chemical Society: Washington, DC, USA, 2009; Volume 1007, p. 314.
6. Irshad, A. Applications of Nanotechnology in Food Packaging and Food Safety (Barrier Materials, Antimicrobials and Sensors). 2013. Available online: http://www.slideshare.net/irshad2k6/applications-of-nanotechnology-in-food-packaging-and-food-safety-barrier-materials-antimicrobials-and-sensors (accessed on 14 June 2017).

7. Regulation (EC) No 1935/2004 of The European Parliament and of The Council of 27 October 2004 on Materials and Articles Intended to Come into Contact with Food and Repealing Directives 80/590/EEC and 89/109/EEC. Available online: https://www.interpack.com/cgi-bin/md_interpack/lib/pub/tt.cgi/Food_Safety.html?oid=55076&lang=2&ticket=g_u_e_s_t (accessed on 5 September 2018).
8. Aliofkhazraei, M. *Nanocoatings: Size Effect in Nanostructured Films*; Springer: Berlin, Germany, 2011; p. 212, ISBN 978-3-642-17966-2.
9. Brody, A.L. Nanocomposite technology in food packaging. *Food Technol.* **2007**, *61*, 80–83.
10. Brody, A.L.; Strupinsky, E.R.; Kline, L.R. *Active Packaging for Food Applications*; Technomic Pub. Co.: Lancaster, PA, USA, 2001; p. 218.
11. Kerry, J.; Butler, P. *Smart Packaging Technologies for Fast Moving Consumer Goods*; John Wiley: Chichester, UK; Hoboken, NJ, USA, 2008; p. 340.
12. De Azeredo, H.M.C. Nanocomposites for food packaging applications. *Food Res. Int.* **2009**, *42*, 1240–1253. [CrossRef]
13. Polyethylene Plastic Packaging Applications—Packaging Blog. Available online: https://packagingblog.org/2014/02/08/polyethylene-plastic-packaging-applications/ (accessed on 5 September 2018).
14. Jin, T.; Zhang, H. Biodegradable Polylactic Acid Polymer with Nisin for Use in Antimicrobial Food Packaging. *J. Food Sci.* **2008**, *73*, 127–134. [CrossRef] [PubMed]
15. Mallegni, N.; Phuong, T.V.; Coltelli, M.-B.; Cinelli, P.; Lazzeri, A. Poly(lactic acid) (PLA) Based Tear Resistant and Biodegradable Flexible Films by Blown Film Extrusion. *Materials* **2018**, *11*, 148. [CrossRef] [PubMed]
16. Xie, J.; Wang, Z.; Zhao, Q.; Yang, Y.; Xu, J.; Waterhouse, G.I.N.; Zhang, K.; Li, S.; Jin, P.; Jin, G. Scale-Up Fabrication of Biodegradable Poly(butylene adipate-co-terephthalate)/Organophilic–clay Nanocomposite Films for Potential Packaging Applications. *ACS Omega* **2018**, *3*, 1187–1196. [CrossRef]
17. ISO/TS27687:2008. This standard has been replaced by ISO/TS 80004-2:2015 ISO/TS 80004-2:2015(en) Nanotechnologies—Vocabulary—Part 2: Nano-objects: Nanotechnologies—Terminology and definitions for nano-objects—Nanoparticle, nanofibre and nanoplate.
18. Schadler, L.S.; Brinson, L.C.; Sawyer, W.G. Polymer Nanocomposites: A Small Part of the Story. Overview Nanocomposite Materials. *J. Mater.* **2007**, *59*, 53–60. [CrossRef]
19. Pedrazzoli, D. Understanding the Effect of Nanofillers on the Properties of Polypropylene and Glass Fiber/Polypropylene Multiscale Composites. Ph.D. Thesis, University of Trento, Rrento, Italy, October 2014. Available online: http://eprints-phd.biblio.unitn.it/1322/1/Thesis_PhD_Pedrazzoli_2014.pdf (accessed on 13 June 2017).
20. Tiwari, J.N.; Tiwari, R.N.; Kim, K.S. Zero-dimensional, one-dimensional, two-dimensional and three-dimensional nanostructured materials for advanced electrochemical energy devices. *Prog. Mater. Sci.* **2012**, *57*, 724–803. [CrossRef]
21. LeBaron, P.C.; Wang, Z.; Pinnavaia, T.J.Ž. Polymer-layered silicate nanocomposites: An overview. *Appl. Clay Sci.* **1999**, *15*, 11–29. [CrossRef]
22. Pavlidou, S.; Papaspyrides, C.D. A review on polymer–layered silicate nanocomposites. *Prog. Polym. Sci.* **2008**, *33*, 1119–1198. [CrossRef]
23. Esposito Corcione, C.; Frigione, M. Characterization of Nanocomposites by Thermal Analysis. *Materials* **2012**, *5*, 2960–2980. [CrossRef]
24. Fernandez, A.; Torres-Giner, S.; Lagaron, J.M. Novel route to stabilization of bioactive antioxidants by encapsulation in electrospun fibres of zein prolamine. *Food Hydrocoll.* **2009**, *23*, 1427–1432. [CrossRef]
25. Reig, C.S.; Lopez, A.D.; Ramos, M.H.; Ballester, V.A.C. Nanomaterials: A Map for Their Selection in Food Packaging Applications. *Packag. Technol. Sci.* **2014**, *27*, 839–866. [CrossRef]
26. Sanchez-Garcia, M.D.; Lopez-Rubio, A.; Lagaron, J.M. Natural micro and nanobiocomposites with enhanced barrier properties and novel functionalities for food biopackaging applications. *Trends Food Sci. Technol.* **2010**, *21*, 528–536. [CrossRef]
27. Siracusa, V.; Rocculi, P.; Romani, S.; Rosa, M.D. Biodegradable polymers for food packaging: A review. *Trends Food Sci. Technol.* **2008**, *19*, 634–643. [CrossRef]
28. Smolander, M.; Chaudhry, Q. Nanotechnologies in Food Packaging. In *Nanotechnologies in Food*; Chaudhry, Q., Castle, L., Watkins, R., Eds.; RSC Publishing: London, UK, 2010; Chapter 6; pp. 86–101, ISBN 13: 978-0854041695, 10: 0854041699.

29. Lvov, Y.; Abdullayev, E. Functional polymer–clay nanotube composites with sustained release of chemical agents. *Prog. Polym. Sci.* **2013**, *38*, 1690–1719. [CrossRef]
30. Dufresne, A. Processing of Polymer Nanocomposites Reinforced with Polysaccharide Nanocrystals. *Molecules* **2010**, *15*, 4111–4128. [CrossRef] [PubMed]
31. Azizi Samir, M.A.S.; Alloin, F.; Sanchez, J.Y.; Dufresne, A. Cross-linked nanocomposite polymer electrolytes reinforced with cellulose whiskers. *Macromolecules* **2004**, *37*, 4839–4844. [CrossRef]
32. Bhattacharya, M. Review Polymer Nanocomposites—A Comparison between Carbon Nanotubes, Graphene, and Clay as Nanofillers. *Materials* **2016**, *9*, 262. [CrossRef] [PubMed]
33. Kny, E. Polymer nanocomposite materials used for food packaging. In *Ecosustainble polymer NANOMATERIALS for Food Packaging. Innovative Solutions, Characterisation Needs, Safety and Environmental Issues*; Silvestre, C., Cimmino, S., Eds.; CRC Press Taylor & Francis Group: Boca Raton, FL, USA, 2013; Chapter 13; pp. 337–375, ISBN 9781138034266.
34. Liu, J.; Boo, W.-J.; Clearfield, A.; Sue, H.-J. Intercalation and Exfoliation: A Review on Morphology of Polymer Nanocomposites Reinforced by Inorganic Layer Structures. *Mater. Manuf. Process.* **2006**, *21*, 143–151. [CrossRef]
35. Oliveira, M.; Machado, A.V. Preparation of Polymer-Based Nanocomposites by Different Routes. 2013. Available online: https://repositorium.sdum.uminho.pt/bitstream/1822/26120/1/Chapter.pdf (accessed on 4 September 2018).
36. Bottom-up Methods for Making Nanotechnology Products. Available online: https://www.azonano.com/article.aspx?ArticleID=1079 (accessed on 4 September 2018).
37. Pal, S.L.; Jana, U.P.; Manna, K.; Mohanta, G.P.; Manava, R. Nanoparticle: An overview of preparation and characterization. *J. Appl. Pharma. Sci.* **2011**, *01*, 228–234.
38. Mittal, V. *In-Situ Synthesis of Polymer Nanocomposites*; Wiley-VCH: Hoboken, NJ, USA, Chapter 1; pp. 1–26. Available online: https://application.wiley-vch.de/books/sample/3527328793_c01.pdf (accessed on 6 September 2018).
39. Ahmad, M.B.; Gharayebi, Y.; Salit, M.S.; Hussein, M.Z.; Shameli, K. Comparison of In Situ Polymerization and Solution-Dispersion Techniques in the Preparation of Polyimide/Montmorillonite (MMT) Nanocomposite. *Int. J. Mol. Sci.* **2011**, *12*, 6040–6050. [CrossRef] [PubMed]
40. Darie, R.N.; Pâslaru, E.; Sdrobis, A.; Pricope, G.M.; Hitruc, G.E.; Poiată, A.; Baklavaridis, A.; Vasile, C. Effect of Nanoclay Hydrophilicity on the Poly(lactic acid)/Clay Nanocomposites Properties. *Ind. Eng. Chem. Res.* **2014**, *53*, 7877–7890. [CrossRef]
41. Reddy, T.R.K.; Kim, H.-J.; Park, J.-W. Bio-nanocomposite Properties and its Food Packaging Applications. In *Ebook: Polymer Science: Research Advances, Practical Applications and Educational Aspects*; Méndez-Vilas, A., Solano, A., Eds.; Formatex Research Center: Badajoz, Spain, 2016; ISBN 978-84-942134-8-9. Available online: http://www.formatex.org/polymerscience1/ (accessed on 26 September 2018).
42. Lagarón, J.-M. (Ed.) *Multifunctional and Nanoreinforced Polymers for Food Packaging*; Woodhead Publishing Limited, Elsevier B.V.: Amsterdam, The Netherlands, 2011; pp. 485–497.
43. Mohanty, F.; Swain, S.K. Bionanocomposites for Food Packaging Applications. In *Nanotechnology Applications in Food. Flavor, Stability, Nutrition and Safety*; Oprea, A.E., Grumezescu, A.M., Eds.; Elsevier B.V.: Amsterdam, The Netherlands, 2018; Chapter 18; ISBN 978-0-12-811942-6363-379.
44. Wróblewska-Krepsztul, J.; Rydzkowski, T.; Borowski, G.; Szczypiński, M.; Klepka, T.; Thakur, V.K. Recent progress in biodegradable polymers and nanocomposite-based packaging materials for sustainable environment. *Int. J. Polym. Anal. Charact.* **2018**, *23*, 383–395. [CrossRef]
45. Horst, M.F.; Quinzani, L.M.; Failla, M.D. Rheological and barrier properties of nanocomposites of HDPE and exfoliated montmorillonite. *Thermoplast. Compos. Mater.* **2014**, *27*, 106–125. [CrossRef]
46. Lagaron, J.M.; Sanchez-Garcia, M.; Gimenez, E. Thermoplastic nanobiocomposites for rigid and flexible food packaging applications. In *Environmentally Compatible Food Packaging*; Chiellini, E., Ed.; Woodhead Publ. Ltd.: Cambridge, UK, 2008; Chapter 3; pp. 63–90, ISBN 9781845691943.
47. Petersson, L.; Oksman, K. Biopolymer based nanocomposites: Comparing layered silicates and microcrystalline cellulose as nanoreinforcement. *Compos. Sci. Technol.* **2006**, *66*, 2187–2196. [CrossRef]
48. Xu, Y.; Ren, X.; Hanna, M.A. Chitosan/clay nanocomposite film preparation and characterization. *J. Appl. Polym. Sci.* **2006**, *99*, 1684–1691. [CrossRef]

49. Marras, S.I.; Kladi, K.P.; Tsivintzelis, I.; Zuburtikudis, I.; Panayiotou, C. Biodegradable polymer nanocomposites: The role of nanoclays on the thermomechanical characteristics and the electrospun fibrous structure. *Acta Biomater.* **2008**, *4*, 756–765. [CrossRef] [PubMed]
50. Jang, W.S.; Rawson, I.; Grunlan, J.C. Layer-by-layer assembly of thin film oxygen barrier. *Thin Solid Films* **2008**, *516*, 4819–4825. [CrossRef]
51. Dean, K.; Yu, L.; Wu, D.Y. Preparation and characterization of melt extruded thermoplastic starch/clay nanocomposites. *Compos. Sci. Technol.* **2007**, *67*, 413–421. [CrossRef]
52. Yu, J.; Yang, J.; Liu, B.; Ma, X. Preparation and characterization of glycerol plasticized-PEA starch/ZnO–carboxymethylcellulose sodium nanocomposite. *Bioresour. Technol.* **2009**, *100*, 2832–2841. [CrossRef] [PubMed]
53. Fortunati, E.; Peltzer, M.; Armentano, I.; Torre, L.; Jiménez, A.; Kenny, J.M. Effects of modified cellulose nanocrystals on the barrier and migration properties of PLA nano-biocomposites. *Carbohydr. Polym.* **2012**, *90*, 948–956. [CrossRef] [PubMed]
54. Sung, S.H.; Chang, Y.; Han, J. Development of polylactic acid nanocomposite films reinforced with cellulose nanocrystals derived from coffee silversk. *Carbohydr. Polym.* **2017**, *169*, 495–503. [CrossRef] [PubMed]
55. Kvien, I.; Tanem, B.S.; Oksman, K. Characterization of cellulose whiskers and their nanocomposites by atomic force and electron microscopy. *Biomacromolecules* **2005**, *6*, 3160–3165. [CrossRef] [PubMed]
56. Chen, Y.; Liu, C.; Chang, P.R.; Cao, X.; Anderson, D.P. Bionanocomposites based on PEA starch and cellulose nanowhiskers hydrolyzed from pea hull fibre: Effect of hydrolysis time. *Carbohydr. Polym.* **2009**, *76*, 607–615. [CrossRef]
57. Svagan, A.J.; Hedenqvist, M.S.; Berglund, L. Reduced water vapour sorption in cellulose nanocomposites with starch matrix. *Compos. Sci. Technol.* **2009**, *69*, 500–506. [CrossRef]
58. Kvien, I.; Oksman, K. Orientation of cellulose nanowhiskers in polyvinyl alcohol. *Appl. Phys. A Mater. Sci. Process.* **2007**, *87*, 641–643. [CrossRef]
59. Martinez-Sanz, M.; Lopez-Rubio, A.; Lagaron, J.M. Cellulose nanowhiskers: Properties and applications as nanofiller in nanocomposites with interest in food packaging applications. In *Ecosustainble Polymer Nanomaterials for Food Packaging. Innovative Solutions, Characterisation Needs, Safety and Environmental Issues;* Silvestre, C., Cimmino, S., Eds.; CRC Press Taylor & Francis Group: Boca Raton, FL, USA, 2013; Chapter 8; pp. 195–219, ISBN 9781138034266.
60. Helbert, W.; Cavaille, C.Y.; Dufresne, A. Thermoplastic nanocomposites filled with wheat straw cellulose whiskers. Part I: Processing and mechanical behaviour. *Polym. Compos.* **1996**, *17*, 604–611. [CrossRef]
61. Chen, G.; Wei, M.; Chen, J.; Huang, J.; Dufresne, A.; Chang, P.R. Simultaneous reinforcing and toughening: New nanocomposites of waterborne polyurethane filled with low loading level of starch nanocrystals. *Polymer* **2008**, *49*, 1860–1870. [CrossRef]
62. Ljungberg, N.; Bonini, C.; Bortolussi, F.; Boisson, C.; Heux, L.; Cavaillé, J.Y. New nanocomposite materials reinforced with cellulose whiskers in atactic polypropylene: Effect of surface and dispersion characteristics. *Biomacromolecules* **2005**, *6*, 2732–2739. [CrossRef] [PubMed]
63. Ljungberg, N.; Cavaillé, J.Y.; Heux, L. Nanocomposites of isotactic polypropylene reinforced with rod-like cellulose whiskers. *Polymer* **2006**, *47*, 6285–6292. [CrossRef]
64. Lu, Y.; Weng, L.; Zhang, L. Morphology and properties of soy protein isolate thermoplastics reinforced with chitin whiskers. *Biomacromolecules* **2004**, *5*, 1046–1051. [CrossRef] [PubMed]
65. Sriupayo, J.; Supaphol, P.; Blackwell, J.; Rujiravanit, R. Preparation and characterization of a-chitin whisker-reinforced chitosan nanocomposite films with or without heat treatment. *Carbohydr. Polym.* **2005**, *62*, 130–136. [CrossRef]
66. De Moura, M.R.; Aouada, F.A.; Avena-Bustillos, R.J.; McHugh, T.H.; Krochta, J.M.; Mattoso, L.H.C. Improved barrier and mechanical properties of novel hydroxypropyl methylcellulose edible films with chitosan/tripolyphosphate nanoparticlses. *J. Food Eng.* **2009**, *92*, 448–453. [CrossRef]
67. Ramanathan, T.; Abdala, A.A.; Stankovich, S.; Dikin, D.A.; Herrera-Alonso, M.; Piner, R.D.; Adamson, D.H.; Schniep, H.C.; Ruoff, R.S.; Nguyen, S.T.; et al. Functionalized graphene sheets for polymer nanocomposites. *Nat. Nanotechnol.* **2008**, *3*, 327–331. [CrossRef] [PubMed]
68. Moraru, C.I.; Panchapakesan, C.P.; Huang, Q.; Takhistov, P.; Liu, S.; Kokini, J.L. Nanotechnology: A new frontier in Food Science. *Food Technol.* **2003**, *57*, 24–29.
69. Brody, A.L. Nano and food packaging technologies converge. *Food Technol.* **2006**, *60*, 92–94.

70. Ling, S.; Chen, W.; Fan, Y.; Ke, Z.; Jin, K.; Haipeng, Y.; Buehler, M.J.; Kaplan, D.L. Biopolymer nanofibrils: Structure, modeling, preparation, and applications. *Prog. Polym. Sci.* **2018**, *85*, 1–56. [CrossRef]
71. Ling, S.; Kaplan, D.L.; Buehler, M.J. Nanofibrils in nature and materials engineering. *Nat. Rev. Mater.* **2018**, *3*, 18016. [CrossRef]
72. Xiong, R.; Grant, A.M.; Ma, R. Naturally-derived biopolymer nanocomposites: Interfacial design, properties and emerging applications. *Mater. Sci. Eng. R Rep.* **2018**, *125*, 1–41. [CrossRef]
73. You, J.; Zhu, L.; Wang, Z.; Zong, L.; Li, M.; Wu, X.; Li, C. Liquid Exfoliated Chitin Nanofibrils for Re-dispersibility and Hybridization of Two-Dimensional Nanomaterials. *Chem. Eng. J.* **2018**, *44*, 498–505. [CrossRef]
74. Barbash, V.A.; Yaschenko, O.V.; Shniruk, O.M. Preparation and Properties of Nanocellulose from Organosolv Straw Pulp. *Nanoscale Res. Lett.* **2017**, *12*, 241–249. [CrossRef] [PubMed]
75. Wulandari, W.T.; Rochliadi, A.; Arcana, I.M. Nanocellulose prepared by acid hydrolysis of isolated cellulose from sugarcane bagasse. *IOP Conf. Ser. Mater. Sci. Eng.* **2016**, *107*, 012045. [CrossRef]
76. Kargarzadeh, H.; Ioelovich, M.; Ahmad, I.; Thomas, S.; Dufresne, A. *Methods for Extraction of Nanocellulose from Various Sources*; Wiley-VCH Verlag GmbH & Co. KGaA: Hoboken, NJ, USA, 2017; Available online: https://application.wiley-vch.de/books/sample/3527338667_c01.pdf (accessed on 25 September 2018).
77. Purdue University. Manufacturing Process Provides Low-Cost, Sustainable Option for Food Packaging. 2018. Available online: https://phys.org/news/2018-06-low-cost-sustainable-option-food-packaging.html#jCp (accessed on 25 September 2018).
78. Rahman, M.; Netravali, A.N. Oriented bacterial cellulose-soy protein based fully 'green' nanocomposites. *Compos. Sci. Technol.* **2016**, *136*, 85–93. [CrossRef]
79. Rahman, M.; Netravali, A.N. High-performance green nanocomposites using aligned bacterial cellulose and soy protein. *Compos. Sci.Technol.* **2017**, *146*, 183–190. [CrossRef]
80. Abdul Khalil, H.P.S.; Davoudpour, Y.; Islam, M.N.; Mustapha, A.; Sudesh, K.; Dungani, R.; Jawaid, M. Production and modification of nanofibrillated cellulose using various mechanical processes: A review. *Carbohydr. Polym.* **2014**, *99*, 649–665. [CrossRef] [PubMed]
81. Martínez-Sanz, M.; Lopez-Rubio, A.; Lagaron, J.M. Optimization of the dispersion of unmodified bacterial cellulose nanowhiskers into polylactide via melt compounding to significantly enhance barrier and mechanical properties. *Biomacromolecules* **2012**, *13*, 3887–3899. [CrossRef] [PubMed]
82. Hoeng, F.; Denneulin, A.; Neuman, C.; Bras, J. Charge density modification of carboxylated cellulose nanocrystals for stable silver nanoparticles suspension preparation. *J. Nanopart. Res.* **2015**, *17*, 244. [CrossRef]
83. Trifol Guzman, J.; Szabo, P.; Daugaard, A.E.; Hassager, O. Hybrid Nanocellulose/Nanoclay Composites for Food Packaging Applications. Danmarks Tekniske Universitet, Kgs. Lyngby, Denmark, 2016. Available online: http://orbit.dtu.dk/files/128126567/Preprint_Jon_Trifol_Guzman.pdf (accessed on 4 September 2018).
84. Trifol, J.; Plackett, D.; Sillard, C.; Szabo, P.; Bras, J.; Daugaard, A.E. Hybrid poly(lactic acid)/nanocellulose/nanoclay composites with synergistically enhanced barrier properties and improved thermomechanical resistance. *Polym. Int.* **2016**, *65*, 988–995. [CrossRef]
85. Lopes, T.A.; Bufalino, L.; Cunha Claro, P.I.; Martins, M.A.; Tonoli, G.H.D.; Mendes, L.M. The effect of surface modifications with corona discharge in pinus and eucalyptus nanofibril films. *Cellulose* **2018**, *25*, 5017–5033. [CrossRef]
86. Singh, A.A.; Wei, J.; Vargas, N.H.; Geng, S.; Oksman, N.K. Synergistic effect of chitin nanocrystals and orientations induced by solid-state drawing on PLA-based nanocomposite tapes. *Compos. Sci. Technol.* **2018**, *162*, 140–145. [CrossRef]
87. Ling, S.; Qin, Z.; Li, C.; Huang, W.; Kaplan, D.L.; Buehler, M.J. Polymorphic Regenerated Silk Fibers Assembled Through Bioinspired Spinning. Available online: https://www.researchgate.net/publication/320959224_Polymorphic_regenerated_silk_fibers_assembled_through_bioinspired_spinning (accessed on 25 July 2018).
88. Araki, J.; Wada, M.; Kuga, S. Steric stabilization of a cellulose microcrystal suspension by poly(ethylene glycol) grafting. *Langmuir* **2001**, *17*, 21–27. [CrossRef]
89. Gopalan, N.K.; Dufresne, A.; Gandini, A.; Belgacem, M.N. Crab shells chitin whiskers reinforced natural rubber nanocomposites, 3. Effect of chemical modification of chitin whiskers. *Biomacromolecules* **2003**, *4*, 1835–1842. [CrossRef] [PubMed]

90. Angellier, H.; Molina-Boisseau, S.; Belgacem, M.N.; Dufresne, A. Surface chemical modification of waxy maize starch nanocrystals. *Langmuir* **2005**, *21*, 2425–2433. [CrossRef] [PubMed]
91. Nogi, M.; Abe, K.; Handa, K.; Nakatsubo, F.; Ifuku, S.; Yano, H. Property enhancement of optically transparent bionanofiber composites by acetylation. *Appl. Phys. Lett.* **2006**, *89*, 233123. [CrossRef]
92. Ifuku, S.; Nogi, M.; Abe, K.; Handa, K.; Nakatsubo, F.; Yano, H. Surface modification of bacterial cellulose nanofibres for property enhancement of optically transparent composites: Dependence on acetyl-group DS. *Biomacromolecules* **2007**, *8*, 1973–1978. [CrossRef] [PubMed]
93. Chen, Y.; Cao, X.; Chang, P.R.; Huneault, M.A. Comparative study on the films of poly(vinyl alcohol)/pea starch nanocrystals and poly(vinyl alcohol)/native pea starch. *Carbohydr. Polym.* **2008**, *73*, 8–17. [CrossRef]
94. Kristo, E.; Biliaderis, C.G. Physical properites of starch nanocrystal reinforced pullulan films. *Carbohydr. Polym.* **2007**, *68*, 146–158. [CrossRef]
95. Bin, Y.; Mine, M.; Koganemaru, A.; Jiang, X.; Matsuo, M. Morphology and mechanical and electrical properties of oriented PVA–VGCF and PVA–MWNT composites. *Polymer* **2006**, *47*, 1308–1317. [CrossRef]
96. Lopez Manchado, M.A.; Valentini, L.; Biagotti, J.; Kenny, J.M. Thermal and mechanical properties of single-walled carbon nanotubes–polypropylene composites prepared by melt processing. *Carbon* **2005**, *43*, 1499–1505. [CrossRef]
97. Zeng, H.; Gao, C.; Wang, Y.; Watts, P.C.P.; Kong, H.; Cui, X.; Yan, D. In situ polymerization approach to multiwalled carbon nanotubes-reinforced nylon 10,10 composites: Mechanical properties and crystallization behavior. *Polymer* **2006**, *47*, 113–122. [CrossRef]
98. Wu, C.L.; Zhang, M.Q.; Rong, M.Z.; Friedrick, K. Tensile performance improvement of low nanoparticles filled-polypropylene composites. *Compos. Sci. Technol.* **2002**, *62*, 1327–1340. [CrossRef]
99. Vladimiriov, V.; Betchev, C.; Vassiliou, A.; Papageorgiou, G.; Bikiaris, D. Dynamic mechanical and morphological studies of isotactic polypropylene/fumed silica nanocomposites with enhanced gas barrier properties. *Compos. Sci. Technol.* **2006**, *66*, 2935–2944. [CrossRef]
100. Tang, S.; Zou, P.; Xiong, H.; Tang, H. Effect of nano-SiO_2 on the performance of starch/polyvinyl alcohol blend films. *Carbohydr. Polym.* **2008**, *72*, 521–526. [CrossRef]
101. Xiong, H.G.; Tang, S.W.; Tang, H.L.; Zou, P. The structure and properties of a starch-based biodegradable film. *Carbohydr. Polym.* **2008**, *71*, 263–268. [CrossRef]
102. Makaremi, M.; Pasbakhsh, P.; Cavallaro, G.; Lazzara, G.; Aw, Y.K.; Lee, S.M.; Milioto, S. Effect of Morphology and Size of Halloysite Nanotubes on Functional Pectin Bionanocomposites for Food Packaging Applications. *ACS Appl. Mater. Interfaces* **2017**, *9*, 17476–17488. [CrossRef] [PubMed]
103. Darie-Niţă, R.N.; Vasile, C. Halloysite Containing Composites for Food Packaging Applications. In *Composites Materials for Food Packaging*; Cirillo, G., Kozlowski, M.A., Spizzirri, U.G., Eds.; Scrivener Publishing LLC: Beverly, MA, USA, 2018; Chapter 2; pp. 73–122, ISBN 978-1-119-16020-5.
104. Gaikwad, K.K.; Singh, S.; Lee, Y.S. High adsorption of ethylene by alkali-treated halloysite nanotubes for food-packaging applications. *Environ. Chem. Lett.* **2018**, *16*, 1055–1062. [CrossRef]
105. Tas, C.E.; Hendessi, S.; Baysal, M.; Unal, S.; Cebeci, F.Ç.; Menceloglu, Y.Z.; Unal, H. Halloysite Nanotubes/Polyethylene Nanocomposites for Active Food Packaging Materials with Ethylene Scavenging and Gas Barrier Properties. *Food Bioprocess Technol.* **2017**, *10*, 789–798. [CrossRef]
106. Meister Meira, S.M.; Zehetmeyer, G.; Scheibel, J.M.; Orlandini Werner, J.; Brandelli, A. Starch-halloysite nanocomposites containing nisin: Characterization and inhibition of Listeria monocytogenes in soft cheese. *LWT Food Sci. Technol* **2016**, *68*, 226–234. [CrossRef]
107. Bugatti, V.; Viscusi, G.; Naddeo, C.; Gorrasi, G. Nanocomposites Based on PCL and Halloysite Nanotubes Filled with Lysozyme: Effect of Draw Ratio on the Physical Properties and Release Analysis. *Nanomaterials* **2017**, *7*, 213. [CrossRef] [PubMed]
108. Kim, H.; Abdala, A.A.; Macosko, C.W. Graphene/Polymer Nanocomposites. *Macromolecules* **2010**, *43*, 6515–6530. [CrossRef]
109. Allahbakhsh, A. High barrier graphene/polymer nanocomposite films. In *Food Packaging*; Academic Press: Cambridge, UK, 2017; Chapter 20; pp. 699–737.
110. Dallasa, P.; Sharma, V.K.; Zborila, R. Silver polymeric nanocomposites as advanced antimicrobial agents: Classification, synthetic paths, applications, and perspectives. *Adv. Colloid Interface Sci.* **2011**, *166*, 119–135. [CrossRef] [PubMed]

111. Llorensa, A.; Lloret, E.; Picouet, P.A.; Trbojevich, R.; Fernandeza, A. Review. Metallic-based micro and nanocomposites in food contact materials and active food packaging. *Trends Food Sci. Technol.* **2012**, *24*, 19–29. [CrossRef]
112. Yang, W.; Owczarek, J.S.S.; Fortunati, E.; Kozanecki, M.; Mazzaglia, A.; Balestra, G.M.M.; Kenny, J.M.M.; Torre, L.; Puglia, D. Antioxidant and antibacterial lignin nanoparticles in polyvinyl alcohol/chitosan films for active packaging. *Ind. Crops Prod.* **2016**, *94*, 800–811. [CrossRef]
113. Dumitriu, R.P.; Stoica, I.; Vasilescu, D.S.; Cazacu, G.; Vasile, C. Alginate/Lignosulfonate Blends with Photoprotective and Antioxidant Properties for Active Packaging Applications. *J. Polym. Environ.* **2018**, *26*, 1100–1112. [CrossRef]
114. Sorrentino, A.; Gorrasi, G.; Vittoria, V. Review, Potential perspectives of bio-nanocomposites for food packaging applications. *Trends Food Sci. Technol.* **2007**, *18*, 84–95. [CrossRef]
115. Kaci, M.; Benhamida, A.; Zaidi, L.; Touati, N.; Remili, C. Photodegradation of Poly(lactic acid)/organo-modified clay nanocomposites under natural weathering exposure. In *Ecosustainble Polymer Nanomaterials for Food Packaging. Innovative solutions, Characterisation Needs, Safety and Environmental Issues*; Silvestre, C., Cimmino, S., Eds.; CRC Press Taylor & Francis Group: Boca Raton, FL, USA, 2013; Chapter 11; pp. 281–315, ISBN 9781138034266.
116. Kozlowski, M.A.; Macyszyn, J. Recycling of nanocomposites. In *Ecosustainble Polymer Nanomaterials for Food Packaging. Innovative Solutions, Characterisation Needs, Safety and Envirinmental Issues*; Silvestre, C., Cimmino, S., Eds.; CRC Press Taylor & Francis Group: Boca Raton, FL, USA, 2013; Chapter 12; pp. 313–335, ISBN 9781138034266.
117. Bora, A.; Mishra, P. Characterization of casein and casein-silver conjugated nanoparticle containing multifunctional (pectin–sodium alginate/casein) bilayer film. *J. Food Sci. Technol.* **2016**, *53*, 3704–3714. [CrossRef] [PubMed]
118. Tharanathan, R.N. Biodegradable films and composite coatings: Past, present and future. *Trends Food Sci. Technol.* **2003**, *14*, 71–78. [CrossRef]
119. Arora, A.; Padua, G.W. Review: Nanocomposites in Food Packaging. *J. Food Sci.* **2010**, *75*, R43–R49. [CrossRef] [PubMed]
120. Sothornvit, R.; Krochta, J.M. Plasticizers in edible films and coatings. In *Innovation in Food Packaging*, 2nd ed.; Han, J.H., Ed.; Elsevier Publishers: New York, NY, USA, 2005; ISBN 9780123948359, Hardcover ISBN 9780123946010.
121. Zhou, J.J.; Wang, S.Y.; Gunasekaran, S. Preparation and characterization of whey protein film incorporated with TiO_2 nanoparticles. *J. Food Sci.* **2009**, *74*, N50–N56. [CrossRef] [PubMed]
122. Shi, L.; Zhou, J.; Gunasekaran, S. Low temperature fabrication of ZnO-whey protein isolate nanocomposite. *Mater. Lett.* **2008**, *62*, 4383–4385. [CrossRef]
123. Chen, P.; Zhang, L. Interaction and properties of highly exfoliated soy protein/montmorillonite nanocomposites. *Biomacromolecules* **2006**, *7*, 1700–1706. [CrossRef] [PubMed]
124. Yu, J.; Cui, G.; Wei, M.; Huang, J. Facile exfoliation of rectorite nanoplatelets in soy protein matrix and reinforced bionanocomposites thereof. *J. Appl. Polym. Sci.* **2007**, *104*, 3367–3377. [CrossRef]
125. Liu, X.; Sun, W.; Wang, H.; Zhang, L.; Wang, J.Y. Microspheres of corn, zein, for an ivermectin drug delivery system. *Biomaterials* **2005**, *26*, 109–115. [CrossRef] [PubMed]
126. Kijchavengkul, T.; Auras, R. Compostability of polymers. *Polym. Int.* **2008**, *57*, 793–804. [CrossRef]
127. Zhao, R.; Torley, P.; Halley, P.J. Emerging biodegradable materials: Starch- and protein-based bio-nanocomposites. *J. Mater. Sci.* **2008**, *43*, 3058–3071. [CrossRef]
128. Armentano, I.; Puglia, D.; Luzi, F.; Arciola, C.R.; Morena, F.; Martino, S.; Torre, L. Nanocomposites Based on Biodegradable Polymers. *Materials* **2018**, *11*, 795. [CrossRef] [PubMed]
129. Lu, D.R.; Xiao, C.M.; Xu, S.J. Starch-based completely biodegradable polymer materials. *Express Polym. Lett.* **2009**, *3*, 366–375. [CrossRef]
130. Castro-Aguirre, E.; Auras, R.; Selke, S.; Rubino, M.; Marsh, T. Impact of Nanoclays on the Biodegradation of Poly(Lactic Acid) Nanocomposites. *Polymers* **2018**, *10*, 202. [CrossRef]
131. Goodwin, D.D.; Boyer, I.; Devahif, T.; Gao, C.; Frank, B.P.; Lu, X.; Kuwama, L.; Gordon, T.B.; Wang, J.; Ranville, J.F.; et al. Biodegradation of Carbon Nanotube/Polymer Nanocomposites using a Monoculture. *Environ. Sci. Technol.* **2018**, *52*, 40–51. [CrossRef] [PubMed]

132. Nanocoatings, Nanowerk. Available online: https://www.nanowerk.com/nanotechnology-news/newsid=47370.php (accessed on 4 September 2018).
133. Egodage, D.P.; Jayalath, H.T.S.; Samarasekara, A.M.P.B. Novel antimicrobial nano coated polypropylene based materials for food packaging systems. In Proceedings of the Moratuwa Engineering Research Conference (MERCon), Moratuwa, Sri Lanka, 29–31 May 2017.
134. Müller, K.; Bugnicourt, E.; Latorre, M.; Jorda, M.; Echegoyen Sanz, Y.; Lagaron, J.M.; Miesbauer, O.; Bianchin, A.; Hankin, S.; Bölz, U.; et al. Review on the Processing and Properties of Polymer Nanocomposites and Nanocoatings and Their Applications in the Packaging, Automotive and Solar Energy Fields. *Nanomaterials* **2017**, *7*, 74. [CrossRef] [PubMed]
135. Bastarrachea, L.J.; Wong, D.E.; Roman, M.J.; Lin, Z.; Goddard, J.M. Active Packaging Coatings. *Coatings* **2015**, *5*, 771–791. [CrossRef]
136. Vasile, C.; Pâslaru, E.; Sdrobis, A.; Pricope, G.; Ioanid, G.E.; Darie, R.N. Plasma assisted functionalization of synthetic and natural polymers to obtain new bioactive food packaging materials in Ionizing Radiation and Plasma discharge Mediating Covalent Linking of Stratified Composites Materials for Food Packaging. In Proceedings of the Co-Ordinated Project: Application of Radiation Technology in the Development of Advanced Packaging Materials for Food Products, Vienna, Austria, 22–26 April 2013; Safrany, A., Ed.; pp. 100–110. Available online: http://www-naweb.iaea.org/napc/iachem/working_materials/F2-22063-CR-1-report.pdf (accessed on 19 June 2017).
137. Cagri, A.; Ustunol, Z.; Ryser, E.T. Antimicrobial edible films and coatings. *J. Food Prot.* **2004**, *67*, 833–848. [CrossRef] [PubMed]
138. Smirnova, V.; Krasnoiarova, O.; Pridvorova, S.; Zherdev, A.; Gmoshinskii, I.; Kazydub, G.; Popov, K.; Khotimchenko, S. Characterization of silver nanoparticles migration from package materials destined for contact with foods. *Voprosy Pitaniia* **2012**, *81*, 34–39. (In Russian) [PubMed]
139. Totolin, M. *Plasma Chemistry and Natural Polymers*; PIM Publisher: Iasi, Romania, 2007; pp. 15–25, ISBN 978-973-716-776-7.
140. Vasile, C.; Stoleru, E.; Munteanu, B.S.; Zaharescu, T.; Ioanid, E.; Pamfil, D. Radiation Mediated Bioactive Compounds Immobilization on Polymers to Obtain Multifunctional Food Packaging Materials. In Proceedings of the International Conference on Applications of Radiation Science and Technology (ICARST' 2017), Vienna, Austria, 23–28 April 2017.
141. Riccardi, C.; Zanini, S.; Tassetti, D. A Polymeric Film Coating Method on a Substrate by Depositing and Subsequently Polymerizing a Monomeric Composition by Plasma Treatment. Patent WO2014191901 A1, 4 December 2014.
142. Vasile, C. General survey of the properties of polyolefins. In *Handbook of Polyolefins*, 2nd ed.; Vasile, C., Ed.; Marcel Dekker: New York, NY, USA, 2000; pp. 401–416, ISBN 13: 978-0824786038, 10: 0824786033.
143. Bastarrachea, L.; Dhawan, S.; Sablani, S.S. Engineering properties of polymeric-based antimicrobial films for food packaging. *Food Eng. Rev.* **2011**, *3*, 79–93. [CrossRef]
144. Butnaru, E.; Stoleru, E.; Rapa, M.; Pricope, G.; Vasile, C. Development and manufacturing of formulations containing as active compounds CS and rosehip seeds oil by emulsion technique. In Proceedings of the 4th Technical Meeting of ActiBiosafe Project, Medias, Romania, 19–20 May 2016.
145. Caruso, R.A.; Antonietti, M. Sol−Gel Nanocoating: An Approach to the Preparation of Structured Materials. *Chem Mater.* **2001**, *13*, 3272–3282. [CrossRef]
146. Vasile, C.; Stoleru, E.; Irimia, A.; Zaharescu, T.; Dumitriu, R.P.; Ioanid, G.E.; Munteanu, B.S.; Oprica, L.; Pricope, G.M.; Hitruc, G.E. Ionizing Radiation and Plasma Discharge Mediating Covalent Linking of Bioactive Compounds onto Polymeric Substrate to Obtain Stratified Composites for Food Packing. In *Report of the 3rd RCM: Application of Radiation Technology in the Development of Advanced Packaging Materials for Food Products*; Safrany, A., Ed.; IAEA: Vienna, Austria, 2016; Chapter 15; pp. 11–15.
147. Silvestre, C.; Cimmino, S.; Stoleru, E.; Vasile, C. Application of Radiation technology to food packaging. In *Applications of Ionizing Radiation in Materials Processing*; Sun, Y., Chmielewski, A., Eds.; Institute of Nuclear Chemistry and Technology: Warsaw, Poland, 2017; Chapter 20; pp. 461–485.
148. Goddard, J.M.; Hotchkiss, J.H. Polymer surface modification for the attachment of bioactive compounds. *Prog. Polym. Sci.* **2007**, *32*, 698–725. [CrossRef]
149. Moy, V.T.; Florin, E.L.; Gaub, H.E. Intermolecular forces and energies between ligands and receptors. *Science* **1994**, *266*, 257–259. [CrossRef] [PubMed]

150. Stoleru, E.; Zaharescu, T.; Hitruc, E.G.; Vesel, A.; Ioanid, E.G.; Coroaba, A.; Safrany, A.; Pricope, G.; Lungu, M.; Schick, C.; et al. Lactoferrin-immobilized surfaces onto functionalized PLA assisted by the gamma-rays and nitrogen plasma to create materials with multifunctional properties. *ACS Appl. Mater. Interfaces* **2016**, *8*, 31902–31915. [CrossRef] [PubMed]
151. Oniz-Magan, A.B.; Pastor-Blas, M.M.; Martin-Martinez, J.M. Different Performance of Ar, O_2 and CO_2 RF Plasmas in the Adhesion of Thermoplastic Rubber to Polyurethane Adhesive. In *Plasma Processes and Polymers*; D'Agostino, R., Favia, P., Oehr, C., Werheimer, M.E., Eds.; Wiley-VCH: Weinheim, Germany, 2005; pp. 177–192.
152. Park, S.J.; Kim, J.S. Influence of Plasma Treatment on Microstructures and Acid–Base Surface Energetics of Nanostructured Carbon Blacks: N2 Plasma Environment. *J Colloid Interface Sci.* **2001**, *244*, 336–341. [CrossRef]
153. Bryjak, M.; Gancarz, I.; Pozniak, G. Surface evaluation of plasma-modified polysulfone (Udel P-1700) films. *Langmuir* **1999**, *15*, 6400–6404. [CrossRef]
154. Strobel, M.; Lyons, C.S.; Mittal, K.L. *Plasma Surface Modification of Polymer: Relevance to Adhesion*; VSP: Utrecht, Germany, 1994; ISBN 90-6764-164-2.
155. Goddard, J.M.; Hotchkiss, J.H. Tailored functionalization of low-density polyethylene surfaces. *J. Appl. Polym. Sci.* **2008**, *108*, 2940–2949. [CrossRef]
156. Stoleru, E.; Dumitriu, R.P.; Munteanu, B.S.; Zaharescu, T.; Tanase, E.E.; Mitelut, A.; Ailiesei, G.-L.; Vasile, C. Novel Procedure to Enhance PLA Surface Properties By Chitosan Irreversible Immobilization. *Appl. Surf. Sci.* **2016**, *367*, 407–417. [CrossRef]
157. Stoleru, E.; Munteanu, S.B.; Dumitriu, R.P.; Coroaba, A.; Drobotă, M.; Fras Zemljic, L.; Pricope, G.M.; Vasile, C. Polyethylene Materials with Multifunctional Surface Properties by Electrospraying Chitosan/Vitamin E Formulation Destined to Biomedical and Food Packaging Applications. *Iran. Polym. J.* **2016**, *25*, 295–307. [CrossRef]
158. Munteanu, B.S.; Dumitriu, R.P.; Profire, L.; Sacarescu, L.; Hitruc, G.E.; Stoleru, E.; Dobromir, M.; Matricala, A.L.; Vasile, C. Hybrid Nanostructures Containing Sulfadiazine Modified Chitosan as Antimicrobial Drug Carriers. *Nanomaterials* **2016**, *6*, 207. [CrossRef] [PubMed]
159. Munteanu, B.S.; Stoleru, E.; Ioanid, E.G.; Zaharescu, T.; Secarescu, L.; Vasile, C. Plasma Discharge and Gamma Irradiation Mediating Covalent Linking of Stratified Composites Materials for Food Packaging. In Proceedings of the XX-th International Conference "Inventica 2016", Iasi, Romania, 30 June–1 July 2016; pp. 213–221.
160. Irimia, A.; Ioanid, G.E.; Zaharescu, T.; Coroabă, A.; Doroftei, F.; Safrany, A.; Vasile, C. Comparative study on Gamma Irradiation and Cold Plasma pretreatment for a Cellulosic substrate modification with phenolic compounds. *Radiat. Phys. Chem.* **2017**, *130*, 52–61. [CrossRef]
161. Pâslaru (Stoleru), E.; Munteanu, B.S.; Vasile, C. Electrospun Nanostructures as Biodegradable Composite Materials for Biomedical Applications. In *Biodegradable Polymeric Nanocomposites Advances in Biomedical Applications*; Depan, D., Ed.; CRC Press, Taylor & Francis Group: Roca Baton, FL, USA, 2015; Chapter 3; pp. 49–73, ISBN 978-1-4822-6052-6.
162. Irimia, A.; Vasile, C. Surface Functionalization of Cellulose Fibers. In *Cellulose and Cellulose Derivatives: Synthesis, Modification, Nanostructure and Applications*; Mondal, I.H., Ed.; Nova Science Publishers, Inc.: New York, NY, USA, 2015; Chapter 5; ISBN 978-1-63483-150-5.
163. Vasile, C.; Darie, R.N.; Sdrobis, A.; Pâslaru, E.; Pricope, G.; Baklavaridis, A.; Munteanu, B.S.; Zuburtikudis, I. Effectiveness of chitosan as antimicrobial agent in LDPE/CS composite films as minced poultry meat packaging materials. *Cell. Chem. Technol.* **2014**, *48*, 325–336.
164. Munteanu, B.S.; Pâslaru, E.; Fras Zemljic, L.; Sdrobis, A.; Pricope, G.M.; Vasile, C. Chitosan coatings applied to polyethylene surface to obtain food-packaging materials. *Cell. Chem. Technol.* **2014**, *48*, 565–575.
165. Paslaru, E.; Fras-Zemljic, L.; Bracic, M.; Vesel, A.; Petrinic, I.; Vasile, C. Stability of a Chitosan Layer Deposited onto a Polyethylene Surface. *J. Appl. Polym. Sci.* **2013**, *130*, 2444–2457. [CrossRef]
166. Savard, T.; Beauliu, C.; Boucher, I.; Champagne, C.P. Antimicrobial action of hydrolyzed chitosan against spoilage yeasts and lactic acid bacteria of fermented vegetables. *J. Food Prot.* **2002**, *65*, 828–833. [CrossRef] [PubMed]
167. Rabea, E.I.; Badawy, M.E.; Stevens, C.V.; Smagghe, G.; Steurbaut, W. Chitosan as antimicrobial agent: Applications and mode of action. *Biomacromolecules* **2003**, *4*, 1457–1465. [CrossRef] [PubMed]

168. Burton, G.W.; Traber, M.G. Vitamin E: Antioxidant activity, biokinetics, and bioavailability. *Annu. Rev. Nutr.* **1990**, *10*, 357–382. [CrossRef] [PubMed]
169. Vasile, C.; Sivertsvik, M.; Mitelut, A.C.; Brebu, M.A.; Stoleru, E.; Rosnes, J.T.; Tanase, E.E.; Khan, W.; Pamfil, D.; Cornea, C.P.; et al. Comparative Analysis of the Composition and Active Property Evaluation of Certain Essential Oils to Assess their Potential Applications in Active Food Packaging. *Materials* **2017**, *10*, 45. [CrossRef] [PubMed]
170. Kong, F.; Hu, Y.F. Biomolecule immobilization techniques for bioactive paper fabrication. *Anal. Bioanal. Chem.* **2012**, *403*, 7–13. [CrossRef] [PubMed]
171. Sdrobiş, A.; Biederman, H.; Kylian, O.; Vasile, C. Modification of cellulose/chitin mix fibers under different cold plasma conditions. *Cellulose* **2012**, *20*, 509–524. [CrossRef]
172. Barish, J.A.; Goddard, J.M. Topographical and chemical characterization of polymer surfaces modified by physical and chemical processes. *J. Appl. Polym. Sci.* **2011**, *120*, 2863–2871. [CrossRef]
173. Carlini, C.; Angiolini, L. Polymers as free radical photoinitiators. In *Synthesis and Photosynthesis*; Springer: Heidelberg, Germany, 1995; pp. 127–214. Available online: https://link.springer.com/bookseries/12 (accessed on 25 September 2018).
174. Hermanson, G.T. *Bioconjugate Techniques*, 3rd ed.; Academic Press: San Diego, CA, USA, 1996; 785p, ISBN 9780123822406. Hardcover ISBN 9780123822390.
175. Maeda, T.; Hagiwara, K.; Hasebe, T.; Hotta, A. *Polymers with DLC Nanocoating by CVD Method for Biomedical Devices and Food Packaging in Comprehensive Guide for Nanocoatings Technology*; Volume 3: Properties and Development; Nova Science Publishers: New York, NY, USA, 2015; pp. 405–433, ISBN 978-1-63482-647-1.
176. Technical Research Centre of Finland (VTT). Nanowerk News A Fully Recyclable Nanocoating for Food and Pharmaceuticals, 24 March 2010. Available online: http://www.nanowerk.com/news/newsid=15499.php (accessed on 23 June 2017).
177. SPALASTM SPALASTM (Spray Assisted Layer-by-Layer Assembly) Coating System. Available online: http://www.agiltron.com/PDFs/SPALASTM.pdf (accessed on 23 June 2017).
178. Bastarrachea, L.J.; Denis-Rohr, A.; Goddard, J.M. Antimicrobial food equipment coatings: Applications and challenges. *Annu. Rev. Food Sci. Technol.* **2015**, *6*, 97–118. [CrossRef] [PubMed]
179. Yang, Y.; Haile, M.; Park, Y.T.; Malek, F.A.; Grunlan, J.C. Super gas barrier of all-polymer multilayer thin films. *Macromolecules* **2011**, *44*, 1450–1459. [CrossRef]
180. Cerkez, I.; Kocer, H.B.; Worley, S.D.; Broughton, R.M.; Huang, T.S. N-Halamine biocidal coatings via a layer-by-layer assembly technique. *Langmuir* **2011**, *27*, 4091–4097. [CrossRef] [PubMed]
181. Bastarrachea, L.J.; McLandsborough, L.A.; Peleg, M.; Goddard, J.M. Antimicrobial N-Halamine modified polyethylene: Characterization, biocidal efficacy, regeneration, and stability. *J. Food Sci.* **2014**, *79*, E887–E897. [CrossRef] [PubMed]
182. Bastarrachea, L.J.; Peleg, M.; McLandsborough, L.A.; Goddard, J.M. Inactivation of Listeria Monocytogenes on a polyethylene surface modified by layer-by-layer deposition of the antimicrobial *N*-Halamine. *Food Eng.* **2013**, *117*, 52–58. [CrossRef]
183. Jokar, M.; Abdul Rahman, R. Study of silver ion migration from melt-blended and layered deposited silver polyethylene nanocomposite into food simulants and apple juice. *Food Addit. Contam. Part A* **2014**, *31*, 734–742. [CrossRef] [PubMed]
184. Sono-Tek Corporation. Ultrasonic Nozzle Systems for Nanotechnology Coating Applications in Nanotechnology Coatings Overview. Available online: http://www.sono-tek.com/nanotechnology-overview/ (accessed on 23 June 2017).
185. Food-Safe, Nano-Coating Appropriate for Food with for Optimal Corrosion Protection and Quality Imprinting on Tubes and Cans. Available online: http://www.plasmatreat.com/industrial-applications/packaging/glass-metal-packages/nano-coating-of-beverage-cans.html (accessed on 2 June 2017).
186. Nobile, M.A.; Cannarsi, M.; Altieri, C.; Sinigaglia, M.; Favia, P.; Iacoviello, G.; D'Agostino, R. Effect of Ag-containing Nano-composite Active Packaging System on Survival of Alicyclobacillus acidoterrestris. *J. Food Sci.* **2004**, *69*, E379–E383. [CrossRef]
187. Mirjalili, M.; Zohoori, S. Review for application of electrospinning and electrospun nanofibers technology in textile industry. *J. Nanostruct. Chem.* **2016**, *6*, 207–213. [CrossRef]

188. Munteanu, B.S.; Aytac, Z.; Pricope, G.M.; Uyar, T.; Vasile, C. Polylactic acid (PLA)/Silver-NP/Vitamin E bionanocomposite electrospun nanofibers with antibacterial and antioxidant activity. *J. Nanopart. Res.* **2014**, *16*, 2643–2646. [CrossRef]
189. Vasile, C.; Darie-Niță, R.; Brebu, M.; Râpă, M.; Ștefan, M.; Stan, M.; Macavei, S.; Barbu-Tudoran, L.; Borodi, G.; Vodnar, D.; et al. New PLA/ZnO:Cu/Ag bionanocomposites for food packaging. *Express Polym. Lett.* **2017**, *11*, 531–544. [CrossRef]
190. Munteanu, B.S.; Ioanid, G.E.; Pricope, G.M.; Mitelut, A.C.; Tanase, E.E.; Vasile, C. Development and manufacturing of capsules with active substances by electrospinning technique. Encapsulated forms. In Proceedings of the 4th Technical Meeting of ActiBiosafe Project, Medias, Romania, 19–20 May 2016.
191. Castro-Mayorga, J.L.; Fabra, M.J.; Cabedo, L.; Lagaron, J.M. On the Use of the Electrospinning Coating Technique to Produce Antimicrobial Polyhydroxyalkanoate Materials Containing In Situ-Stabilized Silver Nanoparticles. *Nanomaterials* **2017**, *7*, 4. [CrossRef] [PubMed]
192. Sanchez-Valdes, S.; Ortega-Ortiz, H.; Ramos-de Valle, L.F.; Medellin-Rodriguez, F.J.; Guedea-Miranda, R. Mechanical and antimicrobial properties of multilayer films with a polyethylene/silver nanocomposite layer. *J. Appl. Polym. Sci.* **2009**, *111*, 953–962. [CrossRef]
193. Garces, L.O.; de la Puerta, C.N. Antimicrobial Packaging Based on the Use of Natural Extracts and the Process to Obtain This Packaging. European Patent EP1657181-B1, 13 January 2010.
194. Lacroix, M. Use of Irradiation, for the Development of Active Edible Coatings, Beads and Packaging to Assure Food Safety and to Prolong Preservation. 2017. Available online: https://media.superevent.com/documents/20170425/1c03bea5d97576aa37769311cb3ae83b/m.-lacroix.pdf (accessed on 25 September 2018).
195. Min, M.; Shi, Y.; Ma, H.; Huang, H.; Shi, J.; Chen, X.; Liu, Y.; Wang, L. Polymer-nanoparticle composites composed of poly(3-hydroxybutyrate-co-3-hydroxyvalerate) and coated silver nanoparticles. *J. Macromol. Sci. Part B* **2015**, *54*, 411–423. [CrossRef]
196. Martínez-Abad, A.; Lagarón, J.M.; Ocio, M.J. Characterization of transparent silver loaded poly(L-lactide) films produced by melt-compounding for the sustained release of antimicrobial silver ions in food applications. *Food Control* **2014**, *43*, 238–244. [CrossRef]
197. George, M.; Shen, W.-Z.; Qi, Z.; Bhatnagar, A.; Montemagno, C. Development and Property Evaluation of Poly (Lactic) Acid and Cellulose Nanocrystals Based Films with Either Silver or Peptide Antimicrobial Agents: Morphological, Permeability, Thermal, and Mechanical Characterization. *IOSR J. Polym. Text. Eng.* **2017**, *4*, 8–24. [CrossRef]
198. Girdthep, S.; Worajittiphon, P.; Leejarkpai, T.; Punyodom, W. Effect of Silver-loaded Kaolinite on Real Ageing, Hydrolytic Degradation, and Biodegradation of Composite Blown Films Based on Poly(lactic acid) and Poly(butylene adipate-co-terephthalate). *Eur. Polym. J.* **2016**, *82*, 244–259. [CrossRef]
199. Mbhele, Z.H.; Salemane, M.G.; van Sitter, C.G.C.E.; Nedeljkov, J.M.; Djokovic, V.; Luyt, A.S. Fabrication and characterization of silver–polyvinyl alcohol nanocomposites. *Chem. Mater.* **2003**, *15*, 5019–5024. [CrossRef]
200. Damm, C.; Munstedt, H.; Rosch, A. Long-term antimicrobial polyamide 6/silver-nanocomposites. *J. Mater. Sci.* **2007**, *42*, 6067–6073. [CrossRef]
201. Damm, C.; Munstedt, H.; Rosch, A. The antimicrobial efficacy of polyamide 6/silver-nano- and microcomposites. *Mater. Chem. Phys.* **2008**, *108*, 61–66. [CrossRef]
202. Cheng, Q.; Li, C.; Pavlinek, V.; Saha, P.; Wang, H. Surface-modified antibacterial TiO_2/Ag+ nanoparticles: Preparation and properties. *Appl. Surf. Sci.* **2006**, *252*, 4154–4160. [CrossRef]
203. Li, H.; Li, F.; Wang, L.; Sheng, J.; Xin, Z.; Zhao, L.; Xiao, H.; Zheng, Y.; Hu, Q. Effect of nano-packing on preservation quality of Chinese jujube. *Food Chem.* **2009**, *114*, 547–552. [CrossRef]
204. Yu, H.; Sun, B.; Zhang, D.; Chen, G.; Yang, X.; Yao, J. Reinforcement of biodegradable poly(3-hydroxybutyrate-co-3-hydroxyvalerate) with cellulose nanocrystal/silver nanohybrids as bifunctional nanofillers. *J. Mater. Chem. B* **2014**, *2*, 8479–8489. [CrossRef]
205. Manso, S.; Becerril, R.; Nerin, C.; Gomez-Lus, R. Influence of pH and temperature variations on vapor phase action of an antifungal food packaging against five mold strains. *Food Control* **2015**, *47*, 20–26. [CrossRef]
206. Valderrama Solano, A.C.; Rojas de Gante, C. Two different processes to obtain antimicrobial packaging containing natural oils. *Food Bioprocess Technol.* **2012**, *5*, 2522–2528. [CrossRef]
207. Minelli, M.; de Angelis, M.G.; Doghieri, F.; Rocchetti, M.; Montenero, A. Barrier properties of organic-inorganic hybrid coatings based on polyvinyl alcohol with improved water resistance. *Polym. Eng. Sci.* **2010**, *50*, 144–153. [CrossRef]

208. Lantano, C.; Alfieri, I.; Cavazza, A.; Corradini, C.; Lorenzi, A.; Zucchetto, N.; Montenero, A. Natamycin based sol-gel antimicrobial coatings on polylactic acid films for food packaging. *Food Chem.* **2014**, *165*, 342–347. [CrossRef] [PubMed]
209. Jokar, M.; Alsing Pedersen, G.; Loeschner, K. Six open questions about the migration of engineered nano-objects from polymer-based food-contact materials: A review. *Food Addit. Contam. Part A* **2017**, *34*, 434–450. [CrossRef] [PubMed]
210. Zhu, Y.; Buonocore, G.G.; Lavorgna, M. Photocatalytic activity of PLA/TiO_2 nanocomposites and TiO_2-active multilayered hybrid coatings. *Ital. J. Food Sci.* **2012**, *24*, 102–106.
211. Makwana, S.; Choudhary, R.; Dogra, N.; Kohli, P.; Haddock, J. Nanoencapsulation and immobilization of cinnamaldehyde for developing antimicrobial food packaging material. *LWT Food Sci. Technol.* **2014**, *57*, 470–476. [CrossRef]
212. Theinsathid, P.; Visessanguan, W.; Kruenate, J.; Kingcha, Y.; Keeratipibul, S. Antimicrobial activity of lauric arginate-coated polylactic acid films against listeria monocytogenes and salmonella typhimurium on cooked sliced ham. *J. Food Sci.* **2012**, *77*, M142–M149. [CrossRef] [PubMed]
213. Guo, M.; Jin, T.Z.; Yang, R. Antimicrobial polylactic acid packaging films against listeria and salmonella in culture medium and on ready-to-eat meat. *Food Bioprocess Technol.* **2014**, *7*, 3293–3307. [CrossRef]
214. Auxier, J.A.; Schilke, K.F.; McGuire, J. Activity retention after nisin entrapment in a polyethylene oxide brush layer. *J. Food Prot.* **2014**, *77*, 1624–1629. [CrossRef] [PubMed]
215. Fernandes, S.C.M.; Sadocco, P.; Causio, J.; Silvestre, A.J.D.; Mondragon, I.; Freire, C.S.R. Antimicrobial pullulan derivative prepared by grafting with 3-aminopropyltrimethoxysilane: Characterization and ability to form transparent films. *Food Hydrocoll.* **2014**, *35*, 247–252. [CrossRef]
216. Muriel-Galet, V.; Talbert, J.N.; Hernandez-Munoz, P.; Gavara, R.; Goddard, J.M. Covalent immobilization of lysozyme on ethylene vinyl alcohol films for nonmigrating antimicrobial packaging applications. *J. Agric. Food Chem.* **2013**, *61*, 6720–6727. [CrossRef] [PubMed]
217. Anthierens, T.; Billiet, L.; Devlieghere, F.; du Prez, F. Poly(butylene adipate) functionalized with quaternary phosphonium groups as potential antimicrobial packaging material. *Innov. Food Sci. Emerg. Technol.* **2012**, *15*, 81–85. [CrossRef]
218. Mackiw, E.; Maka, L.; Sciezynska, H.; Pawlicka, M.; Dziadczyk, P.; Rzanek-Boroch, Z. The impact of plasma-modified films with sulfur dioxide, sodium oxide on food pathogenic microorganisms. *Packag. Technol. Sci.* **2015**, *28*, 285–292. [CrossRef]
219. Pinheiro, A.C.; Bourbon, A.I.; Medeiros, B.G.D.S.; da Silva, L.H.M.; da Silva, M.C.H.; Carneiro-da-Cunha, M.G.; Coimbra, M.A.; Vicente, A.A. Interactions between κ-carrageenan and chitosan in nanolayered coatings-structural and transport properties. *Carbohydr. Polym.* **2012**, *87*, 1081–1090. [CrossRef]
220. Carneiro-da-Cunha, M.G.; Cerqueira, M.A.; Souza, B.W.S.; Carvalhoc, S.; Quintas, M.A.C.; Teixeira, J.A.; Vicente, A.A. Physical and thermal properties of a chitosan/alginate nanolayered PET film. *Carbohydr. Polym.* **2010**, *82*, 153–159. [CrossRef]
221. Medeiros, B.G.D.S.; Pinheiro, A.C.; Teixeira, J.A.; Vicente, A.A.; Carneiro-da-Cunha, M.G. Polysaccharide/protein nanomultilayer coatings: Construction, characterization and evaluation of their effect on "Rocha" pear (*Pyrus communis* L.) shelf-life. *Food Bioprocess Technol.* **2012**, *5*, 2435–2445. [CrossRef]
222. Contini, C.; Álvarez, R.; O'Sullivan, M.; Dowling, D.P.; Gargan, S.O.; Monahan, F.J. Effect of an active packaging with citrus extract on lipid oxidation and sensory quality of cooked turkey meat. *Meat Sci.* **2014**, *96*, 1171–1176. [CrossRef] [PubMed]
223. Contini, C.; Katsikogianni, M.G.; O'Neill, F.T.; O'Sullivan, M.; Boland, F.; Dowling, D.P.; Monahan, F.J. Storage stability of an antioxidant active packaging coated with citrus extract following a plasma jet pretreatment. *Food Bioprocess Technol.* **2014**, *7*, 2228–2240. [CrossRef]
224. Bolumar, T.; Andersen, M.L.; Orlien, V. Antioxidant active packaging for chicken meat processed by high pressure treatment. *Food Chem.* **2011**, *129*, 1406–1412. [CrossRef]
225. Lee, C.H.; An, D.S.; Lee, S.C.; Park, H.J.; Lee, D.S. A coating for use as an antimicrobial and antioxidative packaging material incorporating nisin and α-tocopherol. *J. Food Eng.* **2004**, *62*, 323–329. [CrossRef]
226. Schreiber, S.B.; Bozell, J.J.; Hayes, D.G.; Zivanovic, S. Introduction of primary antioxidant activity to chitosan for application as a multifunctional food packaging material. *Food Hydrocoll.* **2013**, *33*, 207–214. [CrossRef]

227. Struller, C.F.; Kelly, P.J.; Copeland, N.J. Aluminum oxide barrier coatings on polymer films for food packaging applications. *Surf. Coat. Technol.* **2014**, *241*, 130–137. [CrossRef]
228. Chatham, H. Oxygen diffusion barrier properties of transparent oxide coatings on polymeric substrates. *Surf. Coat. Technol.* **1996**, *78*, 1–9. [CrossRef]
229. Shutava, T.G.; Prouty, M.D.; Agabekov, V.E.; Lvov, Y.M. Antioxidant properties of layer-by-layer films on the basis of tannic acid. *Chem. Lett.* **2006**, *35*, 1144–1145. [CrossRef]
230. Arrua, D.; Strumia, M.C.; Nazareno, M.A. Immobilization of Caffeic acid on a polypropylene film: Synthesis and antioxidant properties. *J. Agric. Food Chem.* **2010**, *58*, 9228–9234. [CrossRef] [PubMed]
231. Roman, M.J.; Tian, F.; Decker, E.A.; Goddard, J.M. Iron chelating polypropylene films: Manipulating photoinitiated graft polymerization to tailor chelating activity. *J. Appl. Polym. Sci.* **2014**, *131*, 39948. [CrossRef]
232. Ogiwara, Y.; Roman, M.J.; Decker, E.A.; Goddard, J.M. Iron chelating active packaging: Influence of competing ions and pH value on effectiveness of soluble and immobilized hydroxamate chelators. *Food Chem.* **2016**, *196*, 842–847. [CrossRef] [PubMed]
233. Tian, F.; Decker, E.A.; Clements, D.J.; Goddard, J.M. Influence of non-migratory metal-chelating active packaging film on food quality: Impact on physical and chemical stability of emulsions. *Food Chem.* **2014**, *151*, 257–265. [CrossRef] [PubMed]
234. Tian, F.; Decker, E.A.; Goddard, J.M. Controlling lipid oxidation via a biomimetic iron chelating active packaging material. *J. Agric. Food Chem.* **2013**, *61*, 12397–12404. [CrossRef] [PubMed]
235. Roman, M.J.; Decker, E.A.; Goddard, J.M. Performance of nonmigratory iron chelating active packaging materials in viscous model food systems. *J. Food Sci.* **2015**, *80*, 1965–1973. [CrossRef] [PubMed]
236. Tian, F.; Roman, M.J.; Decker, E.A.; Goddard, J.M. Biomimetic design of chelating interfaces. *J. Appl. Polym. Sci.* **2015**, *132*, 41231–41239. [CrossRef]
237. Wong, D.E.; Dai, M.; Talbert, J.N.; Nugen, S.R.; Goddard, J.M. Biocatalytic polymer nanofibers for stabilization and delivery of enzymes. *J. Mol. Catal. B Enzym.* **2014**, *110*, 16–22. [CrossRef]
238. Johansson, K.; Winestrand, S.; Johansson, C.; Jarnstrom, L.; Jonsson, L.J. Oxygen-scavenging coatings and films based on lignosulfonates and laccase. *J. Biotechnol.* **2012**, *161*, 14–18. [CrossRef] [PubMed]
239. Goddard, J.M.; Talbert, J.N.; Hotchkiss, J.H. Covalent attachment of lactase to low-density polyethylene films. *J. Food Sci.* **2007**, *72*, E36–E41. [CrossRef] [PubMed]
240. Mahoney, K.W.; Talbert, J.N.; Goddard, J.M. Effect of polyethylene glycol tether size and chemistry on the attachment of lactase to polyethylene films. *J. Appl. Polym. Sci.* **2013**, *127*, 1203–1210. [CrossRef]
241. Talbert, J.N.; Goddard, J.M. Influence of nanoparticle diameter on conjugated enzyme activity. *Food Bioprod. Process.* **2013**, *91*, 693–699. [CrossRef]
242. Wong, D.E.; Talbert, J.N.; Goddard, J.M. Layer by layer assembly of a biocatalytic packaging film: Lactase covalently bound to low-density polyethylene. *J. Food Sci.* **2013**, *78*, E853–E860. [CrossRef] [PubMed]
243. Ge, L.; Zhao, Y.; Mo, T.; Li, J.; Li, P. Immobilization of glucose oxidase in electrospun nanofibrous membranes for food preservation. *Food Control* **2012**, *26*, 188–193. [CrossRef]
244. Caseli, L.; Santos, D.S., Jr.; Foschini, M.; Goncalves, D.; Oliveira, O.N., Jr. Control of catalytic activity of glucose oxidase in layer-by-layer films of chitosan and glucose oxidase. *Mater. Sci. Eng. C* **2007**, *27*, 1108–1110. [CrossRef]
245. Winestrand, S.; Johansson, K.; Järnström, L.; Jönsson, L.J. Co-immobilization of oxalate oxidase and catalase in films for scavenging of oxygen or oxalic acid. *Biochem. Eng. J.* **2013**, *72*, 96–101. [CrossRef]
246. Shutava, T.G.; Kommireddy, D.S.; Lvov, Y.M. Layer-by-layer enzyme/polyelectrolyte films as a functional protective barrier in oxidizing media. *J. Am. Chem. Soc.* **2006**, *128*, 9926–9934. [CrossRef] [PubMed]
247. Soares, N.F.F.; Hotchkiss, J.H. Bitterness reduction in grapefruit juice through active packaging. *Packag. Technol. Sci.* **1998**, *11*, 9–18. [CrossRef]
248. Nunes, M.A.P.; Vila-Real, H.; Fernandes, P.C.B.; Ribeiro, M.H.L. Immobilization of naringinase in PVA-alginate matrix using an innovative technique. *Appl. Biochem. Biotechnol.* **2010**, *160*, 2129–2147. [CrossRef] [PubMed]
249. Krenkova, J.; Lacher, N.A.; Svec, F. Highly efficient enzyme reactors containing trypsin and endoproteinase LysC immobilized on porous polymer monolith coupled to MS suitable for analysis of antibodies. *Anal. Chem.* **2009**, *81*, 2004–2012. [CrossRef] [PubMed]

250. Yamada, K.; Iizawa, Y.; Yamada, J.; Hirata, M. Retention of activity of urease immobilized on grafted polymer films. *J. Appl. Polym. Sci.* **2006**, *102*, 4886–4896. [CrossRef]
251. Andersson, M.; Andersson, T.; Adlercreutz, P.; Nielsen, T.; Hornsten, E. Toward an enzyme-based oxygen scavenging laminate. Influence of industrial lamination conditions on the performance of glucose oxidase. *Biotechnol. Bioeng.* **2002**, *79*, 37–42. [CrossRef] [PubMed]
252. Park, H.P.; Braun, P.V. Coaxial Electrospinning of Self-Healing Coatings. *Adv. Mater.* **2010**, *22*, 496–499. [CrossRef] [PubMed]
253. Díez-Pascual, A.M. (Ed.) *Antibacterial Activity of Nanomaterials*; ISBN 978-3-03897-048-4 (Pbk); ISBN 978-3-03897-049-1 (PDF).
254. Wanga, J.; Huang, N.; Pan, C.J.; Kwok, S.C.H.; Yang, P.; Leng, Y.X.; Chen, J.Y.; Sun, H.; Wan, G.J.; Liu, Z.Y.; et al. Bacterial repellence from polyethylene terephthalate surface modified by acetylene plasma immersion ion implantation-deposition. *Surf. Coat. Technol.* **2004**, *186*, 299–305. [CrossRef]
255. Brobbey, K.J.; Saarinen, J.J.; Alakomi, H.-L.; Yang, B.; Toivakka, M. Efficacy Of Natural Plant Extracts In Antimicrobial Packaging Systems. *J. Appl. Packag. Res.* **2017**, *9*, 6.
256. Miteluţ, A.C.; Popa, E.E.; Popescu, P.A.; Popa, M.E.; Munteanu, B.S.; Vasile, C.; Ştefănoiu, G. Research on chitosan and oil coated PLA as food packaging material. Presented at the International Worshop, "Progress in Antimicrobial Materials", Iasi, Romania, 30 March 2017.
257. Busolo, M.A.; Fernandez, P.; Ocio, M.J.; Lagaron, J.M. Novel silver-based nanoclay as an antimicrobial in polylactic acid food packaging coatings. *Food Addit. Contam. Part A* **2010**, *27*, 1617–1626. [CrossRef] [PubMed]
258. Luo, P.G.; Stutzenberger, F.J. Nanotechnology in the detection and control of microorganisms. In *Advances in Applied Microbiology*; Laskin, A.I., Sariaslani, S., Gadd, G.M., Eds.; Elsevier: London, UK, 2008; Volume 63, pp. 145–181, ISBN 9780120026623, 9780080468921.
259. Makwana, S.; Choudhary, R.; Kohli, P. Advances in Antimicrobial Food Packaging with Nanotechnology and Natural Antimicrobials. *Int. J. Food Sci. Nutr. Eng.* **2015**, *5*, 169–175. [CrossRef]
260. Li, Q.; Mahendra, S.; Lyon, D.Y.; Brunet, L.; Liga, M.V.; Li, D.; Alvarez, P.J.J. Antimicrobial nanomaterials for water disinfection and microbial control: Potential applications and implications. *Water Res.* **2008**, *42*, 4591–4602. [CrossRef] [PubMed]
261. Ghaffari-Moghaddam, M.; Eslahi, H. Synthesis, characterization and antibacterial properties of a novel nanocomposite based on polyaniline/polyvinyl alcohol/Ag. *Arab. J. Chem.* **2013**, 7. [CrossRef]
262. Kumar, R.; Munstedt, H. Silver ion release from antimicrobial polyamide/silver composites. *Biomaterials* **2005**, *26*, 2081–2088. [CrossRef] [PubMed]
263. Hannon, J.C.; Cummins, E.; Kerry, J.; Cruz-Romero, M.; Morris, M. Advances and challenges for the use of engineered nanoparticles in food contact materials. *Trends Food Sci. Technol.* **2015**, *43*, 43–62. [CrossRef]
264. Drew, R.; Hagen, T.; ToxConsult Pty Ltd. Nanotechnologies in Food Packaging: An Exploratory Appraisal of Safety and Regulation, Report Prepared for Food Standards Australia New Zealand. May 2016. Available online: https://www.foodstandards.gov.au/publications/Documents/Nanotech%20in%20food%20packaging.pdf (accessed on 10 June 2017).
265. Qi, L.F.; Xu, Z.R.; Jiang, X.; Hu, C.; Zou, X. Preparation and antibacterial activity of chitosan nanoparticles. *Carbohydr. Res.* **2004**, *339*, 2693–2700. [CrossRef] [PubMed]
266. Kang, S.; Pinault, M.; Pfefferle, L.D.; Elimelech, M. Single-walled carbon nanotubes exhibit strong antimicrobial activity. *Langmuir* **2007**, *23*, 8670–8673. [CrossRef] [PubMed]
267. Tian, F.; Decker, E.A.; Goddard, J.M. Controlling lipid oxidation of food by active packaging technologies. *Food Funct.* **2013**, *4*, 669–680. [CrossRef] [PubMed]
268. Gomez-Estaca, J.; Lopez-de-Dicastillo, C.; Hernandez-Munoz, P.; Catala, R.; Gavara, R. Advances in antioxidant active food packaging. *Trends Food Sci. Technol.* **2014**, *35*, 42–51. [CrossRef]
269. Nerin, C.; Tovar, L.; Salafranca, J. Behaviour of a new antioxidant active film versus oxidizable model compounds. *J. Food Eng.* **2008**, *84*, 313–320. [CrossRef]
270. Yemmireddy, V.K.; Farrell, G.D.; Hung, Y.C. Development of Titanium Dioxide (TiO_2) Nanocoatings on Food Contact Surfaces and Method to Evaluate Their Durability and Photocatalytic Bactericidal Property. *J. Food Sci.* **2015**, *80*, N1903–N1911. [CrossRef] [PubMed]

271. Barbiroli, A.; Bonomi, F.; Capretti, G.; Iametti, S.; Manzoni, M.; Piergiovanni, L.; Rollini, M. Antimicrobial activity of lysozyme and lactoferrin incorporated in cellulose-based food packaging. *Food Control* **2012**, *26*, 387–392. [CrossRef]
272. Mendes de Souza, P.; Fernandez, A.; Lopez-Carballo, G.; Gavara, R.; Hernandez-Munoz, P. Modified sodium caseinate films as releasing carriers of lysozyme. *Food Hydrocoll.* **2010**, *24*, 300–306. [CrossRef]
273. Moskovitz, Y.; Srebnik, S. Mean-field model of immobilized enzymes embedded in a grafted polymer layer. *Biophys. J.* **2005**, *89*, 22–31. [CrossRef] [PubMed]
274. Robertson, G.L. *Food Packaging: Principles and Practice*, 2nd ed.; Taylor & Francis/CRC Press: Boca Raton, FL, USA, 2006; p. 550, ISBN 9781439862414.
275. Chan, C.M.; Ko, T.M.; Hiraoka, H. Polymer surface modification by plasmas and photons. *Surf. Sci. Rep.* **1996**, *24*, 3–54. [CrossRef]
276. Jansen, B.; Kohnen, W. Prevention of Biofilm Formation by Polymer Modification. *J. Ind. Microbiol.* **1995**, *15*, 391–396. [CrossRef] [PubMed]
277. Karkhanisa, S.S.; Stark, N.M.; Sabo, R.C.; Matuana, L.M. Water vapor and oxygen barrier properties of extrusion-blown poly(lactic acid)/cellulose nanocrystals nanocomposite films. *Compos. Part A Appl. Sci. Manuf.* **2018**, *114*, 204–211. [CrossRef]
278. López de Dicastillo, C.; Garrido, L.; Alvarado, N.; Romero, J.; Palma, J.L.; Galotto, M.J. Improvement of Polylactide Properties through Cellulose Nanocrystals Embedded in Poly(Vinyl Alcohol) Electrospun Nanofibers. *Nanomaterials* **2017**, *7*, 106. [CrossRef] [PubMed]
279. Chakrabarty, A.; Teramoto, Y. Review Recent Advances in Nanocellulose Composites with Polymers: A Guide for Choosing Partners and How to Incorporate Them. *Polymers* **2018**, *10*, 517. [CrossRef]
280. Ratner, B.D.; Hoffman, A.S. Physichochemical surface modification of materials used in medicine. In *Biomaterials Science: An Introduction to Materials in Medicine*, 3rd ed.; Ratner, B.D., Hoffman, A.S., Schoen, F.J., Lemons, J.E., Eds.; Academic Press: Oxford, UK; Waltham, MA, USA, 2013; pp. 259–276, ISBN 9780123746269, 9780080877808.
281. Wagner, J.R., Jr. *Multilayer Flexible Packaging Technology and Applications for the Food, Personal Care and Over-the-Counter Pharmaceutical Industries*, 1st ed.; Elsevier Science: Oxford, NY, USA, 2010; p. 258, ISBN 9780815520214, 9780815520221.
282. Nistor, M.T.; Vasile, C.; Chiriac, A.P.; Rusu, A.; Zgardan, C.; Nita, L.E.; Neamtu, I. Hybrid Sensitive Hydrogels for Medical Applications. In *Polymer Materials with Smart Properties*; Bercea, M., Ed.; Nova Science Publishers: New York, NY, USA, 2013; Chapter 3; pp. 67–90, ISBN 978-1-62808-876-2.
283. Cheaburu, C.N.; Vasile, C. pH-responsive Hydrogels based on Chitosan and its derivatives. In *Material Science, Synthesis, Properties, Applications, Polyme Yearbook*; Pethrick, R.A., Zaikov, G.E., Eds.; Nova Science Publishers: New York, NY, USA, 2010; Volume 24, ISBN 978-1-60876-872-1.
284. Vasile, C.; Dumitriu, R.P. (Eds.) *Polymeric Materials Responsive to External Stimuli—Smart Polymeric Materials*; PIM Publishng House: Iasi, Romania, 2008; p. 285, ISBN 978-606-520-068-5.
285. Pamfil, D.; Vasile, C. Responsive Polymeric Nanotherapeutics. In *Polymeric Nanomaterials in Nanotherapeutics*; Vasile, C., Ed.; Elsevier: Amstredam, The Netherlands, 2019; Chapter 2; p. 53.
286. Cordeiro, A.L. Stimuli-Responsive Polymer Nanocoatings, Nanotechnologies for the Life Sciences. In *Nanostructured Thin Films and Surfaces*; Challa, S.S., Kumar, R., Eds.; John Wiley and Sons: New York, NY, USA, 2011; Volume 5.
287. Nath, N.; Chilkoti, A. *Creating "Smart" Surfaces Using Stimuli Responsive Polymers, Advanced Materials*; WILEY-VCH Verlag GmbH & Co. KGaA: Weinheim, Germany, 2002.
288. Cohen Stuart, M.; Huck, W.; Genzer, J.; Müller, M.; Ober, C.; Stamm, M.; Sukhorukov, G.B.; Szleifer, I.; Tsukruk, V.V.; Urban, M.; et al. Emerging Applications of Stimuli-Responsive Polymer Materials. *Nat. Mater.* **2010**, *9*, 101–113. [CrossRef] [PubMed]
289. Wei, M.; Gao, Y.; Li, X.; Serpe, M.J. Stimuli-responsive polymers and their applications. *Polym. Chem.* **2017**, *8*, 10–11. [CrossRef]
290. Noonan, G.O.; Whelton, A.J.; Carlander, D.; Duncan, T.V. Measurement methods to evaluate engineered nanomaterial release from food contact materials. *Compr. Rev. Food Sci. Food Saf.* **2014**, *13*, 679–692. [CrossRef]
291. Weiss, J.; Takhistov, P.; Mc Clements, D.J. Functional materials in food nanotechnology. *J. Food Sci.* **2006**, *71*, R107–R116. [CrossRef]

292. Honarvar, Z.; Hadian, Z.; Mashayekh, M. Nanocomposites in food packaging applications and their risk assessment for health. *Electron. Physician* **2016**, *8*, 2531–2538. [CrossRef] [PubMed]
293. Farhoodi, M. Nanocomposite Materials for Food Packaging Applications: Characterization and Safety Evaluation. *Food Eng. Rev.* **2016**, *8*, 35–51. [CrossRef]
294. Lan, T. Nanocomposite Materials for Packaging Applications, ANTEC. 2007. Available online: http://www.nanocor.com/tech_papers/antec-nanocor-tie%20lan-5-07.pdf (accessed on 25 September 2018).
295. BCC Research, The Advanced Packaging Solutions Market Value for 2017 Is Projected to Be Nearly $44.3 Billion. 2015. Available online: http://www.bccresearch.com (accessed on 20 June 2017).
296. Realini, C.E.; Marcos, B. Active and intelligent packaging systems for a modern society. *Meat Sci.* **2014**, *98*, 404–419. [CrossRef] [PubMed]
297. Day, B.P.F. Active packaging of foods. In *Smart Packaging Technologies for Fast Moving Consumer Goods*; Kerry, J.P., Butler, P., Eds.; Wiley & Sons, Ltd.: West Sussex, UK, 2008; pp. 1–18, ISBN 978-0-470-02802-5.
298. Barnes, K.A.; Sinclair, C.R.; Watson, D.H. *Chemical Migration and Food Contact Materials*, 1st ed.; Woodhead Publishing: Amsterdam, The Netherlands, 2007; p. 464, ISBN 9781845692094; Hardcover ISBN 9781845690298.
299. Farris, S.; Introzzi, L.; Piergiovanni, L.; Cozzolino, C.A. Effects of different sealing conditions on the seal strength of polypropylene films coated with a bio-based thin layer. *Ital. J. Food Sci.* **2011**, *23*, 111–114. [CrossRef]
300. Siegrist, M.M.-E.; Cousin, H.; Kastenholz, A.; Wiek, A. Public acceptance of nanotechnology foods and food packaging: The influence of affect and trust. *Appetite* **2007**, *49*, 459–466. [CrossRef] [PubMed]
301. Siegrist, M.; Stampfli, N.; Kastenholz, H.; Keller, C. Perceived Risks and Perceived Benefits of Different Nanotechnology Foods and Nanotechnology Food Packaging. *Appetite* **2008**, *51*, 283–290. [CrossRef] [PubMed]
302. Siegrist, M.; Sütterlin, B. Importance of perceived naturalness for acceptance of food additives and cultured meat. *Appetite* **2017**, *113*, 320–326. [CrossRef] [PubMed]
303. Bearth, A.; Siegrist, M. Are risk or benefit perceptions more important for public acceptance of innovative food technologies: A meta-analysis. *Trends Food Sci. Technol.* **2016**, *49*, 14–23. [CrossRef]
304. Marquis, D.M.; Guillaume, É.; Chivas-Joly, C. Properties of Nanofillers in Polymer. Chap 11. In *Nanocomposites and Polymers with Analytical Methods*; Cuppoletti, J., Ed.; IntechOpen: London, UK, 2011; Chapter 13; pp. 261–284. Available online: http://www.intechopen.com/books/nanocomposites-and-polymers-with-analyticalmethods/properties-of-nanofillers-in-polymer (accessed on 25 September 2018).

8

Custom Design of Packaging through Advanced Technologies: A Case Study Applied to Apples

Lucía Rodríguez-Parada *, Pedro F. Mayuet and Antonio J. Gámez

Department of Mechanical Engineering and Industrial Design, University of Cadiz, 11519 Cadiz, Spain; pedro.mayuet@uca.es (P.F.M.); antoniojuan.gamez@uca.es (A.J.G.)
* Correspondence: lucia.rodriguez@uca.es

Abstract: In the context of food packaging design, customization enhances the value of a product by meeting consumer needs. Personalization is also linked to adaptation, so the properties of the packaging can be improved from several points of view: functional, aesthetic, economic and ecological. Currently, functional and formal properties of packaging are not investigated in depth. However, the study of both properties is the basis for creating a new concept of personalized and sustainable product. In accordance with this approach, a conceptual design procedure of packaging with personalized and adapted geometries based on the digitization of fresh food is proposed in this work. This study is based on the application of advanced technologies for the design and development of food packaging, apples in this work, in order to improve the quality of the packaging. The results obtained show that it is possible to use advanced technologies in the early stages of product design in order to obtain competitive products adapted to new emerging needs.

Keywords: food packaging; customization; product design; personalized design; reverse engineering; computer aid design (CAD); fused deposition modelling (FDM)

---

1. Introduction

During the last decade, the demand for healthy and fresh food, especially fruits and vegetables, was gradually growing in a context where eating and consumption habits were constantly changing due to the society lifestyle. For this reason, the European Union raised new objectives for the food packaging industry: sustainability of raw materials, minimization of waste, reduction of energy consumption during the production process, minimization of environmental impact, recyclability of packaging and littering reduction [1–3].

Besides, consumers demand specific needs, mainly related to the design and adaptability of packaging [4]. Among others, providing more nutritional information on the packaging, greater food security and less risk to health are requested [5–7]. Thus, there are studies that place the packaging of fresh product as the second reason for choosing a product where factors such as comfort, appearance, transparency and texture should be considered [5].

Consequently, regarding the selection of material, thermoplastic polymers comply with all of these conditions, which is why they have been chosen by manufacturers of different products as materials for their packaging [8]. Polyethylene terephthalate (PET) is a polymer whose properties include its mechanical resistance to both impact and chemical products, its transparency, its lightness, the reduced demand for energy in its manufacture and transport, its mouldability and its recyclability [8]. In addition, it is the most recycled plastic in the world, and the European Food Safety Authority has corroborated that PET does not contain bisphenol-A (BPA), phthalates or dioxins [2].

However, PET has been identified as one of the main causes of global environmental degradation and this problem is expected to increase due to growing demand from developing or re-industrializing countries around the world [9–11]. For example, China, which is one of the largest exporters as well as producers, accounted for 24% of the world's demand for PET plastic in 2016. For this reason, understanding the consumption needs of this type of market is critical when examining the prospects for global PET volume growth [12].

In fact, the clear trend towards the increase in consumption of food that needs packaging carries with it the constant development of materials and techniques that improve the performance of this service [13]. In short, several aspects can be highlighted [14]: the improvement of the production times, the reduction of the material and the improvement of its aesthetic and functional properties.

Thus, the fresh food packaging industry is interested in developing efficient and innovative solutions to ensure the quality of the products by taking into account their sustainability, especially at the environmental level [12]. For this reason, the design evaluation in the development of sustainable products should include aspects related to the design, manufacture and use of the packaging [15,16].

In this context, the packaging design process implies the consideration of aspects associated with its cost, appearance, usability, manufacturing, sustainability, standards or competitiveness [15]. Therefore, the selection of the manufacturing process of the food package is extremely important and, in this case, one of most commonly used is the thermoforming process [17]. Previous research on the packaging design process in thermoforming focused on the study of the moulds and the materials used in the process [10,18–20]. Also, recent research implements functions for intelligent packaging development through which sensitive labels control the condition of the food [21]. These initiatives are aimed at improving food health and freshness, achieving in this way its better preservation [4].

On the other hand, the packaging sector demands customized solutions in terms of shapes, sizes and colours, so that each packaging solution is unique [22]. In turn, it is intended to pursue product differentiation by means of sustainable products and new designs [23]. In such a way, it is also possible to respond to demographic changes and consumption habits [24–26]. Thus, packaging can be personalised in design, brand and/or size, among other characteristics. Also, it can be understood that an adapted packaging is the one that adapts itself to the inner shape and size of the product [27].

However, despite the needs that were determined in previous research and after having analysed the fresh food packaging which is currently commercialised on the market, it is observed that the majority of them are standard and just a few include customization in terms of forms [28]. In addition, there is no adaptation to specific sizes according to calibres in these cases.

For this reason, this work studies the personalization and adaptation in packaging design for the protection of the product in order to provide designers with a tool for the packaging 4.0 generation. This is linked to the sustainability of both the interior of the product and the material expense. A case study with apples is shown in order to validate the use of advanced technologies as part of the unconventional design of packaging, increasing the sustainability and functionality of these products. Thus, a technique that allows the design of personalized packaging to optimize the cost of packaging material, the technological resources and to improve the functionality of packaging is developed. The research also focuses on obtaining packaging that offers brands the possibility of differentiating themselves by means of personalization and in accordance with sales expectations and consumer perception.

2. Materials and Methods

2.1. Tools and Materials Used to Obtain the Customized Packaging

To carry out the experimental development, apples were used as the target product because their size facilitates the acquisition of measurements, studying the adaptation to packaging depending on two different calibres: category I and category II, according to [29].

For the development of the experimental procedure, 10 units of each calibre were used. To carry out the design-adapted concept from the computer-aided design (CAD), the apples were digitized by means of 3D techniques [30], Figure 1. To this end, 8 images were captured for each unit to create a three-dimensional model, Figure 2. An SLS-1 David® 3D scanner (David Vision Systems GmbH, Koblenz, Germany) of Hewlett-Packard (HP inc.) was used, according to [31]. These scanned elements have, as their main purpose, the creation of adapted geometries through the generation of curves based on the scanned elements.

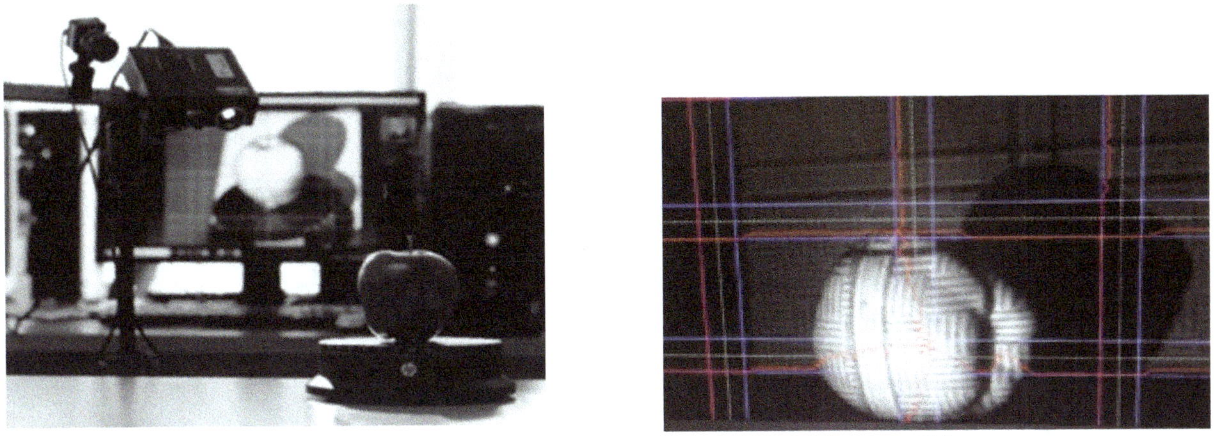

(**a**) (**b**)

Figure 1. (**a**) Scanning procedure using the SLS-1 David® V5 scanner, (**b**) Visualization of the apple through the software (HP 3D Scan David®, version Pro V5, HP inc., Palo Alto, CA, US).

Figure 2. Example of image capture, which the 3D scanner performs, of an apple of calibre 1.

Thus, the concept development was carried out through CAD software, Solidworks® (2016 version, Dassault Systèmes SE, Velizy-Villacoublay, France). As a result, the development of packaging design has been simplified in time and has allowed us to generate more complex and personalized forms for the food.

The evaluation of the design proposals was validated through the generation of reliable prototypes. The mould design and The Standard Triangle Language (STL) file of the mould were generated by Solidworks® and were then parameterized using software for 3D printing, Simplify3D® (Cincinnati, OH, USA). A Gcode file was generated for printing in the FDM machine BQ Witbox (Mundo Reader S.L., Madrid, Spain), that uses a diameter filament of 1.75 mm. A PET sheet of 500 μm thick was thermoformed to generate the prototype with Formech 450DT using the FDM mould, Figure 3.

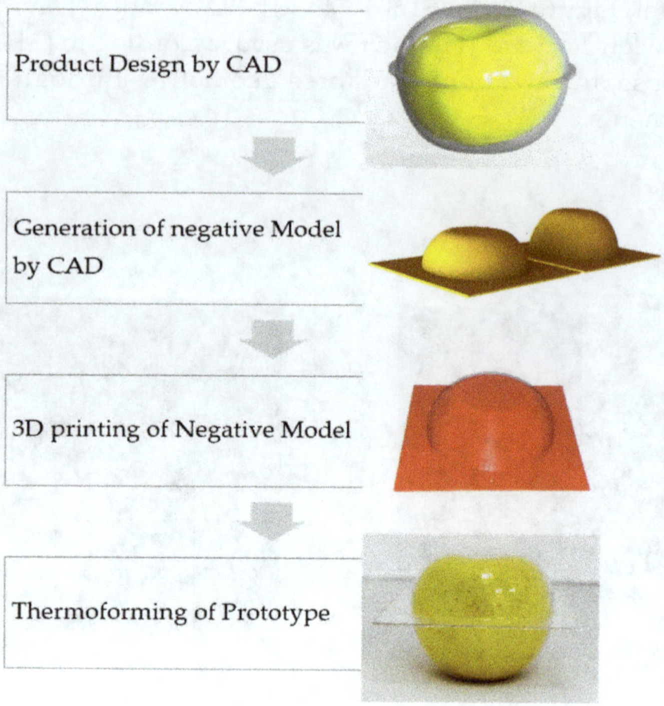

Figure 3. Generation of prototypes in thermoforming with moulds created by FDM.

2.2. *Parameterization of the Packaging Design*

Figure 4 details the procedure carried out for the design and development of customised and adapted packaging. Once the conceptual design of the package has been carried out for this case study, the properties of these products are studied through the application to scanned apples, in order to define the main parameters to be taken into account. These data were used for two purposes: packaging design using CAD tools and parameterization of the final geometry. Then, the parameterized package was validated by prototypes made with FDM moulds.

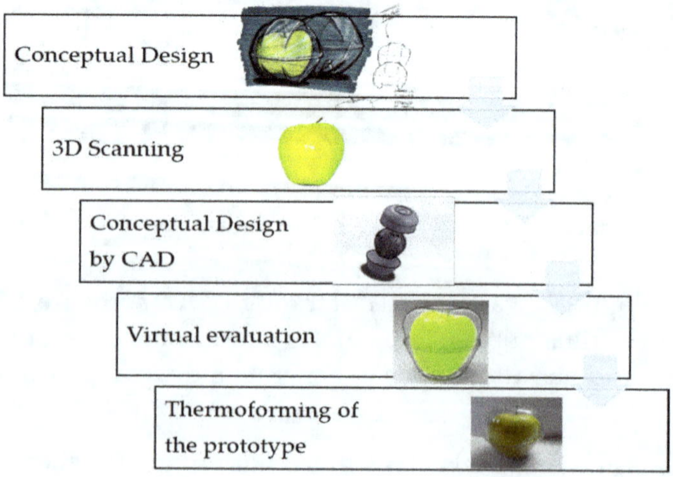

Figure 4. General procedure used to obtain the final package.

Also, several functional measures of each natural product were evaluated to analyse the differences between them, also using Solidworks® software. These dimensions were selected according to the parameters established for the calibres: the largest diameter, Dm, and the maximum height, H. According to this, two measurements were collected for each of the dimensions studied: L01 and

L02 for Dm and L03 and L04 for H, Figure 5. It should be noted that L01/L02 were made in the two directions of the maximum diameter of the fruit and L03/L04 for the two highest recorded heights. This allows the calculation of the variation of the measurement between pieces of fruit.

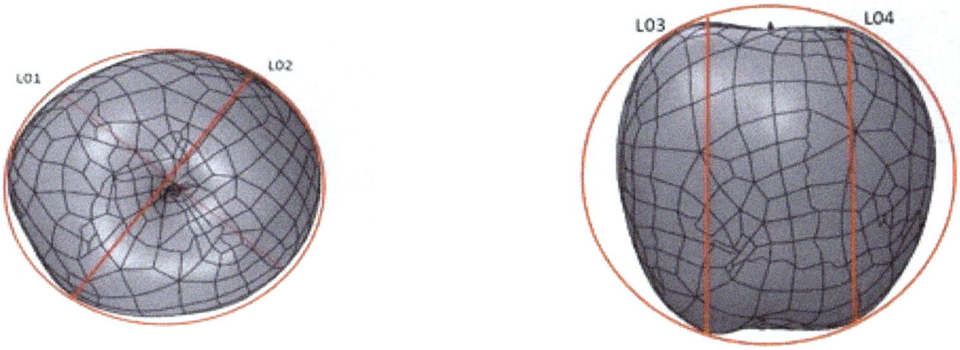

(a) (b)

Figure 5. Measurements collected on the digitized products: (**a**) width measurements, (**b**) high measurements.

One of the objectives of digitizing these elements, because of the normal variations of a natural product, is to define the range of measures that present representative variations that must be taken into account for the design of the packaging. As a result, comparative tables were obtained and the dimensional range was defined, which will serve to obtain the adaptation parameters on the design of the packaging, which is the object of study.

Once the concept was generated, the design was developed in Solidworks® using the 10 scanned elements as a means of generating the construction curves. In accordance with this, the set of lines and tangent arcs, which together form the design of the idea previously conceptualized, were defined. Likewise, the numerical relations between the different container geometries were defined, using as relation parameters the maximum height, H and the maximum diameter, Dm, of the container.

Then, when the final design was developed in Solidworks®, the relationship equations, defined above, were introduced in order to evaluate the degree of adaptability through this type of digital tool. Then, the variable measures of the packaging were defined in order to carry out the adaptation to the two categories of size of apple studied according to [31].

Finally, the adaptability range of the design created for this practical case was established and, in addition, the degree of adaptation of the dimensions obtained with each of the digitalized units for the two calibres studied were evaluated. Figure 6 shows an example of the results obtained.

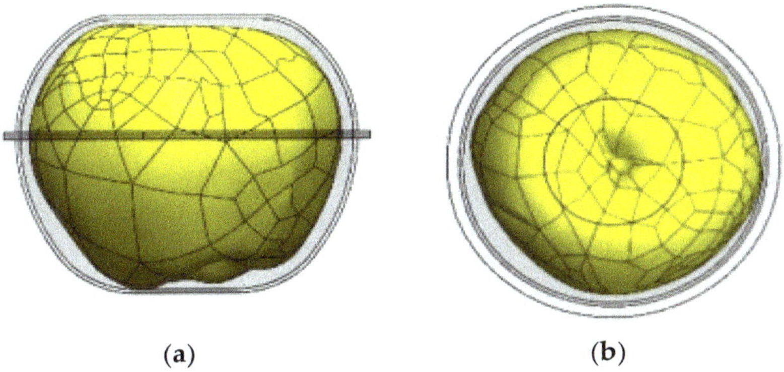

(a) (b)

Figure 6. Example of the result of the adaptation parameters with respect to the digitized product: (**a**) evaluation of the high packaging designed, (**b**) evaluation of the width of the packaging designed.

The two calibres studied were evaluated through the creation of a reliable physical prototype using FDM moulds for thermoforming of the sheet. The developed prototypes were then used to validate the dimensions obtained by introducing real apples. In total, 20 apples were used for each calibre at this stage and the selection of each of them was made taking into account shapes and size variations. Also, different types of varieties were included.

3. Results

3.1. Virtual Evaluation of Digital Elements

As described above, the study was conducted for two types of calibres, according to [29]. As seen, the calibre refers to the predominant size within the packaging and is defined according to the maximum equatorial diameter [32]. Thus, the apple samples were scanned to generate a parametric model of each one to make the measurements according to the methodology. The data obtained from the samples studied are detailed in Figure 7.

From the results obtained from the study of the morphology of apples it is determined that the dominant geometry of the fruit is slightly oval. Therefore, a major axis, L01, and a minor axis, L02, can be defined in order to name the maximum dimensions of the digitized samples. It is worth mentioning again that L01 and L02 correspond to the average dimensions of the width of the apple.

On the other hand, L03 and L04 are the height measurements collected on the digitized elements. Figure 7 shows that, generally speaking, calibre 1 (category Extra) is larger than calibre 2 (category I). For calibre 1, parameter L01 lies between 83 mm and 85.5 mm, while L02 varies between 80.2 mm and 83.75 mm, Figure 8a. This shows the disparity of measurement for apples of the same calibre, intrinsically affecting their standardisation for subsequent parameterisation.

Regarding the height of the apple, for parameters L03 and L04, the aim is to find the maximum per calibre of the apples. In this case, L03 is the larger dimension with measures between 77.7 mm and 88.16 mm.

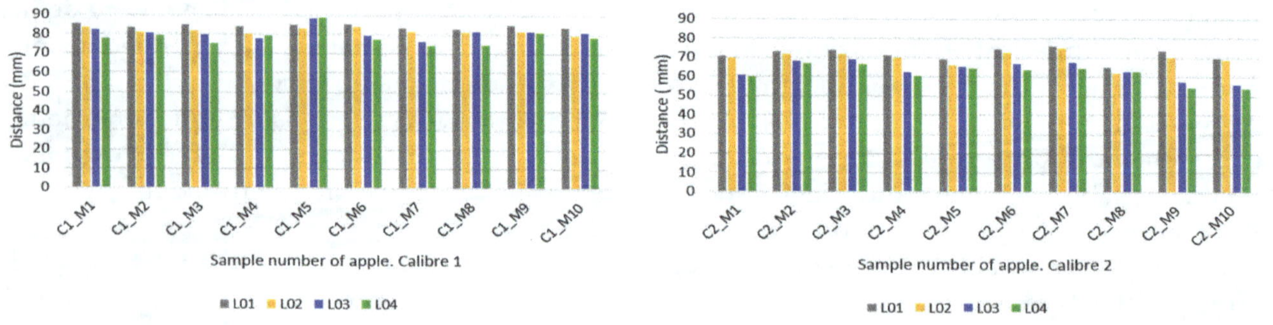

(**a**) (**b**)

Figure 7. Measurements collected from scanned parametric models of apple samples: (**a**) calibre 1, (**b**) calibre 2.

As for calibre 2, in Figure 8b, the measurement intervals tend to increase according to standard (CE) N° 85/2004 [29]: 5% for calibre 1 and 10% for calibre 2. As for the measurement results and because parameter L01 is predominant from the point of view of packaging design, a measurement range between 69.4 mm and 75.9 mm is observed. Likewise, as far as the maximum height is concerned, parameter L03 also plays a part in the design and varies between 56 mm and 68.5 mm.

On the other hand, Figure 8 shows the mean data for the four parameters with their standard dispersion. In both cases, the measurements show a greater dispersion for the parameters L03 and L04, being slightly higher for the apples of calibre 2. L01, which has a greater influence in the design of the package, exhibit more homogeneity than the other dimensions.

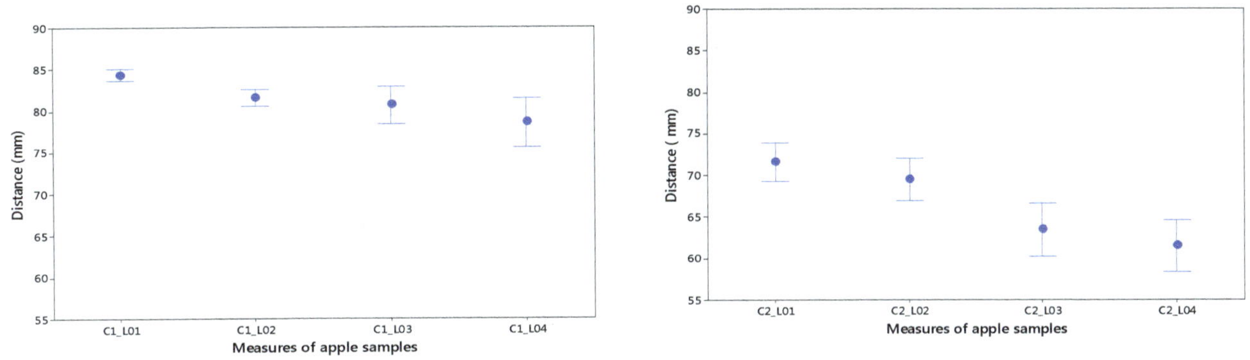

(**a**) (**b**)

Figure 8. Mean data for parameters L01, L02, L03 and L04 with measurement dispersion: (**a**) calibre 1; (**b**) calibre 2.

Analysing the data presented so far, it can be deduced that the sizes of the packages of a certain calibre can be grouped in a dimension that encompasses all sizes of apple inside a category, including the difference between all the fruits of the same type and calibre. Therefore, the variations in the samples that affect the design of the packaging are the maximum width and height obtained from grouping the digital models.

For calibre 1, the maximum measurement within the calibre established according to the standard and with a tolerance of 5% [29] was 85 mm. For virtual measurements, a maximum diameter of 86.7 mm was obtained, Figure 9a. For calibre 2, with a maximum nominal size of 73 mm and a permitted tolerance of 10% according to [29], 78 mm was obtained as maximum diameter for the virtual measurements, Figure 9b. These dimensions served as starting points for the dimensional study of the container, although these dimensions could be reduced due to the irregular geometries.

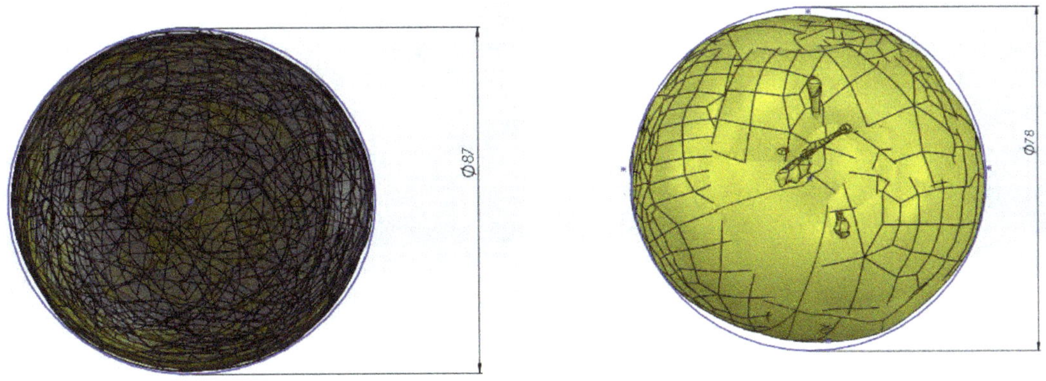

(**a**) (**b**)

Figure 9. Overlap of the 10 apple units to establish a common diameter based on the sample: (**a**) larger size 1: (**b**) size 2 with dimensions between 63 and 73 mm.

Although all apples are included in the same diameter, the plant shape is slightly oval, as discussed previously. Thus, the linear dimension on the plant in one of the sides is lower with respect to its perpendicular components, L02 and L01 respectively. According to this, a dimensional relationship, Dr, was established for the diameter, according to Equation (1):

$$\mathrm{Dr} = \frac{\mathrm{L01}}{\mathrm{L02}}, \quad (1)$$

Thus, from the data obtained in the 20 case studies corresponding to calibres 1 and 2, a dimensional relationship was established between both distances of 0.97 and 0.98, respectively, Figure 10. This relationship was used for the design and parameterization of the packaging by CAD, giving rise to the final geometry of a non-cylindrical container.

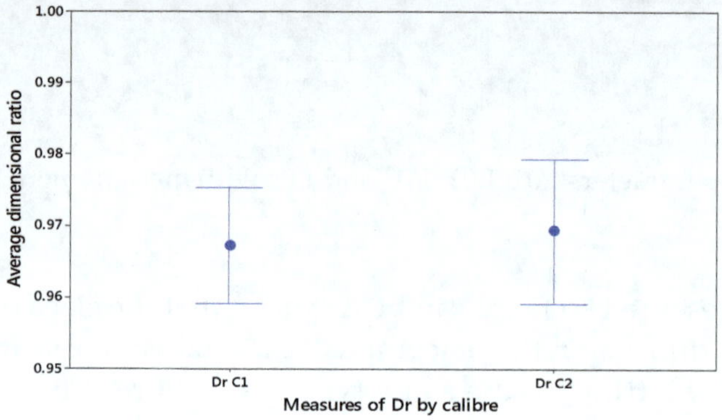

Figure 10. Average dimensional ratio between dimensions L01 and L02 for calibre 1 and 2.

It is important to bear in mind that these dimensions were studied at an experimental level. The aim is to propose a method in which, using scanned elements, designs can be generated. To this end, a clearance coefficient, that can be applied to the dimension of a given calibre, Cd, was proposed.

Analysing the dimensional results of the three-dimensional models of the scanned apples and the measurements according to the norm, a particular Cd can be defined for the design of a packaging. This coefficient was defined as the ratio between the observed maximum dimension per calibre according to the norm, and the virtual dimensions studied. Consequently, this definition makes it possible to ensure that the size of the containers is adapted to all geometries included within a calibre.

As an example, if, for calibre 1, the maximum diameter allowed is 89.25 mm including tolerance, and the maximum diameter that appears as a measure of nominal calibre is 85 mm, then Cd could lie between 1.05 and 1.06. However, because of the differences that naturally arise in fresh foods, 1.06 was considered for use to ensure proper functioning. Thus, the validation of this coefficient was carried out by implementing this coefficient in the development of the adaptation of calibre 2.

Therefore, the calculation of packaging dimensions can be done according to the following equation:

$$\mathrm{Dm} = \mathrm{Cd} \times \mathrm{Cm}, \quad (2)$$

where Cm is the maximum nominal size of the calibre, Dm is the width of the packaging and Cd is the clearance coefficient.

Then, for designing the packaging for calibre 1, Dm is 90 mm; this size is obtained by multiplying 1.06 and 85 mm. In the case of calibre 2, Dm is 77.4 mm, obtained by the multiplication 1.06 and 73 mm. These data were validated during the parametric design, which is explained in the following section.

The norm does not specify apple heights by calibre. Therefore, from the measurements made on the scanned samples, a ratio, Rd, was established between the average dimensions of width (L02) and height (L03) of the apples, Equation (3).

$$\mathrm{Rd} = \mathrm{L02}/\mathrm{L03}, \quad (3)$$

Therefore, the Rd for calibre 1 corresponds to 1.05 and 1.07 for calibre 2. This means that, compared to calibre 1, calibres 2 have lower height in relation to their diameter. Then, the total height H of the package in this case study is given by Equation (4).

$$H = Dm/Rd, \tag{4}$$

Thus, in the parametric design of the package, Cd is implicitly included in all the dimensions of the package from the Dm obtained in Equation (4). It should be stressed that this ratio, Rd, can be modified to obtain containers with different height in relation to the width.

3.2. Parameterization of the Conceptual Proposal

Once the design of the packaging was carried out, two variables, the maximum height of the container, H, and the maximum width, Dm, were used to define a series of parameters that characterise the package. Figure 11a shows a diagram of the variables that affect the sizing of the container.

From the initial geometry extracted from the concept design of the apple, the packaging was parameterized according to a series of equations described below. Figure 11b shows a diagram with the dimensions that affect the mould when thermoforming the designed package. Thus, the equations that affect the overall dimension of the packaging correspond to:

$$Hb = H \times 0.57, \tag{5}$$

$$Ht = H \times 0.43, \tag{6}$$

$$Dn = Dm \times 0.97, \tag{7}$$

where Hb is the height of the bottom half of the designed packaging and Ht is the height of the top half. In the same way, the projected dimensions of the container are given by the greater width, Dm, and the smaller width, Dn. In this case study, an oval geometry has been created.

As mentioned above, the design created in this study consists of a series of arches that are tangentially joined, in plan and profile, and a flat surface at the ends with a circular shape. The curves that define these geometries, Figure 11, could be related by means of equations from the initially created CAD design. Thus, the relationship between H and the slightly oval curved geometry that makes up the package design in the profile view is given by the equations:

$$Rb = H \times 0.27, \tag{8}$$

$$Rt = H \times 0.18, \tag{9}$$

$$db = Dm \times 0.3, \tag{10}$$

$$dt = Dm \times 0.45, \tag{11}$$

$$Rp = (Dm + 10)/2, \tag{12}$$

These equations define the radius of curvature of the upper part, Rt, and lower, Rp. In addition, the adaptation of the dimension of the flat part, so that the packaging can be easily supported, is given by the diameter in both halves of the container, db and dt, and was related to the parameter dm. On the other hand, the radius Rp is the parameter that encompasses the overall geometry of the container in plant, and the shape was constructed by sweeping through the vertical curves given by Rb and Rt.

The rest of the tangent arcs that make up the packaging design are automatically adapted from the curves generated with the above equations.

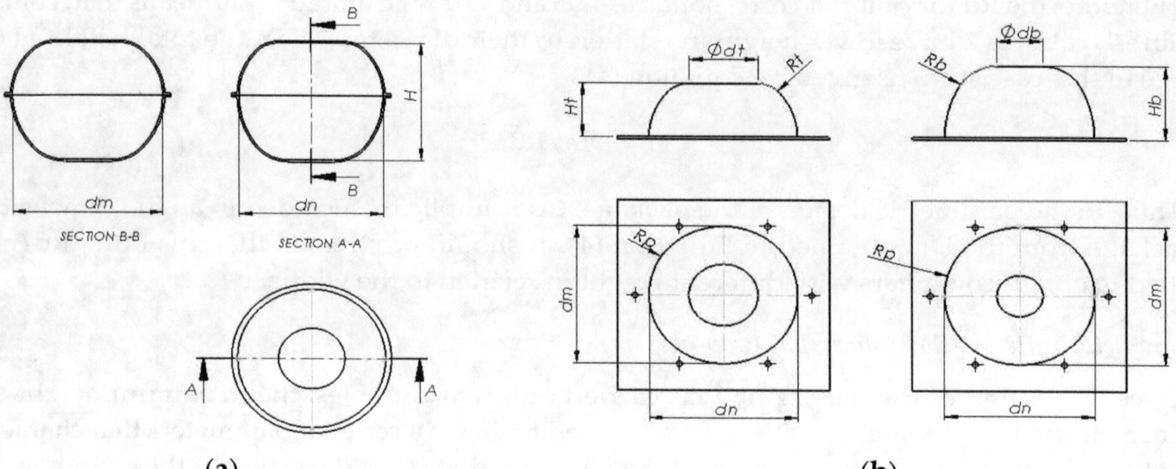

(a) (b)

Figure 11. Dimensions of the container and the mould: (**a**) Plan of the variables that affect the container and therefore the upper and lower mould; (**b**) Dimensions of the moulds, on the left upper half and on the right lower half.

Then, the generated equations were introduced in the parametric design software, Solidworks®. The benefit of the parametric design of a packaging is the customization of the geometry according to the need of adaptation to the product to be contained. In addition, the design can be evaluated in real time by means of the digitized fruit samples. As a result, all construction operations such as roundings, tangent arcs, etc. were related. The number of operations obtained was a total of 12 for each part of the container. These equations and variables serve to quickly modify the geometry of an object.

For the lower part, according to the equations defined in the previous section, a relationship was defined between the curves that make up the container and, therefore, the mould. These equations were related to the sketches made in the parametric design program. In the same way, the equations of the upper part were parameterized.

In short, by modifying one of the measures, the packaging is automatically adapted. This is one of the first steps for the generation adapted to a specific packaging design in the context of industry 4.0.

Another result obtained in this case study is the maximum and minimum ratio (Rd) of measures between H and Dm, obtaining a range of adaptation measures. The maximum Rd is 1.32 and the minimum is 0.24. This range of measures means that the packaging created can also be used for other types of fresh food or measurements. Figure 12 shows an example of the variations in function of the maximum and minimum ratio.

(a) (b)

Figure 12. Example of maximum and minimum ratio between Dm and H for H equal to 85 mm: (**a**) Maximum ratio for Dm = 137 mm; (**b**) Minimum ratio for Dm = 20 mm.

Also, if Dm is kept constant in the two parts, top and bottom, of the parametric packaging, Ht and Hb can be adapted independently, in the CAD file, to obtain intermediate H measurements by modifying the H value only in one of the parts. Also, with this modification, Rt and Rb also adapt automatically according to Ht and Hb, respectively. This parameter offers greater versatility if possible combinations for containers using fewer moulds are taken into account.

Once the packaging was parameterized, a series of configurations were established in Solidworks® software to evaluate the appropriate sizes for each of the samples studied and with which the theoretically established Cd was validated. The configurations allow the packaging to adapt automatically to the measures established, making it easier to adapt to calibres and measures according to specific needs.

3.3. Evaluation of Results

In order to analyse the viability of the coefficient obtained, Cd, using the 3D scanned fruit models, the correct arrangement of the apple was evaluated with respect to the dimensions of the packaging. Then, after analysing the dimensions with the configurations of the packaging, it was observed that the Cd corresponds adequately, although the optimum height for each sample varies, Appendix A.

Bearing this in mind, it can be concluded that, depending on the design needs for the same calibre, several containers with different H could be constructed. This may be possible thanks to new technologies such as additive manufacturing, which allows low-cost moulds to be made in order to optimise the maximum performance of the container. However, if the design requirements allow it, with this methodology it is possible to establish an optimal design that encompasses all food units of the same topology.

On the other hand, generating a custom packaging for a unit is also possible. This example could have multiple applications that could be extended to other types of products.

Thus, the final dimensions, which were adapted to the size obtained according to the Cd calculated in the measurement part, correspond to calibre 1 with the ratio Rd of 1.05, with Dm of 89.3 mm and H of 85 mm. Although, as it was mentioned, some of the samples studied could modify the H for the use of the dimensions having a range between 80 mm and 85 mm. Similarly, for calibre 2, the dimensions correspond to the ratio Rd of 1.07 with Dm and H being 75 mm and 70 mm, respectively, and, according to the samples studied, with a range of H between 67 mm and 70 mm. For more information see Appendix A.

Finally, the results obtained for the two calibres studied were validated by means of a physical prototype. The FDM moulds created for the two calibres are shown in Figure 13.

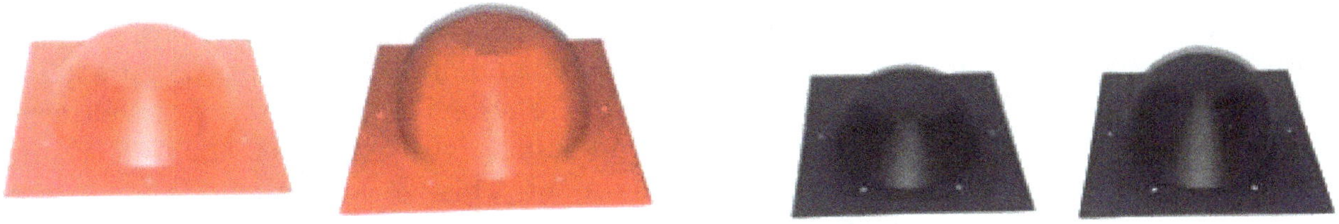

(a) (b)

Figure 13. Parts, inferior and superior, of the mould to generate the physical prototypes: (**a**) Moulds for calibre 1; (**b**) moulds for calibre 2.

The design was validated using the physical prototypes and 20 apples for each calibre studied, Figure 14. These apples were selected taking into account shapes and size variations, and different varieties were included in the units studied. For more information see Appendix B.

Once the results were analysed, it can be said that the final solution obtained is positive as all the units of the same calibre correspond to the dimensions of the packaging. However, according to the studies in the parametric software, it can be seen that, in several of the cases studied, the dimensions are not adjusted in their entirety, causing parts of the packaging to be empty. This is not a functional problem for the design because these variations are given by the wide range of measurements that are included within a calibre. Specifically, these dimensional deviations are accentuated in calibre 2, which is a smaller calibre and comprises a larger range of measurements. In the case of calibre 1, which belongs to category I according to [28], lower dimensional deviations are observed. However, the morphological inequalities between the different apple varieties are larger.

(a) (b) (c)

Figure 14. Images of the evaluation carried out with the prototype and commercial fruits: (**a**) Pacific rose apple, (**b**) Golden apple, (**c**) Jonagold apple.

In this situation, personalisation and adaptation could be increased by reducing the spectrum of fruit types that can be introduced in the same type of packaging. Therefore, depending on the type of product to be packed, adaptation and personalization can be considered with a greater degree of accuracy.

4. Discussion

The main idea of reverse engineering is to synthesize a fruit model so that it can be used and measured as part of the design process. The studies carried out show that it is possible to generate personalized designs and that they can also be adapted according to the specifications required by a specific product. Digitization and flexible designs make it easier to customize and test concepts in real time to help design teams make faster decisions and with greater reliability [33].

Then, the introduction of scanned elements in the design process facilitates the adaptation of the size and shape of the packaging to the type of fruit contained, allowing only the quantity of plastic material necessary for its sale and transport to be used. In this way, the product is optimised, which means a reduction in environmental and economic impact. This fact is relevant because, currently, thousands of tons of plastic and food are discarded daily [12], and great efforts are made to mitigate this impact, generating stricter and stricter regulations in the withdrawal and recycling of food packaging [34]. The huge quantity of packaging manufactured means that a small reduction in each container has a great influence on the environment. This working methodology promotes the reduction of material that is so necessary in a strategic sector such as the food industry. Thus, by adapting the packaging to the size and shape of the food, less raw material is wasted. In addition, the technique presented in this paper for designing efficient packaging also offers the possibility of creating packaging with less food, adapted to emerging consumption needs, which translates into less food waste and, therefore, in the protection of the environment. In addition, this tool can be used with any type of sustainable raw material, i.e., PLA (Polylactic Acid) sheet, although the material currently on the market has been used in this work.

As noted above, much of the food waste is produced by using containers with large amounts of food that end up deteriorating in homes [35]. Thus, the possibility offered by this methodology for the realization of custom packaging by type of fruit and category can lead to the reduction of food waste due to the possibility of reducing the amount of product inside and by improving the preservation of specific properties of each food [36].

The application of digital elements in the design process was validated as part of the creation of a method that includes advanced technologies in its procedure, as was researched in other fields of engineering [37,38]. In the study, two main parameters related to the functional measures for the containers were selected, the maximum height and width, to adapt a packaging to a given calibre. However, the geometry of a fresh food is irregular so four measurements at the top and bottom were taken to calculate the height and width. It should be added that these measurements had been carried out to analyse the dimensions of the product studied at laboratory level. The main objective of this work was to evaluate the direct application of the scanned models for the generation of personalized packages for a range of measures established within a calibre.

This study proposed a 3D scanning scheme for fresh food to support custom packaging design [31]. Then, the three-dimensional model of the fruits approximates the real geometry. Thus, the application of 3D fruits facilitates the realization of personalized and adapted designs, as well as the evaluation in real time of design proposals. Moreover, the parametric design of the package according to the two parameters studied, gives rise to the possibility of generating packages adapted to the dimensions and needs of the food according to its established range of measures. In short, the result was the creation of a packaging, fully defined by equations, which is capable of adapting to the measurements in a given range.

Based on these results, computational design is aimed at the parameterization to favour customization by means of optimal configurations of the variables, favouring the adaptation of the design to specific needs [39,40]. Computer programs facilitate the realization of personalized and flexible designs to adapt to specific needs by means of the parameterization and digitalization of elements [41–43]. Product customization also serves as an engine to improve sustainability throughout the product life cycle [44]. Custom design enhances product design by meeting the specific needs of users [45].

Furthermore, according to new trends and competitiveness, the design and development of packaging need solutions that streamline the working procedure. In this context, based on the approaches made on the customization of packaging, there is also a latent need for quick and flexible solutions. Specifically, the process of manufacturing moulds by conventional methods usually delays the validation time of the final prototype of the product, so designers cannot make changes or explore quick and reliable options [46]. Facing this situation, the application of additive manufacturing techniques can provide design teams with a fast and economical tool that can thus be used from the early stages of product design.

Summarizing, it is possible to build reliable prototypes by additive manufacturing. The similarity of the prototypes generated thanks to 3D printing and thermoforming technologies shows the possibility of creating prototypes that provide greater reliability in a simple way. Then, the evaluation by means of these prototypes provides a trustful tool for the validation of the mentioned designs. In short, it was proven that it is possible to thermoform geometries with different shapes. Therefore, it is possible to obtain products with better performance and, consequently, competition. This can also affect the life cycle of the product as there is a significant improvement from the point of view of the social, economic and environmental impact.

Finally, this work presents several future lines of action, the most important of which is the advanced study of new packaging designs by means of topological optimisation and the extension of the study of virtual environments for early evaluation.

5. Conclusions

The following conclusions can be drawn from the research work carried out in this study:

It was possible to digitize fresh food by means of 3D scanning techniques, obtaining reliable digital elements that could be used in advanced technologies. In this sense, it was possible to define a procedure for the reverse engineering of different types of food, detailing the specific parameters according to size and finish.

It was possible to use the digitized fruits during the conceptual phase in the packaging design process. Thus, custom designs were developed using these elements as a reference during computer-aided design, thus validating the proposed methodology. In addition, the evaluation of the ideas generated was also favoured by the possibility of checking dimensions in real time using these digital products. On the other hand, the adaptation according to the specifications required by a specific product was also improved. Therefore, the application of 3D fruits facilitated the development of customized and adapted designs.

It was possible to completely parameterize the geometry of the package to create custom and, in turn, automatically adaptable designs. In short, the creation of a packaging, fully defined by equations, was able to adapt to measurements in a given range. The parameterized design was made possible by means of the virtual evaluation of digitized fruits.

Finally, by adapting the containers to the size and shape of the food content, less raw material used for the container was wasted. Likewise, the technique presented in this work to design efficient containers also offered the possibility of creating containers with less food, adapted to emerging consumption needs, which lead to less food waste and therefore a better protection of the environment. In addition, this tool could be used with any type of sustainable raw material, although in this study the material currently on the market was used.

Author Contributions: Conceptualization, L.R.-P.; methodology, L.R.-P. and P.F.M.; software, L.R.-P.; validation, L.R.-P., P.F.M. and A.J.G.; formal analysis, L.R.-P., P.F.M. and A.J.G.; investigation, L.R.-P. and P.F.M.; resources, A.J.G.; data curation, L.R.-P., P.F.M. and A.J.G.; writing—original draft preparation, L.R.-P.; writing—review and editing, P.F.M. and A.J.G.; supervision, P.F.M. and A.J.G. All authors read and approved the final manuscript.

Appendix

Table A1. Images of virtual evaluation for both calibres.

Sample	Calibre 1		Calibre 2	
	Front View	Top View	Front View	Top View
M1	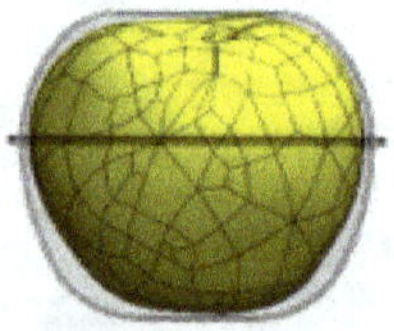	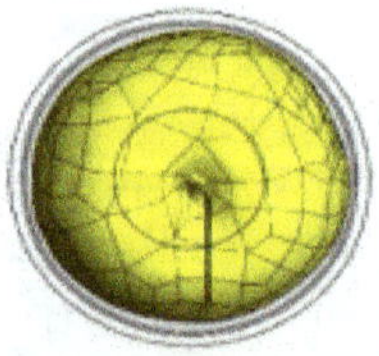		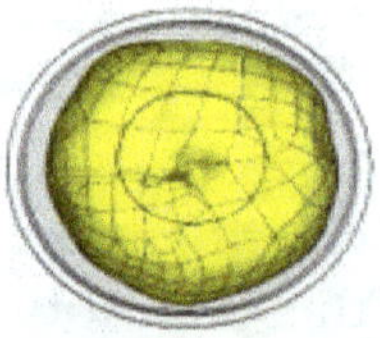
M2		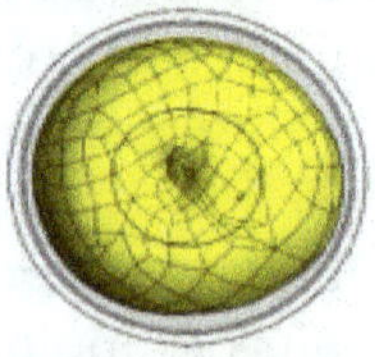	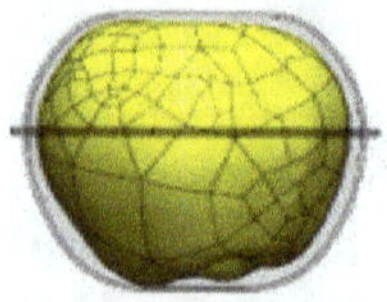	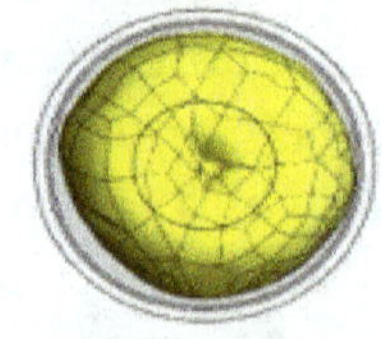
M3	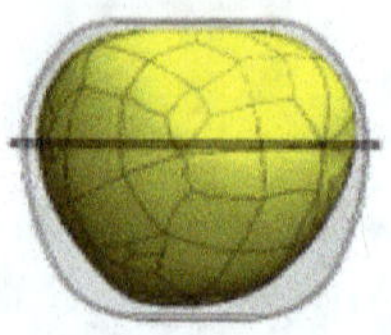	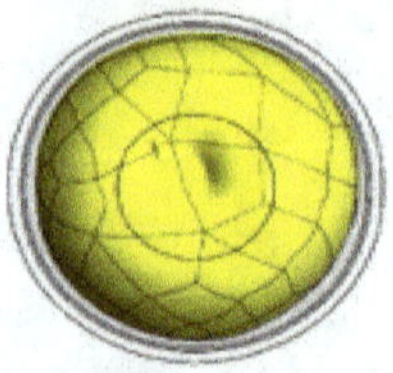		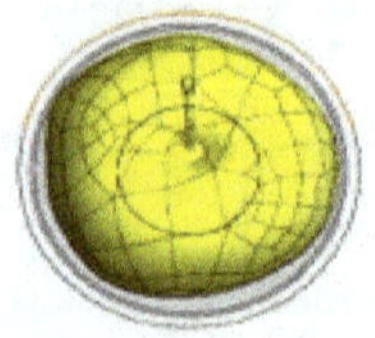

Table A1. *Cont.*

Sample	Calibre 1		Calibre 2	
	Front View	Top View	Front View	Top View
M4	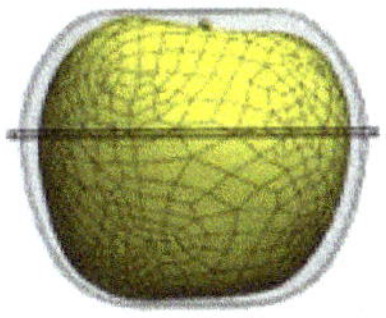	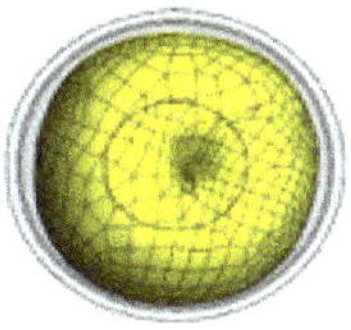	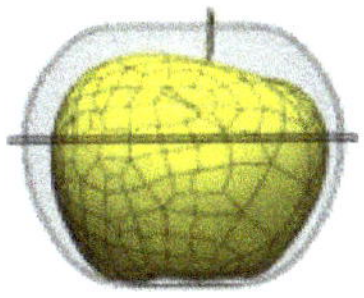	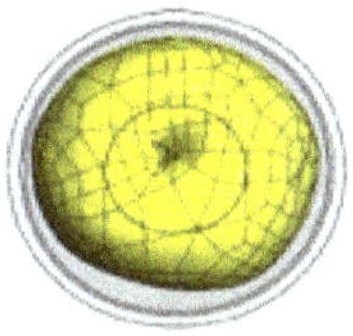
M5	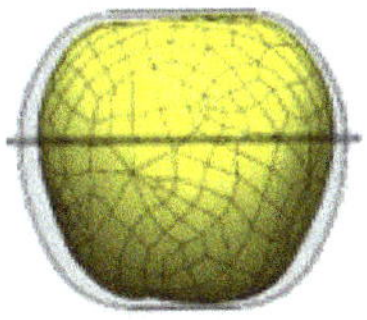	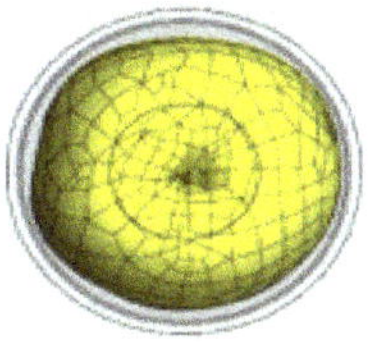		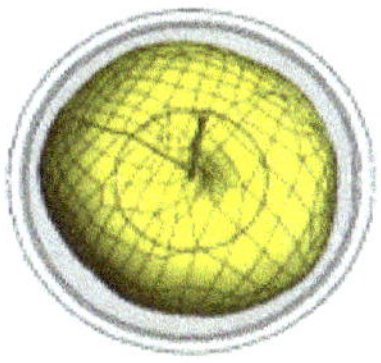
M6	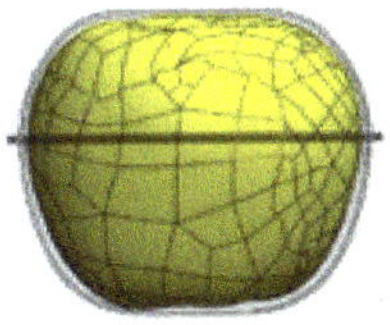	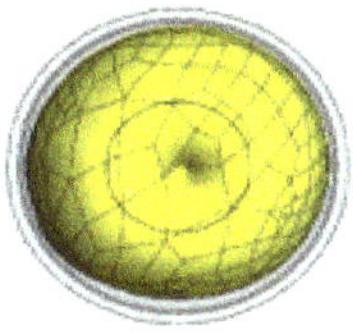	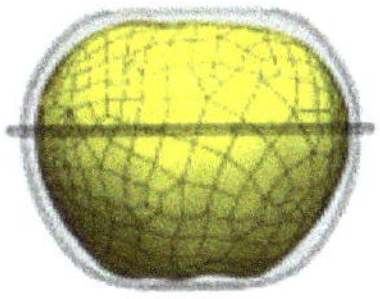	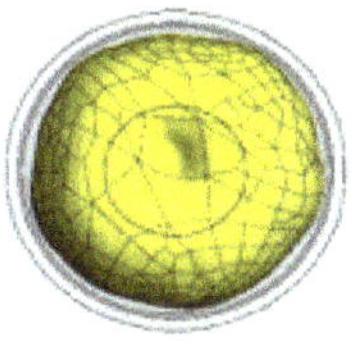
M7		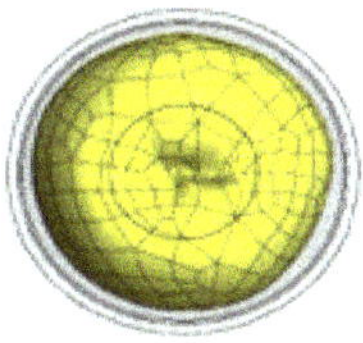	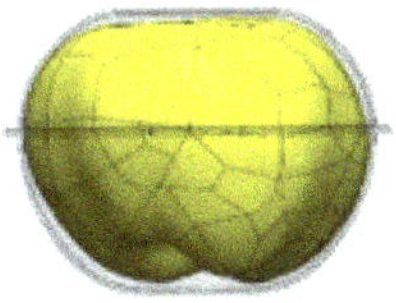	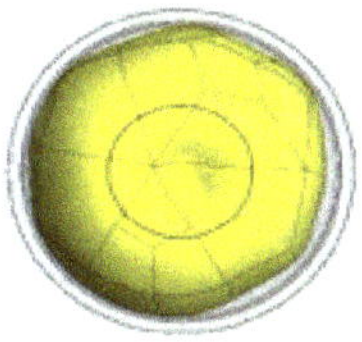
M8	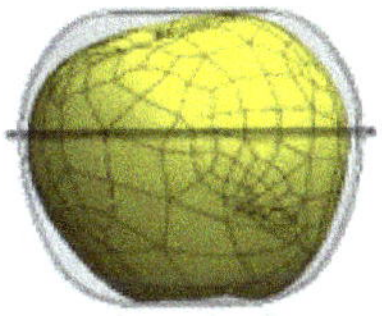	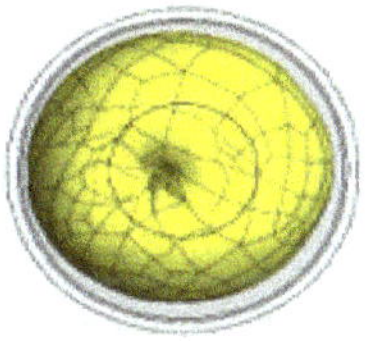		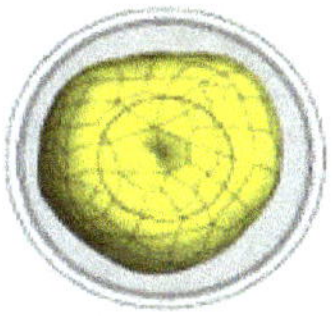
M9	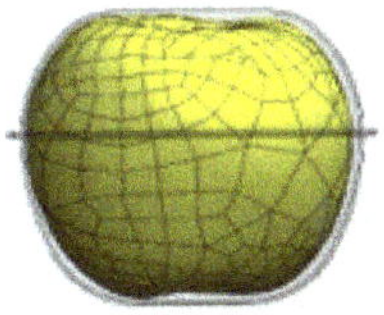	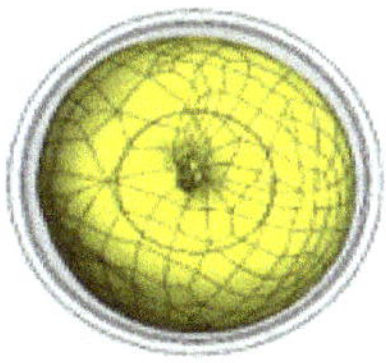	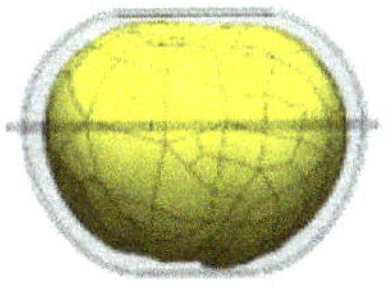	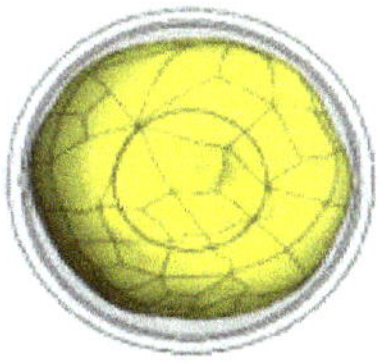
M10	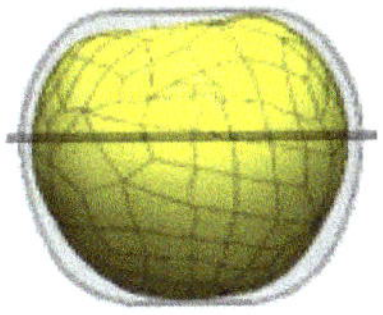	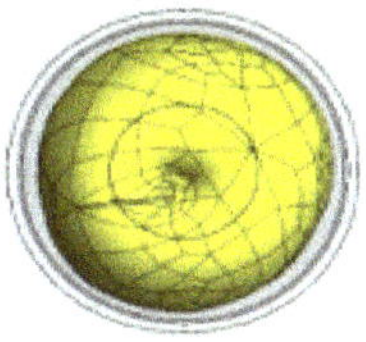		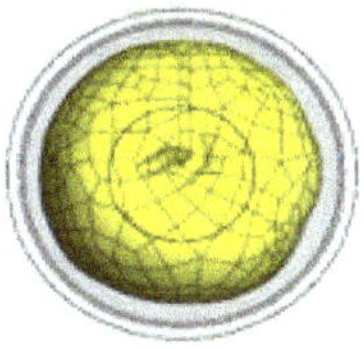

Appendix

Table A2. Images from the validation study with prototype of the custom packaging for Calibre 1.

Diferent apples inside of designed packaging for Calibre 1

Table A3. Images from the validation study with prototype of the custom packaging for Calibre 2.

Diferent apples inside of designed packaging for Calibre 2

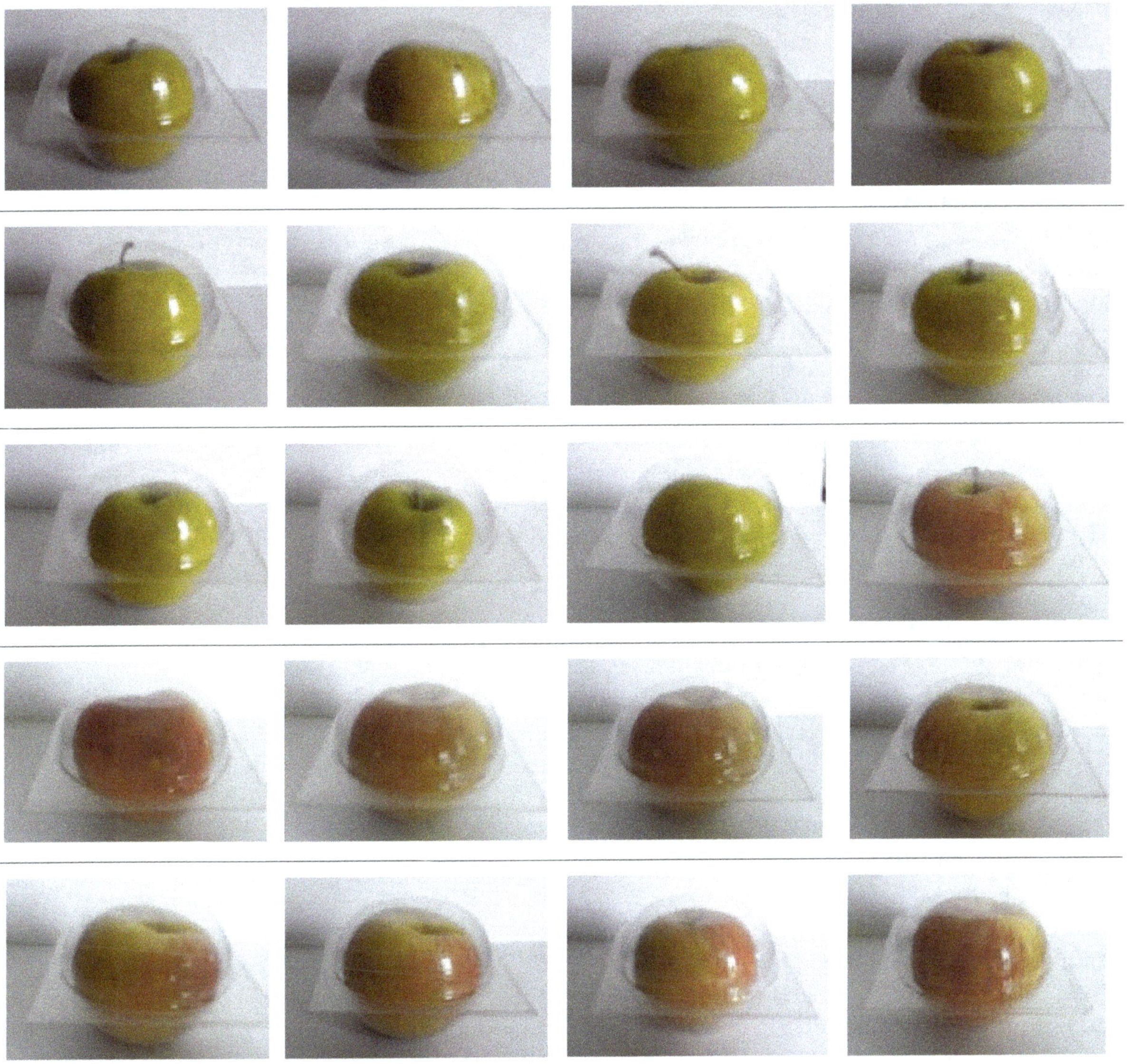

References

1. Marsh, K.; Bugusu, B. Food packaging—Roles, materials, and environmental issues. *J. Food Sci.* **2007**, *72*, R39–R55. [CrossRef] [PubMed]
2. Schweitzer, J.; Gionfra, S.; Pantzar, M.; Mottershead, D.; Watkins, E.; Petsinaris, F.; ten Brink, P.; Ptak, E.; Lacey, C.; Janssens, C. Plastic Packaging and Food Waste—New Perspectives on a Dual Sustainability Crisis. 2018. Available online: https://ieep.eu/publications/plastic-packaging-and-food-waste-new-perspectives-on-a-dual-sustainability-crisis (accessed on 11 July 2018).
3. Vodnar, D.C.; Pop, O.L.; Dulf, F.V.; Socaciu, C. Antimicrobial Efficiency of Edible Films in Food Industry. *Not. Bot. Horti Agrobot. Cluj-Napoca* **2015**, *43*, 302–312. [CrossRef]
4. Santeramo, F.G.; Carlucci, D.; De Devitiis, B.; Seccia, A.; Stasi, A.; Viscecchia, R.; Nardone, G. Emerging trends in European food, diets and food industry. *Food Res. Int.* **2018**, *104*, 39–47. [CrossRef] [PubMed]
5. Baselice, A.; Colantuoni, F.; Lass, D.A.; Nardone, G.; Stasi, A. Trends in EU consumers' attitude towards fresh-cut fruit and vegetables. *Food Qual. Prefer.* **2017**, *59*, 87–96. [CrossRef]
6. Gómez, E. Convenience Food, Tendencia en Alimentación. 2017. Available online: https://www.ainia.es/tecnoalimentalia/consumidor/convenience-food-tendencia-en-alimentacion/ (accessed on 20 November 2018). (In Spanish)

7. Ainia. Las 7 Claves de un Envase Sostenible. Available online: https://www.ainia.es/noticias/prensa/las-7-claves-de-un-envase-sostenible/ (accessed on 17 September 2018). (In Spanish)
8. Farris, S. Main Manufacturing Processes for Food Packaging Materials. In *Reference Module in Food Science*; Elsevier: Amsterdam, The Netherlands, 2016. [CrossRef]
9. Nerin, C. Plastics and Polymers for Food Packaging Manufacturing. In *Reference Module in Food Science*; Elsevier: Amsterdam, The Netherlands, 2016. [CrossRef]
10. Klaiman, K.; Ortega, D.L.; Garnache, C. Perceived barriers to food packaging recycling: Evidence from a choice experiment of US consumers. *Food Control* **2017**, *73*, 291–299. [CrossRef]
11. Baldwin, C.J. *The 10 Principles of Food Industry Sustainability*; John Wiley & Sons: Chichester, UK, 2015; ISBN 9781118447697.
12. Siracusa, V.; Rosa, M.D. Sustainable Packaging. In *Sustainable Food Systems from Agriculture to Industry*, 1st ed.; Galanakis, C., Ed.; Elsevier: London, UK, 2018; pp. 275–307. ISBN 9780128119358.
13. Sundbo, J. Food scenarios 2025: Drivers of change between global and regional. *Futures* **2016**, *83*, 75–87. [CrossRef]
14. Selke, S.; Culter, J. *Plastics Packaging*, 3rd ed.; Hanser: Munich, Germany, 2016; ISBN 9781569904435.
15. Wikström, F.; Williams, H.; Venkatesh, G. The influence of packaging attributes on recycling and food waste behaviour—An environmental comparison of two packaging alternatives. *J. Clean. Prod.* **2016**, *137*, 895–902. [CrossRef]
16. Martinho, G.; Pires, A.; Portela, G.; Fonseca, M. Factors affecting consumers' choices concerning sustainable packaging during product purchase and recycling. *Resour. Conserv. Recycl.* **2015**, *103*, 58–68. [CrossRef]
17. Klein, P. Fundamentals of Plastics Thermoforming. In *Synthesis Lectures on Materials Engineering*; M&C: Williston, ND, USA, 2009; ISBN 9781598298840.
18. Van Mieghem, B.; Desplentere, F.; Van Bael, A.; Ivens, J. Improvements in thermoforming simulation by use of 3D digital image correlation. *Express Polym. Lett.* **2015**, *9*, 119–128. [CrossRef]
19. Schüller, C.; Panozzo, D.; Grundhöfer, A.; Zimmer, H.; Sorkine, E.; Sorkine-Hornung, O. Computational thermoforming. *ACM Trans. Graph.* **2016**, *35*, 43. [CrossRef]
20. Sreedhara, V.S.M.; Mocko, G. Control of thermoforming process parameters to increase quality of surfaces using pin-based tooling. In Proceedings of the ASME Design Engineering Technical Conference, Boston, MA, USA, 2–5 August 2015; Volume 4, p. V004T05A016. [CrossRef]
21. Vanderroost, M.; Ragaert, P.; Devlieghere, F.; De Meulenaer, B. Intelligent food packaging: The next generation. *Trends Food Sci. Technol.* **2014**, *39*, 47–62. [CrossRef]
22. Herbes, C.; Beuthner, C.; Ramme, I. Consumer attitudes towards biobased packaging—A cross-cultural comparative study. *J. Clean. Prod.* **2018**, *194*, 203–218. [CrossRef]
23. Deng, X.; Srinivasan, R. When Do Transparent Packages Increase (or Decrease) Food Consumption? *J. Mark.* **2013**, *77*, 104–117. [CrossRef]
24. Klimchuk, M.R.; Krasovec, S.A. *Packaging Design: Successful Product Branding from Concept to Shelf*; John Wiley & Sons: Hoboken, NJ, USA, 2012; ISBN 1118358546.
25. Otero, C.; Valentini, P.; Fischer, X. Behavioural Modelling for Design. In *Research in Interactive Design*; Springer International Publishing: Cham, Switzerland, 2016; Volume 4, pp. 151–220. [CrossRef]
26. Wang, P.; Hamila, N.; Boisse, P. Thermoforming simulation of multilayer composites with continuous fibres and thermoplastic matrix. *Compos. Part B Eng.* **2013**, *52*, 127–136. [CrossRef]
27. Trinetta, V. Definition and Function of Food Packaging. In *Reference Module in Food Science*; Elsevier: Amsterdam, The Netherlands, 2016. [CrossRef]
28. Simmonds, G.; Spence, C. Thinking inside the box: How seeing products on, or through, the packaging influences consumer perceptions and purchase behaviour. *Food Qual. Prefer.* **2017**, *62*, 340–351. [CrossRef]
29. Marketing Standards for Fruit and Vegetables. Available online: http://www.juntadeandalucia.es/agriculturaypesca/productos/info_comercial.html (accessed on 10 December 2018). (In Spanish)
30. Van Boeijen, A.; Daalhuizen, J.; Van der Schoor, R.; Zijlstra, J. *Delft Design Guide: Design Methods*; BIS: Amsterdam, The Netherlands, 2014; ISBN 9063693273.
31. Rodríguez-Parada, L.; Pardo Vicente, M.A.; Mayuet Ares, P.F. Digitizing fresh food using 3D scanning for custom packaging design. *DYNA Ing. E Ind.* **2018**, *93*, 681–688. [CrossRef]
32. Frutas y Verduras-HortiqualityHortiquality. Available online: http://hortiquality.com/frutas-y-verduras-espana/ (accessed on 26 July 2018). (In Spanish)

33. Becker, L.; van Rompay, T.J.L.; Schifferstein, H.N.J.; Galetzka, M. Tough package, strong taste: The influence of packaging design on taste impressions and product evaluations. *Food Qual. Prefer.* **2011**, *22*, 17–23. [CrossRef]
34. Mwanza, B.G.; Mbohwa, C.; Telukdarie, A. Strategies for the Recovery and Recycling of Plastic Solid Waste (PSW): A Focus on Plastic Manufacturing Companies. *Procedia Manuf.* **2018**, *21*, 686–693. [CrossRef]
35. Tulaphol, N. Food Industry 4. A New Era of Consumer Empowerment. Economic Intelligence Center (EIC), 2016. Available online: https://www.scbeic.com/en/detail/product/2916 (accessed on 4 April 2018).
36. Hellström, D.; Olsson, A. *Managing Packaging Design for Sustainable Development: A Compass for Strategic Directions*; Wiley: Chichester, UK, 2017; ISBN 1119150930.
37. Aromaa, S.; Väänänen, K. Suitability of virtual prototypes to support human factors/ergonomics evaluation during the design. *Appl. Ergon.* **2016**, *56*, 11–18. [CrossRef]
38. Holhorst, F.W.B.; van Rompay, T.J.L.; ten Klooster, R.; Roukema, M. Evaluating Shelf-ready Packaging Designs in a VR Environment. In Proceedings of the EuroVR 2014 conference, Bremen, Germany, 8–10 December 2014; pp. 113–117. [CrossRef]
39. Yassine, A.A. Parametric design adaptation for competitive products. *J. Intell. Manuf.* **2012**, *23*, 541–559. [CrossRef]
40. Harding, J.E.; Shepherd, P. Meta-Parametric Design. *Des. Stud.* **2017**, *52*, 73–95. [CrossRef]
41. Vitković, N.; Stojković, M.; Majstorović, V.; Trajanović, M.; Milovanović, J. Novel design approach for the creation of 3D geometrical model of personalized bone scaffold. *CIRP Ann.* **2018**, *67*, 177–180. [CrossRef]
42. Zhang, J.; Gu, P.; Peng, Q.; Hu, S.J. Open interface design for product personalization. *CIRP Ann.* **2017**, *66*, 173–176. [CrossRef]
43. Aish, R.; Hanna, S. Comparative evaluation of parametric design systems for teaching design computation. *Des. Stud.* **2017**, *52*, 144–172. [CrossRef]
44. Kaneko, K.; Kishita, Y.; Umeda, Y. Toward Developing a Design Method of Personalization: Proposal of a Personalization Procedure. *Procedia CIRP* **2018**, *69*, 740–745. [CrossRef]
45. Chu, C.-H.; Wang, I.-J.; Wang, J.-B.; Luh, Y.-P. 3D parametric human face modeling for personalized product design: Eyeglasses frame design case. *Adv. Eng. Inform.* **2017**, *32*, 202–223. [CrossRef]
46. Kleespies, H.S., III; Crawford, R.H. Vacuum forming of compound curved surfaces with a variable geometry mold. *J. Mannuf. Syst.* **1998**, *17*, 325–337. [CrossRef]

9

Formulation and Characterization of New Polymeric Systems Based on Chitosan and Xanthine Derivatives with Thiazolidin-4-One Scaffold

Sandra Madalina Constantin [1], Frederic Buron [2], Sylvain Routier [2], Ioana Mirela Vasincu [1], Maria Apotrosoaei [1], Florentina Lupașcu [1], Luminița Confederat [3], Cristina Tuchilus [3], Marta Teodora Constantin [4], Alexandru Sava [5] and Lenuța Profire [1,*]

1 Department of Pharmaceutical Chemistry, Faculty of Pharmacy, University of Medicine and Pharmacy "Grigore T. Popa", 16 University Street, 700115 Iasi, Romania; constantin.sandra@umfiasi.ro (S.M.C.); ioanageangalau@yahoo.com (I.M.V.); apotrosoaei.maria@umfiasi.ro (M.A.); florentina-geanina.l@umfiasi.ro (F.L.)

2 Institut de Chimie Organique et Analytique, Univ Orleans, CNRS, ICOA, UMR 7311, F-45067 Orléans, France; frederic.buron@univ-orleans.fr (F.B.); sylvain.routier@univ-orleans.fr (S.R.)

3 Department of Microbiology, Faculty of Pharmacy, University of Medicine and Pharmacy "Grigore T. Popa", 16 University Street, 700115 Iasi, Romania; luminita.confederat@umfiasi.ro (L.C.); cristina.tuchilus@umfiasi.ro (C.T.)

4 University of Kent, School of Physical Sciences, Canterbury CT2 7NH, UK; marta.constantin97@gmail.com

5 Department of Analytical Chemistry; Faculty of Pharmacy, University of Medicine and Pharmacy "Grigore T. Popa", 16 University Street, 700115 Iasi, Romania; alexandru.i.sava@umfiasi.ro

* Correspondence: lenuta.profire@umfiasi.ro

Abstract: In the past many research studies have focused on the thiazolidine-4-one scaffold, due to the important biological effects associated with its heterocycle. This scaffold is present in the structure of many synthetic compounds, which showed significant biological effects such as antimicrobial, antifungal, antioxidant, anti-inflammatory, analgesic, antidiabetic effects. It was also identified in natural compounds, such as actithiazic acid, isolated from *Streptomyces* strains. Starting from this scaffold new xanthine derivatives have been synthetized and evaluated for their antibacterial and antifungal effects. The antibacterial action was investigated against Gram positive (*Staphyloccoccus aureus* ATCC 25923, *Sarcina lutea* ATCC 9341) and Gram negative (*Escherichia coli* ATCC 25922) bacterial strains. The antifungal potential was investigated against Candida spp. (*Candida albicans* ATCC 10231, *Candida glabrata* ATCC MYA 2950, *Candida parapsilosis* ATCC 22019). In order to improve the antimicrobial activity, the most active xanthine derivatives with thiazolidine-4-one scaffold (XTDs: 6c, 6e, 6f, 6k) were included in a chitosan based polymeric matrix (CS). The developed polymeric systems (CS-XTDs) were characterized in terms of morphological (aspect, particle size), physic-chemical properties (swelling degree), antibacterial and antifungal activities, toxicity, and biological functions (bioactive compounds loading, entrapment efficiency). The presence of xanthine-thiazolidine-4-one derivatives into the chitosan matrix was confirmed using Fourier transform infrared (FT-IR) analysis. The size of developed polymeric systems, CS-XTDs, ranged between 614 μm and 855 μm, in a dry state. The XTDs were encapsulated into the chitosan matrix with very good loading efficiency, the highest entrapment efficiency being recorded for CS-6k, which ranged between 87.86 ± 1.25% and 93.91 ± 1.41%, depending of the concentration of 6k. The CS-XTDs systems showed an improved antimicrobial effect with respect to the corresponding XTDs. Good results were obtained for CS-6f, for which the effects on *Staphylococcus aureus* ATCC 25923 (21.2 ± 0.43 mm) and *Sarcina lutea* ATCC 9341 (25.1 ± 0.28 mm) were comparable with those of ciprofloxacin (25.1 ± 0.08 mm/25.0 ± 0.1 mm), which were used as the control. The CS-6f showed a notable antifungal effect, especially on *Candida parapsilosis* ATCC 22019 (18.4 ± 0.42 mm), the effect being comparable to those of nystatin (20.1 ± 0.09 mm), used as the control. Based on the obtained results these polymeric systems, consisting of thiazolidine-4-one derivatives loaded with chitosan microparticles, could have important applications in the food field as multifunctional (antimicrobial, antifungal, antioxidant) packaging materials.

Keywords: thiazolidine-4-one scaffold; chitosan; polymeric systems; antibacterial activity

1. Introduction

In the last time the xanthine scaffold was widely investigated based on its high biological potential. Some of the most important derivatives are caffeine, theophyline, theobromine, and paraxanthine [1]. These compounds are known for important biological effects such as bronchodilator, diuretic, anti-inflammatory and lipolytic [2,3]. New biological active molecules with a xanthine structure have been developed and evaluated for potential applications in asthma [4], diabetes mellitus [5], cancer [6], microbial infections [7] and neurodegenerative diseases such as Parkinson [8] and Alzheimer [2].

In the medicinal chemistry the thiazolidine heterocycle is also known as an important scaffold that is used to modulate the classical structures in order to improve their biological effects and to induce other new ones. For the first time this structure was identified as an actithiazic acid, a natural compound isolated from *Streptomyces spp.*, which showed high and specific action against *Mycobacterium tuberculosis* [9]. Later, many thiazolidine derivatives were synthesized and evaluated for their antimicrobial effect [10–12]. It was shown that the antimicrobial effects are in close correlation with the nature and position of substituents on the heterocycle. For example, the presence of arylazo, phenylhydrazono or sulfamoylphenylazo on the heterocycle was associated with improved antimicrobial effects [13].

In a previous study the researchers focused their studies on micro and nanotechnology as new strategies for increasing solubility, improving the biological properties, and decreasing the toxicity of the drugs. Some of the most important applications are in the regenerative medicine or tissue engineering area, as biosensors or as controlled drug release systems. The polymeric systems such as nano and microparticels are largely used as drug delivery systems due to their small particle size and large surface area [14]. Lately, chitosan, a natural polymer derived from chitin, has attracted the interest of many researchers due to its important and specific properties, such as biocompatibility, biodegradability, and reduced toxicity [15]. This biopolymer is considered the largest biomaterial after cellulose, having important applications in both the pharmaceutical and the food industry [16,17]. It is a linear policationic polysaccharide, copolymer of D-glucosamine and N-acetyl-D-glucosamine linked in 1-4 positions [16]. The amino groups on chitosan allow it to react with varied types of reagents to introduce different functional groups to chitosan. This polymer has important biological effects such as antimicrobial [18,19], antidiabetic [20–22], antitumor [23], antioxidant [24,25], anti-inflammatory [24], antiulceros [26], and not least, hypocholesterolemic [27]. The antimicrobial effect is the result of the interaction between the negatively charged bacterial cell membranes and the positively charged amino groups of chitosan, being directly proportional to the molecular weight and inversely proportional to the pH value. The studies showed also that *Gram-positive* bacterial strains are more susceptible to chitosan than *Gram-negative* ones [28].

Referring to the food industry, it is known that the packaging components have an important role for preservation of food as well as for environment protection, in a previous study the researchers focused on the use of renewable material and on the development of active packaging in order to increase the quality and safety of food [29]. The embedded food technology based on antimicrobial agents represents nowadays a challenge for researchers being strengthened also by the few marked products [30]. To obtain active packaging the antimicrobial agents could be included in the polymer matrix or the polymer surface could be covered with different antimicrobial agents [31,32]. Knowing that the microbial growth occurs mainly on the food surface, this new strategy involves slow diffusion of the antimicrobial agent from the package material, such as polymer films [33]. The controlled release packaging is a new approach for the food industry. The literature reported various packaging systems based on chitosan or embedded chitosan with different antimicrobial agents [34–37]. The studies highlight the important role chitosan plays as a plasticizer and it is frequently used in different composite systems with potential application in the food industry in order to increase the mechanical strength of the developed structure [37].

In this study we report the preparation, physic-chemical characterization, and biological evaluation of new polymeric systems based on chitosan, which have been loaded with new xanthine derivatives with a thiazolidine-4-one scaffold. Based on the unique properties of chitosan (biodegradability, biocompatibility) and of the biological effects of xanthine derivatives with a thiazolidine-4-one scaffold, the developed polymeric systems could have important applications in the food industry as active packaging materials and also in the medical field.

2. Materials and Methods

2.1. Materials

Chitosan (CS, molecular weight 190–310 kDa, 75–85% deacetylation degree), pentasodium tripolyphosphate (TPP), disodium hydrogen phosphate, sodium dihydrogen phosphate, hydrochloric acid 37%, acetic acid (min. 99,8%, p.a. ACS reagent), dimethyl sulfoxide (DMSO, p.a., ACS regent), Mueller–Hinton agar medium (Oxoid), Sabouraud agar medium were purchased from Sigma Aldrich. Xanthine derivatives with thiazolidine-4-one scaffold (XTDs: 6c, 6e, 6f, 6k) have been synthesized by our research group and were reported in a previous paper [38]. Bacterial strains (*Staphyloccoccus aureus* ATCC 25923, *Sarcina lutea* ATCC 9341, *Escherichia coli* ATCC 25922) and yeast strains (*Candida albicans* ATCC 10231, *Candida glabrata* ATCC MYA 2950, *Candida parapsilosis* ATCC 22019) were obtained from the Department of Microbiology, "Grigore T. Popa" University of Medicine and Pharmacy, Iasi, Romania. Ampicillin (30 μg/disc), ciprofloxacin (25 μg/disc) and nystatin (100 μg/disc) were used as positive controls for antibacterial and antifungal activity respectively. Swiss mice were purchased from the Biobase of "Grigore T. Popa" University of Medicine and Pharmacy, Iasi, Romania.

2.2. Development of New Polymeric Systems Based on Chitosan

2.2.1. Preparation of Chitosan Microparticles

Chitosan microparticles were prepared by the ionic gelation method. Briefly, chitosan (CS, 1.2–1.7 g) was dispersed in 100 mL of acetic acid 1% (*v*/*v*) and the mixture was left under stirring overnight. 3 mL of chitosan solution were dropped, using a syringe needle (26 G; 0.45 mm × 16 mm), into 20 mL of TPP (2%–3%), under light stirring and the mixture was left at room temperature, under stirring for 4–8 h in order to achieve an efficient reticulation. After 24 h the formed chitosan beads were separated from the TPP solution, washed three times with distillated water, and air-dried at room temperature.

2.2.2. Preparation of Chitosan Microparticles Loaded with Xanthine Derivatives with Thiazolidine-4-One Scaffold (CS-XTDs)

The xanthine derivatives with thiazolidine-4-one scaffold (XTDs: 6c, 6e, 6f, 6k) were loaded into chitosan microparticles by the ionic gelation method [5]. Briefly, different amounts of XTD (9 mg, 12 mg, 15 mg) were dissolved in 0.3 mL of DMSO and mixed with 3 mL of chitosan solution (1.2%–1.7%, *v*/*v*) in order to obtain three different concentrations (3 mg/mL; 4 mg/mL, 5 mg/mL) for each derivative (6c, 6e, 6f, 6k). The resulted mixtures were dropped, using a syringe needle (26 G; 0.45 mm × 16 mm), into 20 mL of TPP (2%–3%), under light stirring and the mixture was left at room temperature, under stirring for 4–8 h to achieve an efficient reticulation. After 24 h the formed chitosan loaded beads (CS-XTDs) were separated fromTPP solution, washed with distillated water three times, and air-dried at room temperature.

2.3. Characterization of Chitosan Microparticles Loaded with Xanthine Derivatives with Thiazolidine-4-one Scaffold (CS-XTDs)

2.3.1. Particle Size and Morphology

The CS-XTDs microparticle size (in wet and dry state) was measured using a Leica DM750 microscope (Wetzlar, Germany) equipped with a video model ICC50 W0366. The values were recorded with 10× objective by LAS EZ program. The morphology of the microparticles was studied using a scanning electronic microscope (SEM), Vega II SBH model, produced by the Tescan Company (Brno, Czech Republic). All measurements were performed in triplicate and the results are expressed as mean size ± standard deviation.

2.3.2. Swelling Degree (SD)

The swelling studies were carried out using distilled water and simulated gastric fluid (SGF, pH 2.2) at 37 °C (in a thermostated water bath) according to the literature references [5,39]. Briefly, an exact amount of CS-XTDs microparticles was immersed into the media. At different times microparticles were removed from media (water and SGF respectively), dried quickly with filter paper, and weighted (W_1). At the end of the study the microparticles were dried again and weighted (W_2). All tests were carried out in triplicate and the results are expressed as mean SD ± standard deviation. The analysis of Variance (ANOVA) was used to analyze the experimental data. Statistical significance was set to p value ≤ 0.05.

The swelling degree (SD) at different times was calculated using the following formula:

$$\text{SD (\%)} = W_1 - W_2/W_2 \times 100 \tag{1}$$

where: W_1—the weight of the swollen CS-XTDs microparticles; W_2—the weight of the dried CS-XTDs microparticles.

2.3.3. Drug Loading and Entrapment Efficiency

The loading efficiency of XTDs (6c, 6e, 6f, 6k) into the chitosan matrix was analyzed using a GBC Cintra 2010 UV-VIS spectrophotometer (Madrid, Spain) equipped with Cintral Software, according to the literature references [40] with slight modifications. The content of the XTDs into TPP solution was evaluated spectrophotometrically at different wavelengths, corresponding to each derivative, as follows: $\lambda = 276$ nm for 6k, $\lambda = 277$ nm for 6c and 6e, $\lambda = 280$ nm for 6f. The recorded values were used to calculate the non-loaded and loaded amount of each XTD. A standard curve for each XTD with a specific correlation coefficient ($y = 21.23x - 0.0112$; $R^2 = 1$ for 6c, $y = 27.606x - 0.0151$; $R^2 = 1$ for 6e, $y = 28.89x + 0.0219$; $R^2 = 0.9996$ for 6f, $y = 20.914x + 0.0406$; $R^2 = 1$ for 6k) was used.

The drug loading (DL) was calculated using the following formula [40]:

$$\text{DL}(\%) = \frac{W_3}{W1} \times 100\% \tag{2}$$

where: W_1 = the amount of CS-XTD microparticles, after drying (mg); W_3 = the amount of XTD loaded into the chitosan matrix (mg).

The entrapment efficiency (EE) was calculated using the following formula [41]:

$$\text{EE}(\%) = \frac{W_3}{W_2} \times 100\% \tag{3}$$

where: W_2 = the mount of XTD used in study (mg); W_3 = the amount of XTD loaded into chitosan matrix(mg).

All tests were performed in triplicate and the results are expressed as mean DL/EE ± standard deviation.

2.3.4. In Vitro Release

In vitro release of XTDs from polymeric system based chitosan (CS-XTDs: CS-6c, CS-6e, CS-6f, CS-6k) was performed using simulated gastric fluid (SGF, pH 1.2) and simulated intestinal fluid (SIF, pH 6.8), according to the literature references, with slight modifications [42–44]. The polymeric systems (CS-XTDs) were firstly placed in SGF for two hand then were moved to SIF. Briefly, a weighed amount of CS-XTDs microparticles was placed in a flask with 2 mL of media and incubated at 37 ± 0.1 °C under stirring (100 rpm). Every 30 min, a sample of 1.8 mL was collected from the media and replaced with an equal volume (1.8 mL) of fresh media. The concentration of the XTDs in the solution was evaluated spectrophotometrically at the specific wavelengths, corresponding to each derivative, using the standard curve previously obtained.

The Drug release (DR) was calculated using the following formula [42]:

$$\mathrm{DR}(\%) = \frac{\mathrm{C}_1}{\mathrm{C}_0} \times 100\% \tag{4}$$

where: C_1 = concentration of the XTD (mg/mL) in the release media at different times; C_0 = concentration of the XTD (mg/mL) in the CS-XTD microparticles.

All tests were performed in triplicate and the results are expressed as mean DR ± standard deviation.

2.3.5. Fourier Transform Infrared (FTIR) Spectroscopy

FT-IR spectra of chitosan, XTDs (6c, 6e, 6f, 6k) and of CS-XTDs (CS-6c, CS-6e, CS-6f, CS-6k) were recorded using ABB-MB3000 FT-IR MIRacleTM Single Bounce ATR (Zürich, Switzerland) at a resolution of 4 cm^{-1}, after 16 scans, in the 4000–650 cm^{-1} range and processed with the Horizon MBTM FT-IR Software (Horizon MB 3.1.29.5, LabCognition GmbH & Co. KG, Cologne, Germany).

2.4. Biological Evaluation

2.4.1. Acute Toxicity Assay

The acute toxicity of the XTDs (6c, 6e, 6f, 6k) was evaluated by determining the lethal dose of 50 (LD_{50}). The Swiss mice (22–35 g) were housed in polyethylene cages, at constant conditions: temperature of 24 ± 2 °C, humidity of 40–70%, and cycle of 12 h light and 12 h dark. The acclimatization of the animals to laboratory conditions was performed 7 days before the experiment, receiving standard food and water ad libitum. The animals were kept fasting for 24 h before starting the experiment. Each group (six mice/group) received XTDs, orally (p.o.), as suspensions in Tween 80 in doses ranging between 1500–6500 mg/kg b.w. The survival rate was determined at 1 day, 2 days, 3 days, 7 days, and 14 days after administration.

The lethal dose 50 (LD_{50}) was determined by the Karber method using the following formula [45,46]:

$$LD_{50} = LD_{100} - \frac{\Sigma(a \times b)}{n} \tag{5}$$

where: a = the difference between two successive doses of the XTDs; b = the arithmetic average of the mice died from two successive series; n = the number of animals from each group; LD_{100} = the 100% lethal dose.

This study was performed according to the ethics guidelines on laboratory animal studies (Law no. 206/May 27, 2004, EU/2010/63-CE86/609/EEC) and with agreement (no 17826/2016) of the Ethics Committee for Animal Research of "Grigore T. Popa" University of Medicine and Pharmacy Iasi.

2.4.2. Antibacterial/Antifungal Tests

The antibacterial and antifungal activity of XTDs and CS-XTDs[3] (5 mg/mL) were evaluated using the agar disc diffusion and broth micro-dilution methods.

The Agar Disc Diffusion Method.

Antibacterial and antifungal activity of XTDs and CS-XTDs, expressed as the diameter of inhibition area, were evaluated using the standard disk diffusion assay according to literature reference [47] with slight modifications. Prior to the test, the bacterial and yeast strains were diluted in NaCl (0.9 %) in order to achieve a turbidity equivalent of 10^6 CFU/mL (McFarland standard no 5). The suspensions were diluted 1:10 in Mueller Hinton agar (bacteria) and Sabouraud agar (yeasts) and then spread on sterile Petri plates (25 mL/Petri plate). Sterile stainless steel cylinders (10 mm height and 5 mm internal diameter) were placed on the agar surface in Petri dishes. In each cylinder 200 μL of XTDs, solutions in DMSO (20 mg/mL) and an equivalent concentration of CS-XTDs were added. Commercial discs containing ampicillin (25 μg/disc), ciprofloxacin (30 μg/disc), and nystatin (100 μg/disc) were used as a positive control. DMSO was used as a negative control. The plates were incubated for 24 h at 37 °C (bacteria) and for 48 h at 24 °C (yeasts). After incubation the diameters of inhibition were measured. All tests were performed in triplicate and the results are expressed as mean diameter ± standard deviation.

The Broth Micro-Dilution Method.

The minimum inhibitory concentration (MIC) and the minimum bactericidal/fungicidal concentration (MBC/MFC) of XTD were determined using the standard two-fold dilution method standardized by [47], with slight modifications. The strains were inoculated in the specific agar medium and incubated at 37 °C for 24 h for bacteria, respectively 48 h at 24 °C for *Candida* strains. After the incubation, the culture media were diluted in order to obtain a final concentration of 10^6 CFU/mL. In a 96-well microplate, different dilutions were prepared from the stock solution of the XTDs (20 mg/mL), in order to obtain a volume of 100 μL in each well with final concentrations in the range of 10 mg/mL, 5 mg/mL, 2.5 mg/mL, 1.25 mg/mL, 0.625 mg/mL, 0.312 mg/mL, 0.156 mg/mL, 0.078 mg/mL, 0.039 mg/mL, 0.0195 mg/mL, 0.009 mg/mL, and 0.0048 mg/mL. A volume of 100 μL strain suspension was inoculated onto the wells and the microplates were incubated for 24 h at 37 °C (bacterial strains) and at 24 °C for (fungal strains). The MIC value was established as the lowest concentration, which determined the visual inhibition of the strain's growth. For MBC/MFC determination, 1 μl from each visually complete inhibition was transferred onto a plate with solid media and incubated for 24 h at 37 °C (bacterial strains) and at 24 °C for (fungal strains). The MBC/MFC values were considered the lowest concentration, which killed 99.9% of the tested strains. All tests were performed in triplicate.

3. Results and Discussions

The synthesis of new xanthine derivatives with thiazolidine-4-one scaffold (6a–k) (Figure 1) were presented in our previous paper [38]. Some of these compounds showed good antioxidant effects in terms of the antiradical scavenging effect (DPPH, ABTS) and phosphomolybdenum reducing antioxidant power [38]. In this study the compounds were evaluated for antibacterial and antifungal effects and the most active of them were formulated as new polymeric systems based on chitosan in order to increase their biological effects.

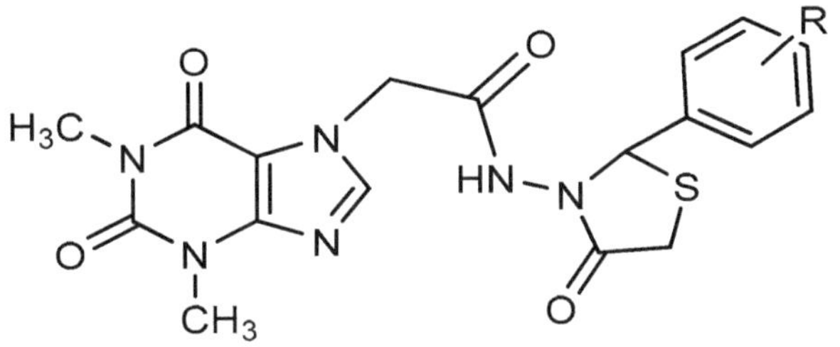

6a-k

6a: R= H; 6b: R= 4-Br; 6c: R= 4-Cl; 6d: R= 4-F; 6e: R= 2-OCH_3; 6f: R= 4-OCH_3; 6g: R= 3-OCH_3; 6h: R= 2,3-diOCH_3; 6i: R= 2,4-diOCH_3; 6j: R= 3,5-diOCH_3; 6k: R= 4-CH_3.

Figure 1. The structure of the xanthine derivatives with thiazolidine-4-one scaffold (XTDs: 6a-k).

3.1. New Polymeric Systems Based on Chitosan

The chitosan microparticles could be prepared using chemical and physical methods. The chemical method is based on the covalent bond formation between the functional groups of chitosan while the physical process involves electrostatic hydrophobic interactions or hydrogen bonds formation in the polymer matrix [48,49]. It is considered that the last type of polymeric system seems to have increased biocompatibility and are more tolerated than the chemical ones [50,51]. The most used physical method is ionic gelation, which is based on the ionic complexes formation between positively charged amino groups and anions such as sulphate, citrate, and phosphate. The most used cross-linking agent is pentasodiumtripolyphosphate, which interacts with the amino groups of chitosan by electrostatic forces [52].

Four XTDs (6c, 6e, 6f, 6k) were selected based on their biological effects and were inglobated into a polymeric matrix based on chitosan using the ionic gelation method. In order to optimize the preparation procedure of chitosan microparticles loaded with xanthine derivatives with thiazolidine-4-one scaffold (CS-XTDs) different parameters were applied: chitosan concentration (1.2%–1.7%), TPP concentration (2%–3%), and reticulation time (4–8 h). The most stable CS-XTDs systems were obtained using the following parameters: CS concentration of 1.7%, TPP concentration of 2%, and reticulation for 8 h. Using three concentrations (3 mg/mL, 4 mg/mL, 5 mg/mL) for each XTD, twelve CS-XTD systems were developed and characterized. The mass ration between CS, TPP, and XTD, based on the concentration of used XTD (9 mg, 12 mg, 15 mg) was as follows:5.7:44:1, 4.25:33:1 and 3.4:27:1.

3.1.1. Particle Size and Morphology

The size of the CS-XTDs ranged between 858–953 μm (wet state) and 611–855 μm (dry state) and it was higher than CS (825 μm—wet state; 611 μm—dry state), thus confirming the loading process (Table 1). The data showed that the size increases with the concentration of the compounds and are in good agreement with similar data presented in the literature [53].

Table 1. Microparticles size of CS and CS-XTDs (CS-6c, CS-6e, CS-6f, CS-6k) at different concentrations.

CS/CS-XTDs	Size in Wet State (μm)	Size in Dry State (μm)	CS-XTDs	Size in Wet State (μm)	Size in Dry State (μm)
CS-6c^1	904 ± 4.2	638 ± 3.1	CS-6f^1	887 ± 2.1	614 ± 3.1
CS-6c^2	897 ± 7.2	762 ± 2.5	CS-6f^2	921 ± 4.0	675 ± 5.0
CS-6c^3	951 ± 3.1	855 ± 2.1	CS-6f^3	947 ± 3.2	751 ± 6.4
CS-6e^1	858 ± 3.1	632 ± 3.5	CS-6k^1	929 ± 6.0	639 ± 7.0
CS-6e^2	923 ± 3.5	634 ± 4.7	CS-6k^2	912 ± 5.0	688 ± 4.0
CS-6e^3	953 ± 2.9	646 ± 4.0	CS-6k^3	918 ± 2.1	728 ± 4.7
CS	825 ± 3.5	611 ± 3.1	-	-	-

^{1} = 3 mg/mL, ^{2} = 4 mg/mL, ^{3} = 5 mg/mL.

The SEM showed that chitosan microparticles (CS) have a spherical and regular shape, regular outline and a slightly rough surface. Upon loading of XTD, a deformation of microparticles was observed, which is intensified by the increasing concentration described in other studies [5]. The most intense deformation was observed in the case of CS-6c^3 (5 mg/mL), for which the spherical shape is lost and pronounced flattening was recorded (Figure 2).

The results showed that a higher concentration of the cross-linking agent was associated with decreased stability of microparticles during the gelation process. Also, it was highlighted that the deformation of CS-XTDs microparticles during the air-drying process was directly proportional to the concentration of the TPP solution, which resulted in strongly flattened microparticles being obtained for a value of 3% TPP. The reticulation time is also an important parameter for the stability of microparticles; increasing the time was associated with the increased stability of the developed polymeric systems [54,55].

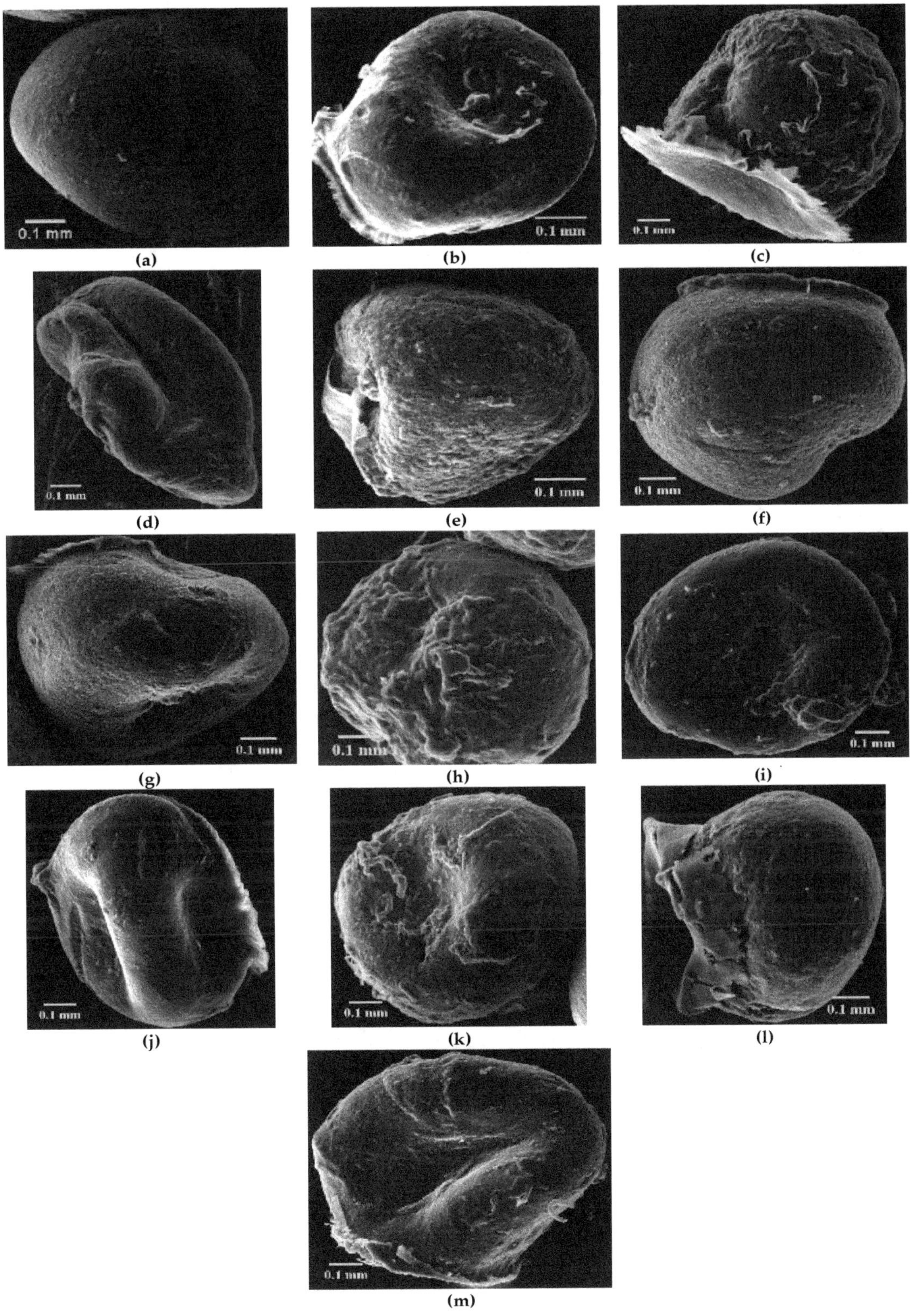

Figure 2. Scanning electron microscope (SEM) images for CS and CS-XTDs microparticles (**a**: CS, **b**: CS-6c^1, **c**: CS-6c^2, **d**: CS-6c^3, **e**: CS-6e^1, **f**: CS-6e^2, **g**: CS-6e^3, **h**: CS-6f^1, **i**: CS-6f^2, **j**: CS-6f^3, **k**: CS-6k^1, **l**: CS-6k^2, **m**: CS-6k^3); ^{1} = 3 mg/mL, ^{2} = 4 mg/mL, ^{3} = 5 mg/mL.

3.1.2. Swelling Degree (SD)

The swelling degree is an important parameter in the release of the drug from the polymer matrix, having an important influence on the drug's bioavailability. The results for CS-XTDs systems recorded in distilled water and simulated gastric fluid (SGF, pH 2.2) are presented in Figures 3–6. For distilled water, the swelling degree values of CS-XTDs was comparable with that of CS. The CS-XTDs absorbed the highest amount of water in the first 10 mins of the experiment, with the SD (%) value ranging around 150% (Figures 3 and 4). This phenomenon could be explained by the cross-linking agent which, by dissolution in aqueous medium, released the amino groups of chitosan, increasing the hydrophilic character of the polymer matrix.

The thermodynamic equilibrium state was reached after different times depending on the structure of XTDs loaded into the polymer matrix. At the end of the experiment the swelling degree values ranged between 190 ± 2.6 (CS-6e[3]) and 206 ± 2.9 (CS-6k[1]). In similar conditions the recorded value for CS was 209 ± 3.5 (Table 2).

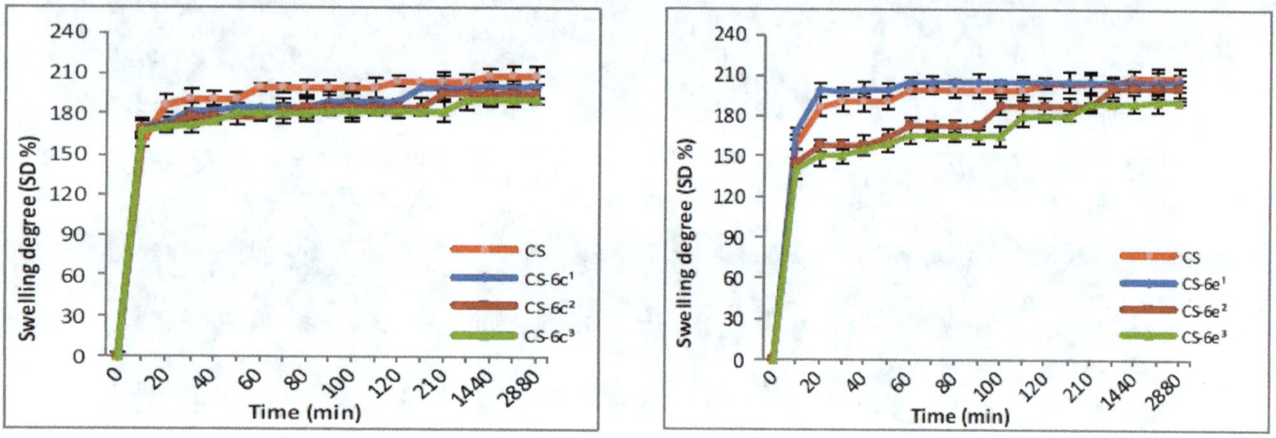

Figure 3. Swelling degree of CS and CS-XTDs (CS-6c; CS-6e) in distilled water at different concentrations (^{1} = 3 mg/mL, ^{2} = 4 mg/mL, ^{3} = 5 mg/mL).

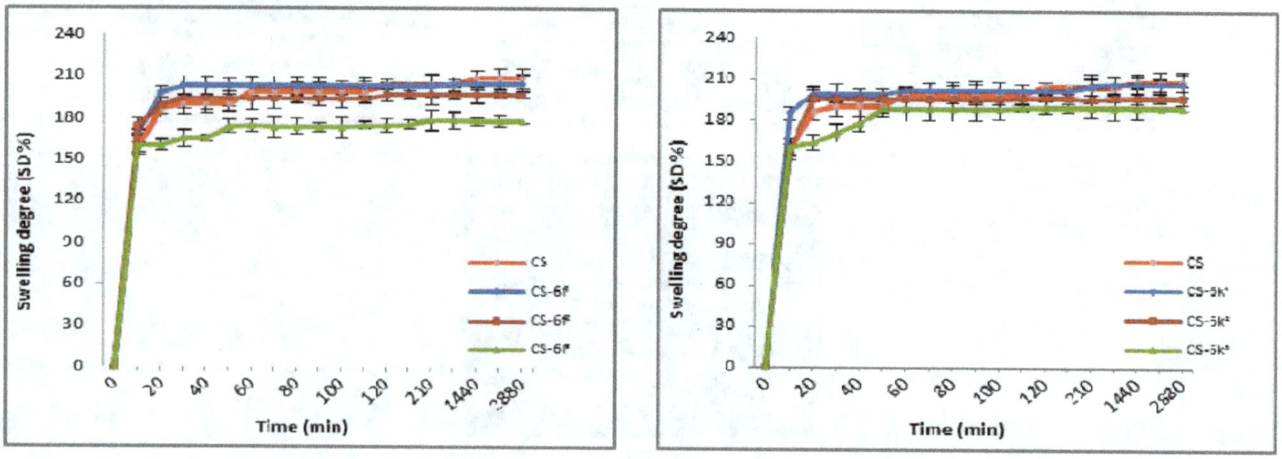

Figure 4. Swelling degree of CS and CS-XTDs (CS-6f; CS-6k) in distilled water at different concentrations (^{1} = 3 mg/mL, ^{2} = 4 mg/mL, ^{3} = 5 mg/mL).

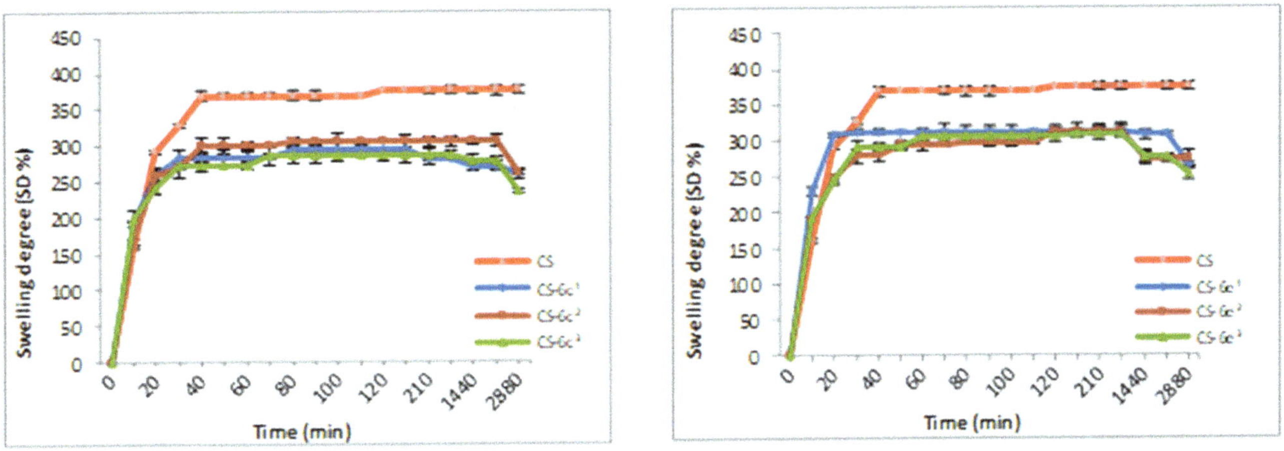

Figure 5. Swelling degree of CS and CS-XTDs (CS-6c; CS-6e) in SGF at different concentrations (^{1} = 3 mg/mL, ^{2} = 4 mg/mL, ^{3} = 5 mg/mL).

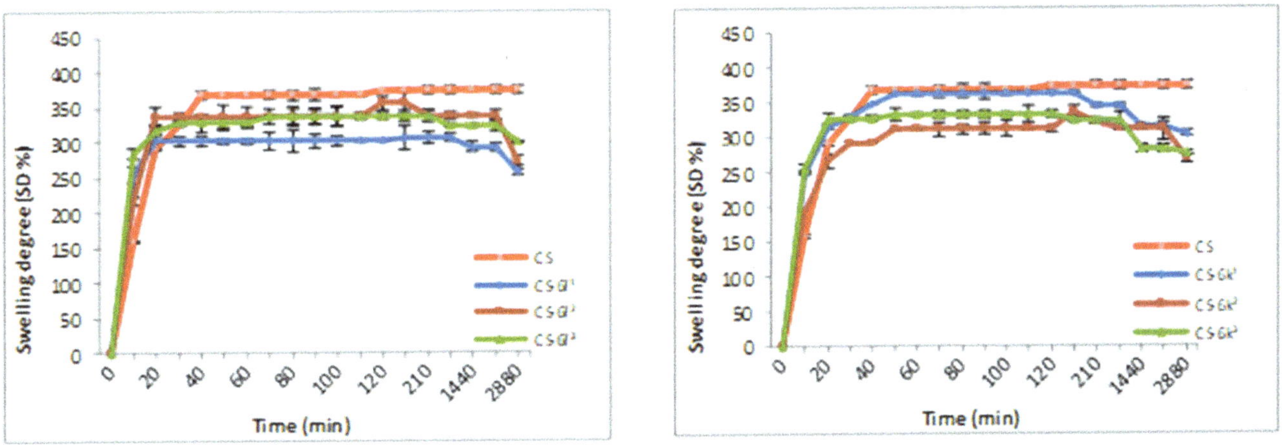

Figure 6. Swelling degree of CS and CS-XTDs (CS-6f; CS-6k) in SGF at different concentrations (^{1} = 3 mg/mL, ^{2} = 4 mg/mL, ^{3} = 5 mg/mL).

Table 2. Swelling degree of CS and CS-XTD at thermodynamic equilibrium state.

CS-XTDs	SD (%)	CS-XTDs	SD (%)	CS-XTDs	SD (%)	CS-XTDs	SD (%)
CS-6c^1	200 ± 2.7 *	CS-6e^1	205 ± 3.6	CS-6f^1	205 ± 4.5	CS-6k^1	206 ± 2.9
CS-6c^2	195 ± 2.2 *	CS-6e^2	201 ± 1.7 *	CS-6f^2	197 ± 3.6 *	CS-6k^2	196 ± 5.2 *
CS-6c^3	190 ± 4.1 *	CS-6e^3	190 ± 2.6 *	CS-6f^3	178 ± 3.1 *	CS-6k^3	189 ± 3.1 *
CS	209 ± 3.5	-	-	-	-	-	-

* significant different ($p < 0.05$) in reference with CS; ^{1} = 3 mg/mL, ^{2} = 4 mg/mL, ^{3} = 5 mg/mL.

The analysis of the SD results recorded in SGF highlighted that the SD of CS-XTDs was lower than of CS (375%) (Figures 5 and 6); the results could be explained based on the hydrophobic character of the XTD loaded into the polymer matrix, a character which results in decreasing the chitosan matrix permeability and thus the swelling degree. The presence of functional groups such as halogen (6c) or methyl (6k), attached to the aromatic ring of the thiazolidine-4-one scaffold increases the hydrophobic character of the XTD. Our observation is in agreement with other literature data [56]. A high absorption degree (over 300%) at the beginning of the experiment was observed, after which it decreased in intensity. For CS the thermodynamic equilibrium state was reached after 2 h and maintained until the end of the experiment. In similar conditions, for CS-XTDs the thermodynamic equilibrium state was reached after 40 min–2 h, depending on the kind of XTD loaded into the polymer matrix, which could result in a shorter period.

3.1.3. Drug Loading and Entrapment Efficiency

The results obtained for drug loading (DL) and entrapment efficiency (EE) parameters are presented in Figure 7 (CS-6c, CS-6e) and Figure 8 (CS-6f, CD-6k). The study highlighted that the entrapment efficiency ranged between 63.71% (CS-6e[1]) and 93.91% (CS-6k[3]) and is directly proportional to the concentration. The highest entrapment efficiency, at all three concentrations (3 mg/mL, 4 mg/mL, 5 mg/mL), was recorded for CS-6k, the values being: 87.86 ± 1.25% (3 mg/mL), 91.58 ± 0.55% (4 mg/mL) and 93.91 ± 1.41% (5 mg/mL). These results could be explained based on the structure of the XTD (6k), which has hydrophobic *methyl* group at *para* position of the aromatic ring. The XTD with hydrophilic *methoxy* group attached at the *ortho* (6e) and *para* (6f) position on the aromatic ring have showed lower values of entrapment efficiency.

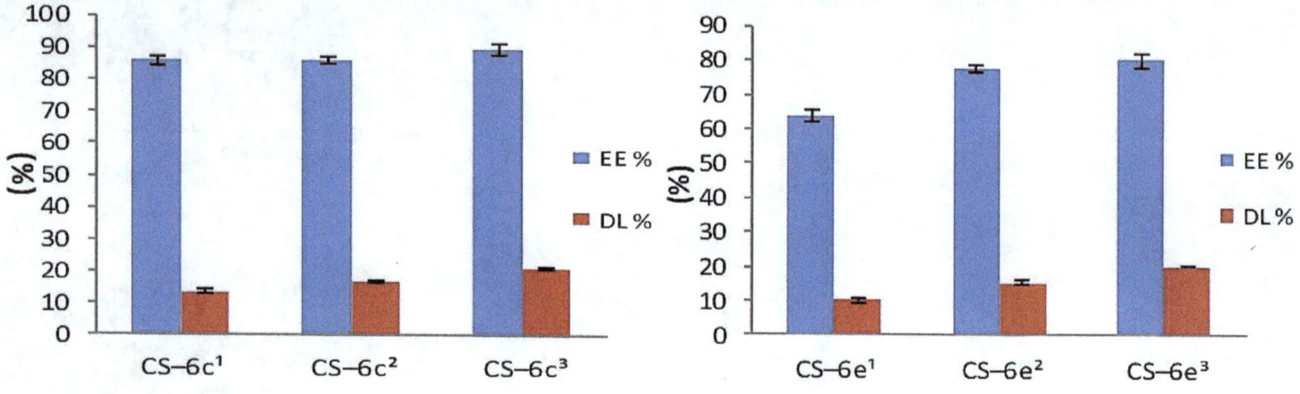

Figure 7. The Entrapment efficiency (EE%) and Drug loading (DL%) for CS–6c andCS-6e ([1] = 3 mg/mL, [2] = 4 mg/mL, [3] = 5 mg/mL).

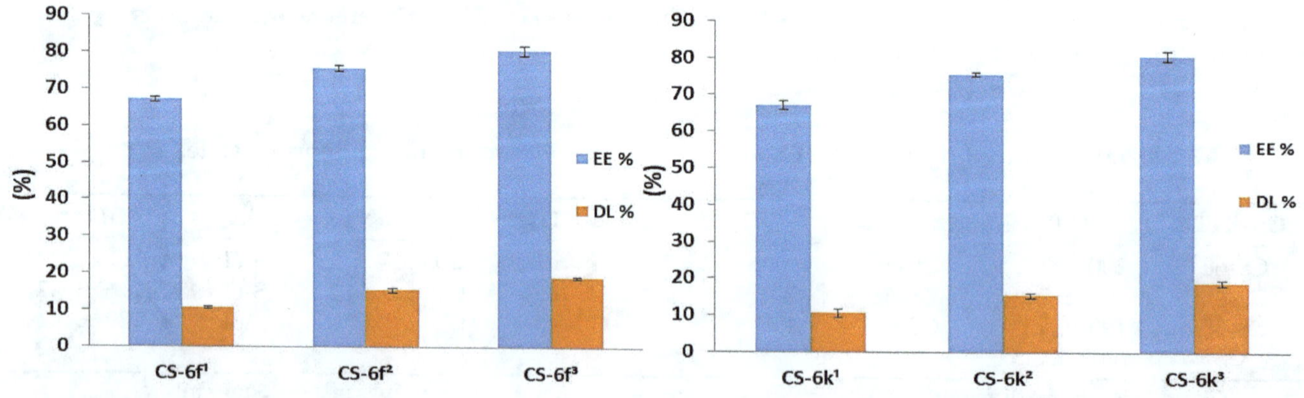

Figure 8. The Entrapment efficiency (EE%) and Drug loading (DL%) for CS–6f and CS-6k ([1]= 3 mg/mL, [2] = 4 mg/mL, [3] = 5 mg/mL).

Concerning the DL of XTDs into the polymeric matrix, it was noted that this parameter increases with the concentration. The obtained values ranged between 10.04 ± 0.49% (CS-6e[1]) and 22.02 ± 0.74% (CS-6k[3]). As a previous parameter, the highest values were obtained for derivative 6k, which has a *methyl* group at the *para* position of the aromatic ring as follows: 14.32 ± 1.12% (CS-6k[1]), 18.90 ± 0.67% (CS-6k[2]), and 22.02 ± 0.74% (CS-6k[3]). A good correlation was also observed between DL and EE values, which proves the accuracy of the study. Thus, it has been demonstrated that the optimal CS-XTD formulation was obtained for 5 mg/mL of XTD.

3.1.4. In Vitro Drug Release

The literature describes two methods for releasing the drugs from the polymer matrix: by direct and indirect diffusion, the last one being based on the dissolution of the drug into the matrix, followed by the release through membrane pores. In turn, the membrane permeability is influenced by the method used to prepare the polymeric systems, by the morphology of microparticles, and not least by the chemistry structure of the drug. The oral route is the most used for drug administration. After the absorption of drugs through gastric or intestinal mucosa, it will pass to the blood system and then will arrive at the site of action. Normally, the gastric pH ranges between 1 and 1.5 and will contain 99% water and only 1% different organic or inorganic substances, while the small intestine has a pH, which ranges between 4.8 to 8.2.

In the SGF, the drug release from CS-XTD was ranging between 51.34% and 98.42%, while the drug release in SFI was lower, being around 5%, excepting the CS-6k for which the drug release in SIF was 17% (Figures 9 and 10). The lowest cumulative release was observed for CS-6f (CS-6f^1 = 58.10%, CS-6f^2 = 51.34%, CS-6f^3 = 54.52%), while the highest values were recorded for CS-6c (CS-6c^1 = 98.42%, CS-6c^2 =89.92%, CS-6c^3 = 97.44%).

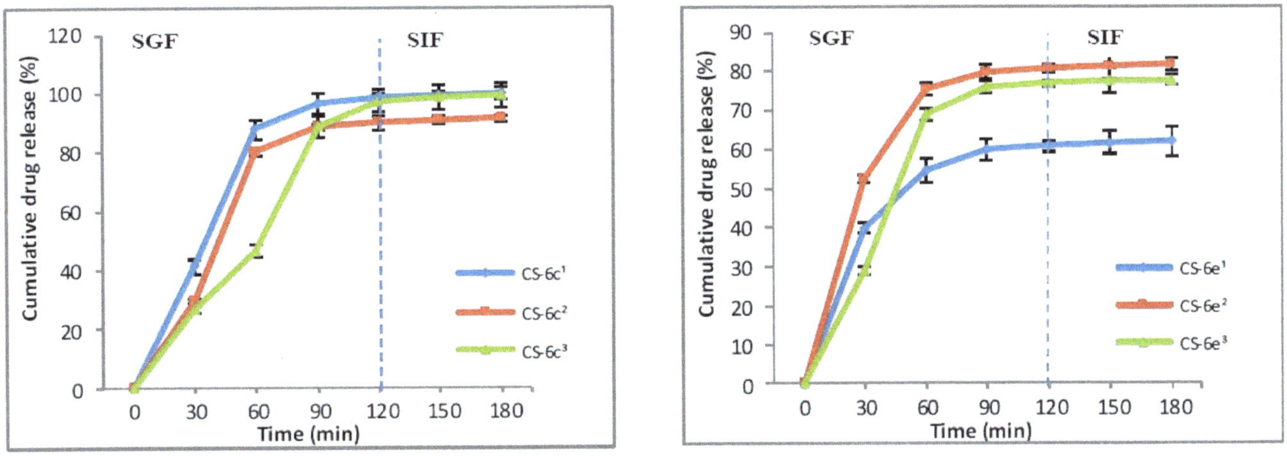

Figure 9. The release profile of compound 6c and 6e from CS-6c and CS-6e respectively (^{1} = 3 mg/mL, ^{2} = 4 mg/mL, ^{3} = 5 mg/mL).

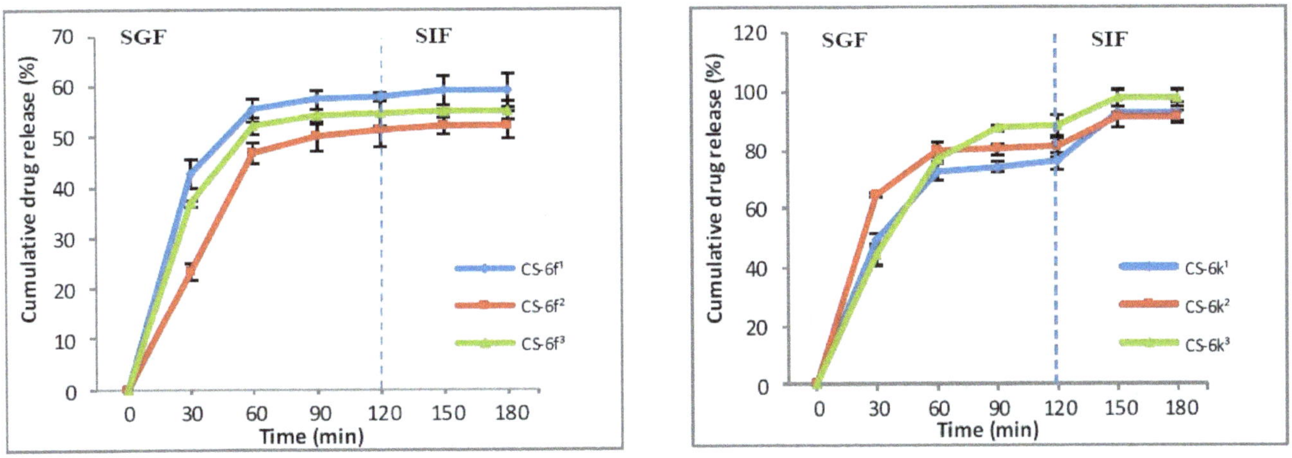

Figure 10. The release profile of compound 6f and 6k from CS-6f and CS-6k respectively (^{1} = 3 mg/mL, ^{2} = 4 mg/mL, ^{3} = 5 mg/mL).

3.1.5. FTIR Analysis

The confirmation of the loading process was performed by highlighting in FTIR spectra the CS-XTDs of the specific functional groups of the components of the polymeric matrix: CS, XTDs, and TPP. The FTIR spectra of CS-XTDs (CS-6c, CS-6e, CS-6f and CS-6k) at different concentrations in reference with the FTIR spectra of CS and XTDs (6c, 6e, 6f and 6k) are presented in Figure 11.

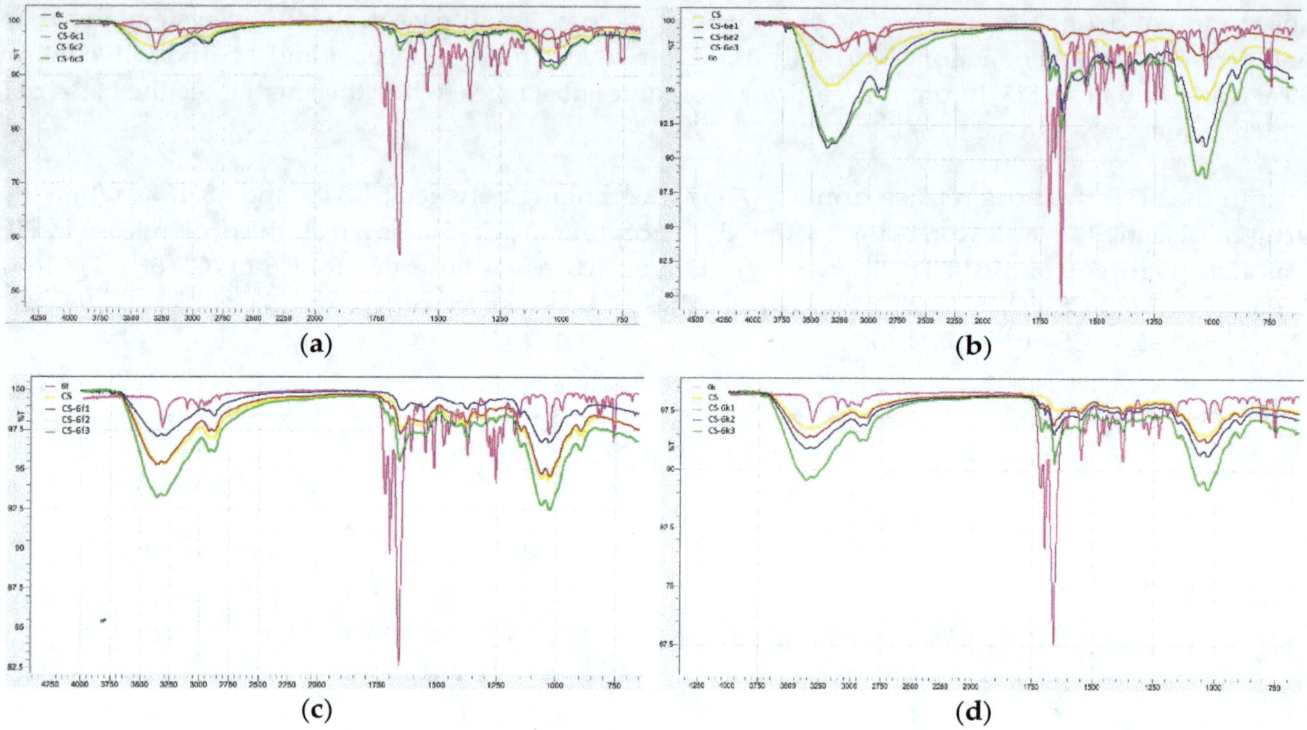

Figure 11. The Fourier transform infrared (FTIR) spectra of CS, XTDs (6c, 6e, 6f, 6k) and CS-XTDs (CS-6c, CS-6e, CS-6f, CS-6k) (^{1} = 3 mg/mL, ^{2} = 4 mg/mL, ^{3} = 5 mg/mL).

In acetic acid solution, chitosan has a pKa of 6.3 and a polycationic structure with positively charged amino groups. In water, TPP dissociates to form both hydroxyl and phosphoric ions. In the crosslinking process the cationic amino groups of chitosan react with negatively charged TPP to form ionic complexes by electrostatic bonds. A secondary mechanism based on hydrogen-hydrogen bonds between the hydroxyl groups of CS and of TPP is also possible [50]. The XTDs are physically and uniformly dispersed into the polymer matrix. The concentration of CS and XTDs are very important parameters, because they influence the physic-chemical properties of the CS-XTD systems [49].

The specific vibrations of functional groups of CS have been identified in the following regions: 3362–3285 cm^{-1} (OH, NH), 2981–2872 cm^{-1}, 1431–1414 cm^{-1} and 1377–1375 cm^{-1} ($-CH_2-CH_3$), 1661–1647 cm^{-1} (C=O), 1267–1250 cm^{-1} (C–N), 1034–1026 cm^{-1} (C–O–C). The presence of XTDs (6c, 6e, 6f and 6k) into the polymer matrix was confirmed by identification of the specific absorption bands: NH bond (3362–3285 cm^{-1}), thiazolidine-4-one ring: 2981–2872 cm^{-1} ($-CH_2-$), 1724–1711 cm^{-1} (C=O) and 671–660 cm^{-1} (C–S), aromatic ring: 3362–3285 cm^{-1}, 810–746 cm^{-1} (=C–H), 1558–1549 cm^{-1} and 1494–1452 cm^{-1} (C=C), amide group, which are overlapping with valence vibrations of the same group in the chitosan structure, and not the substituents on the aromatic ring at 1063–1059 cm^{-1} (C–O from $-OCH_3$ in case of 6e and 6f), 822–818 cm^{-1} (–Cl in case of 6c) and 1377–1375 cm^{-1} ($-CH_3$ in case of 6e, 6f and 6k). The reticulation agent, TPP, was identified by specific absorption bands in the 1151–1149 cm^{-1} region (P=O) and the 895–893 cm^{-1} region (P–O–P).

3.2. Biological Evaluation

3.2.1. Toxicity Degree

The toxicological screening highlighted that all tested XTDs (6c, 6e, 6f, 6k) have a low toxicity, with the following LD_{50} values: 2125 mg/kg bw (6c), 1687.5 mg/kg bw (6e), 1937.5 mg/kg bw (6f) and 1312.5 mg/kg bw (6k). These compounds proved to be less toxic than theophylline, which is used as a starting reagent in their synthesis (LD_{50} = 332 mg/kg bw) [57] and supports the favorable influence of thiazolidine-4-one scaffold upon the chemical modulation of xanthine structure.

3.2.2. Antibacterial/Antifungal Study

The data presented in Table 3 showed that XTDs (6a–k) and CS-XTDs^3 (CS-6c^3, CS-6e^3, CS-6f^3, CS-6k^3) are active on both bacterial and fungal strains, the diameter of the inhibition area being in correlation with structure of the XTD and also with bacterial and yeasts strains, respectively.

Table 3. Antibacterial/antifungal inhibition area (mm) of XTDs (6a–k) and CS-XTDs.

Sample	Diameter of Inhibition Area*(mm)					
	Bacterial Strains			Yeasts Strains		
	SA	SL	EC	CA	CG	CP
6a	15.2 ± 0.23	20.1 ± 0.17	12.2 ± 0.35	11.1 ± 0.36	12.2 ± 0.22	12.2 ± 0.24
6b	15.3 ± 0.40	19.1 ± 0.28	11.1 ± 0.23	10.1 ± 0.21	10.8 ± 0.37	10.3 ± 0.31
6c	14.9 ± 0.21	19.0 ± 0.13	10.4 ± 0.15	15.2 ± 0.21	12.1 ± 0.42	11.2 ± 0.24
6d	17.1 ± 0.24	18.1 ± 0.42	11.2 ± 0.47	10.0 ± 0.11	11.0 ± 0.46	11.3 ± 0.32
6e	14.0 ± 0.38	18.8 ± 0.29	11.1 ± 0.14	10.1 ± 0.28	10.9 ± 0.26	14.2 ± 0.12
6f	15.0 ± 0.50	20.1 ± 0.43	11.2 ± 0.35	12.1 ± 0.07	10.0 ± 0.06	15.2 ± 0.11
6g	15.5 ± 0.27	20.0 ± 0.31	12.1 ± 0.26	11.3 ± 0.23	16.1 ± 0.28	15.4 ± 0.31
6h	15.4 ± 0.34	19.2 ± 0.09	12.0 ± 0.06	15.2 ± 0.26	19.1 ± 0.17	12.0 ± 0.14
6i	15.4 ± 0.25	19.1 ± 0.11	12.1 ± 0.04	12.5 ± 0.10	15.0 ± 0.24	12.1 ± 0.23
6j	15.2 ± 0.14	19.1 ± 0.17	12.2± 0.26	12.2 ± 0.26	15.3 ± 0.09	12.1 ± 0.24
6k	14.3 ± 0.19	17.8 ± 0.23	11.0 ± 0.31	10.3 ± 0.23	13.2 ± 0.11	11.0 ± 0.13
CS-6c^3	16.4 ± 0.41	21.0 ± 0.23	12.3 ± 0.21	17.4 ± 0.29	14.3 ± 0.38	13.2 ± 0.31
CS-6e^3	15.2 ± 0.23	20.4 ± 0.21	13.4 ± 0.25	12.5 ± 0.21	12.6 ± 0.21	15.4 ± 0.26
CS-6f^3	21.2 ± 0.43	25.1 ± 0.28	14.7 ± 0.38	16.7 ± 0.42	12.3 ± 0.51	18.4 ± 0.42
CS-6k^3	17.1 ± 0.32	22.4 ± 0.18	15.2 ± 0.18	14.6 ± 0.21	16.3 ± 0.38	17.2 ± 0.35
CS	12 ± 0.35	11 ± 0.26	9 ± 0.41	-	-	-
C^a	25.1 ± 0.08	25.0 ± 0.1	28.9 ± 0.18	-	-	-
A^b	27.1 ± 0.12	31.8 ± 0.15	21.0 ± 0.21	-	-	-
N^c	-	-	-	20.1 ± 0.11	21.0 ± 0.14	20.1 ± 0.09

* mean values (n = 3) ± standard deviation. SA = *Staphylococcus aureus* ATCC 25923; SL = *Sarcina lutea* ATCC 9341; EC = *Escherichia coli* ATCC 25922; CA = *Candida albicans* ATCC 10231; CG = *Candida glabrata* ATCC MYA 2950; ^{3} = 5 mg/mL; CP = *Candida parapsilosis* ATCC 22019; C^a= Ciprofloxacin (30 μg/disc); A^b = Ampicillin (25 μg/disc); N^c = Nystatin (100 μg/disc).

Referring to the XTDs it was noted the most active compound on *Staphylococcus aureus* ATCC 25923 was 6d, for which the diameter of the inhibition area was 17.1 ± 0.24 mm. The compounds 6a, 6f and 6g showed improved antibacterial effects on *Sarcina lutea* ATCC 9341, their effect (6a: 20.1 ± 0.17 mm, 6f: 20.1 ± 0.43 mm, 6g: 20.0 ± 0.31 mm) being higher than the effect of CS and comparable with ciprofloxacin (25.0 ± 0.1 mm). The effect of XTDs on Gram negative bacterial strain *Escherichia coli* ATCC 25922 was reduced in reference to Gram positive bacterial strains (*Staphylococcus aureus, Sarcina lutea*), the diameter of the inhibition area ranging between 10.4 mm and 12.2 mm.

Appreciable antifungal effects were noted for compounds 6c and 6h on *Candida albicans* ATCC 10231 (15.2 ± 0.21 mm and 15.2 ± 0.26 mm respectively), for 6h and 6g on *Candida glabrata* ATCC MYA 2950 (19.1 ± 0.17 mm and 16.1 ± 0.28 mm respectively) and for 6f and 6g on *Candida parapsilosis* ATCC 22019 (15.2 ± 0.11 mm and 15.4 ± 0.31 mm respectively).

As we expected, the loading of the XTDs into the polymer matrix based on chitosan was associated with increasing the antimicrobial effects, all CS-XTD systems (CS-6c, CS-e, CS-6f, CS-6k) show improved antibacterial and antifungal effects in reference to corresponding XTD (6c, 6e, 6f, 6k) (Table 3).

The antibacterial effect of 6d was confirmed by the broth micro-dilution assay, for which the minimum inhibitory concentration (MIC) and minimum bactericidal concentration (MBC) were 0.3125 mg/mL and 10 mg/mL, respectively (Table 4). The similar results were obtained also for 6a. It was also noted that, in most cases, the value of MBC was 10 mg/mL, higher than the value of MIC, which means that the compounds act, especially, by the inhibition of microbial growth that can then act as bactericidal agents.

Table 4. Antimicrobial activity expressed as MIC and MBC values (mg/mL) of XTDs (6a–k).

Sample	*Staphylococcus Aureus* ATCC 25923		*Escherichia Coli* ATCC 25922		*Candida Albicans* ATCC 90028	
	MIC *	MBC *	MIC *	MBC *	MIC *	MBC *
6a	0.625	10	1.25	1.25	1.25	1.25
6b	2.5	10	2.5	2.5	1.25	1.25
6c	2.5	10	1.25	1.25	1.25	2.5
6d	0.3125	10	1.25	2.5	2.5	2.5
6e	1.25	10	1.25	1.25	1.25	2.5
6f	0.625	10	1.25	2.5	2.5	2.5
6g	0.625	10	1.25	2.5	2.5	2.5
6h	1.25	5	2.5	2.5	2.5	2.5
6i	0.625	10	2.5	2.5	2.5	2.5
6j	0.625	10	1.25	2.5	2.5	2.5
6k	1.25	10	0.625	2.5	0.625	1.25
A	1 [1]	2 [1]	2 [1]	4 [1]	-	-
N	-	-	-	-	8 [1]	16 [1]

* mean values (n = 3) ± standard deviation; [1] μg/mL.

4. Conclusions

In this study new polymeric systems based on chitosan (CS) loaded with new xanthine derivatives with thiazolidine-4-one scaffold (XTDs) were developed and characterized. The success of the loading process was proved by highlighting in FTIR spectra of the CS-XTDs of the specific functional groups of XTD. The optimized polymeric systems were evaluated in terms of particle size, morphology, swelling degree, drug loading and entrapment efficiency. The results demonstrated a good swelling degree for CS-XTDs and the entrapment efficiency of the XTD into polymer matrix was between 63% and 94%. In the simulated biological fluids (SGF, SIF), an increased cumulative release, between 54.52% and 97.44% was observed. The data supports also improved antimicrobial effects for CS-XTDs in reference with XTDs. Based on the obtained results the developed polymeric systems could have important applications in the food field as active multifunctional (antimicrobial, antifungal) packaging materials and in medical and pharmaceutical fields.

Author Contributions: S.M.C., L.P., F.B., and S.R. initiated and designed all of the experiments. S.M.C., I.M.V., M.A., F.L., L.C., C.T., A.S. and M.T.C. performed the experiments. S.M.C., L.P., F.B. and S.R. analyzed the data, S.M.C and L.P. wrote the manuscript.

Abbreviations

ABTS	2,2′-Azino-bis(3-ethylbenzthiazoline-6-sulfonic acid)
bw	Body weight
CFU	Colony Forming Unit
CS	Chitosan
CS-XTDs	Chitosan microparticles loaded with xanthine derivatives with thiazolidine-4-one scaffold
DL	Drug loading
DMSO	Dimethyl sulfoxide
DPPH	2,2-diphenyl-1-picryl-hydrazyl-hydrate
DR	Drug release
EE	Entrapment efficiency
FT-IR	Fourier transform infrared
LD_{50}	Lethal dose 50
MBC	Minimum bactericidal concentration
MFC	Minimum fungicidal concentration
MIC	Minimum inhibitory concentration
SD	Swelling degree
SEM	Scanning electron microscopy
SGF	Simulated gastric fluid
SIF	Simulated intestinal fluid
TPP	Pentasodium tripolyphosphate
XTDs	Xanthine derivatives with thiazolidine-4-one scaffold

References

1. Martinez-Lopez, S.; Sarria, B.; Gomez-Juaristi, M.; Goya, L.; Mateos, R.; Bravo-Clemente, L. Theobromine, caffeine, and theophylline metabolites in human plasma and urine after consumption of soluble cocoa products with different methylxanthine contents. *Food Res. Int.* **2014**, *63*, 446–455. [CrossRef]
2. Singh, N.; Shreshtha, A.K.; Thakur, M.S.; Patra, S. Xanthine scaffold: Scope and potential in drug development. *Heliyon* **2018**, *4*, 00829. [CrossRef] [PubMed]
3. Monteiro, J.P.; Alves, M.G.; Oliveira, P.F.; Silva, B.M. Structure-Bioactivity Relationships of Methylxanthines: Trying to Make Sense of All the Promises and the Drawbacks. *Molecules* **2016**, *21*, 974. [CrossRef] [PubMed]
4. Cazzola, M.; Calzetta, L.; Barnes, P.J.; Criner, G.J.; Martinez, F.J.; Papi, A.; Matera, M.G. Efficacy and safety profile of xanthines in COPD: A network meta-analysis. *Eur. Respir. Rev.* **2018**, *27*, 180010. [CrossRef] [PubMed]
5. Lupascu, F.G.; Dash, M.; Samal, S.K.; Dubruel, P.; Lupusoru, C.E.; Lupusoru, R.V.; Dragostin, O.; Profire, L. Development, optimization and biological evaluation of chitosan scaffold formulations of new xanthine derivatives for treatment of type-2 diabetes mellitus. *Eur. J. Pharm. Sci.* **2015**, *77*, 122–134. [CrossRef] [PubMed]
6. Ma, Q.S.; Yao, Y.; Zheng, Y.C.; Feng, S.; Chang, J.; Yu, B.; Liu, H.M. Ligand-based design, synthesis and biological evaluation of xanthine derivatives as LSD1/KDM1A inhibitors. *Eur. J. Med. Chem.* **2019**, *162*, 555–567. [CrossRef]
7. Ruddarraju, R.R.; Murugulla, A.C.; Kotla, R.; Chandra Babu Tirumalasetty, M.; Wudayagiri, R.; Donthabakthuni, S.; Maroju, R.; Baburao, K.; Parasa, L.S. Design, synthesis, anticancer, antimicrobial activities and molecular docking studies of theophylline containing acetylenes and theophylline containing 1,2,3-triazoles with variant nucleoside derivatives. *Eur. J. Med. Chem.* **2016**, *123*, 379–396. [CrossRef]
8. Wang, X.; Han, C.; Xu, Y.; Wu, K.; Chen, S.; Hu, M.; Wang, L.; Ye, Y.; Ye, F. Synthesis and Evaluation of Phenylxanthine Derivatives as Potential Dual A2AR Antagonists/MAO-B Inhibitors for Parkinson's Disease. *Molecules* **2017**, *22*, 1010. [CrossRef]
9. Borde, R.M.; Gaikwad, M.A.; Waghmare, R.A.; Munde, A.S. Design, Synthesis and In-Vitro Anti-inflammatory, Antimicrobial Activities of Some Novel 2,3-Disubstituted-1,3-Thiazolidin-4-One Derivatives Containing Thiazole Moiety. *J. Ultra Chem.* **2018**, *14*, 104–114. [CrossRef]

10. Da Silva, I.M.; da Silva Filho, J.; Santiago, P.B.; do Egito, M.S.; de Souza, C.A.; Gouveia, F.L.; Ximenes, R.M.; de Sena, K.X.; de Faria, A.R.; Brondani, D.J.; et al. Synthesis and antimicrobial activities of 5-Arylidene-thiazolidine-2,4-dione derivatives. *Biomed. Res. Int.* **2014**, *2014*, 316082. [CrossRef]
11. Samadhiya, P.; Sharma, R.; Srivastava, S.K.; Srivastava, S.D. Synthesis and biological evaluation of 4-thiazolidinone derivatives as antitubercular and antimicrobial agents. *Arab. J. Chem.* **2014**, *7*, 657–665. [CrossRef]
12. Ramachandran, S.; Bharathi, B.; Lavanya, R.; Nandhini, R.; Sivaranjani, R.; Sundhararajan, R. Synthesis, characterisation, antimicrobial evaluation of 2-hydroxy phenyl thiazolidine-4-one derivative. *Int. J. Pharm. Pharm. Sci.* **2017**, *6*, 278–283. [CrossRef]
13. Solankee, A. Synthesis, Characterization and in vitro antimicrobial activity of some new 2,3-disubstituted-4-thiazolidinone and 2,3-disubstituted-5-methyl-4-thiazolidinone derivatives as a biologically active scaffold. *Int. J. Pharm. Pharm. Sci.* **2016**, *6*, 386–399.
14. Ray, M.; Pal, K.; Anis, A.; Banthia, A.K. Development and Characterization of Chitosan-Based Polymeric Hydrogel Membranes. *Des. Monomers Polym.* **2010**, *13*, 193–206. [CrossRef]
15. Ferreira Tomaz, A.; Sobral de Carvalho, S.M.; Cardoso Barbosa, R.; L. Silva, S.M.; Sabino Gutierrez, M.A.; B. de Lima, A.G.; L. Fook, M.V. Ionically Crosslinked Chitosan Membranes Used as Drug Carriers for Cancer Therapy Application. *Materials* **2018**, *11*, 2051. [CrossRef] [PubMed]
16. Pillai, C.K.S.; Paul, W.; Sharma, C.P. Chitin and chitosan polymers: Chemistry, solubility and fiber formation. *Prog. Polym. Sci.* **2009**, *34*, 641–678. [CrossRef]
17. Xie, J.; Li, A.; Li, J. Advances in pH-Sensitive Polymers for Smart Insulin Delivery. *Macromol. Rapid. Commun.* **2017**, *38*, 1700413. [CrossRef]
18. Lizardi-Mendoza, J.; Argüelles-Monal, W.M.; Goycoolea, F.M. Chemical Characteristics and Functional Properties of Chitosan. In *Chitosan in the Preservation of Agricultural Commodities*; Bautista-Baños, S., Romanazzi, G., Jiménez-Aparicio, A., Eds.; Academic Press: Cambridge, MA, USA, 2016; pp. 3–31.
19. O'Callaghan, K.A.M.; Kerry, J.P. Preparation of low- and medium-molecular weight chitosan nanoparticles and their antimicrobial evaluation against a panel of microorganisms, including cheese-derived cultures. *Food Control* **2016**, *69*, 256–261. [CrossRef]
20. Xing, R.; He, X.; Liu, S.; Yu, H.; Qin, Y.; Chen, X.; Li, K.; Li, R.; Li, P. Antidiabetic Activity of Differently Regioselective Chitosan Sulfates in Alloxan-Induced Diabetic Rats. *Mar. Drugs* **2015**, *13*, 3072–3090. [CrossRef]
21. Barakat, N.S.; Almurshedi, A.S. Design and development of gliclazide-loaded chitosan microparticles for oral sustained drug delivery: in-vitro/in-vivo evaluation. *J. Pharm. Pharmacol.* **2011**, *63*, 169–178. [CrossRef]
22. Perumal, V.; Manickam, T.; Bang, K.S.; Velmurugan, P.; Oh, B.T. Antidiabetic potential of bioactive molecules coated chitosan nanoparticles in experimental rats. *Int. J. Biol. Macromol.* **2016**, *92*, 63–69. [CrossRef] [PubMed]
23. Zhao, L.; Zhang, S.; An, F.; Ma, S.; Cheng, Y.; Liu, W.; Xu, F.; Li, M. Water soluble chitosans shows anti-cancer effect in mouse H22 liver cancer by enhancing the immune response. *Int. J. Clin. Exp. Med.* **2016**, *9*, 164–171.
24. Kim, S. Competitive Biological Activities of Chitosan and Its Derivatives: Antimicrobial, Antioxidant, Anticancer, and Anti-Inflammatory Activities. *Int. J. Polym. Sci.* **2018**, *2018*, 1708172. [CrossRef]
25. Ghosh, S.; Bhattacharyya, S.; Rashid, K.; Sil, P.C. Curcumin protects rat liver from streptozotocin-induced diabetic pathophysiology by counteracting reactive oxygen species and inhibiting the activation of p53 and MAPKs mediated stress response pathways. *Toxicol. Rep.* **2015**, *2*, 365–376. [CrossRef] [PubMed]
26. Kalam, M.A. Development of chitosan nanoparticles coated with hyaluronic acid for topical ocular delivery of dexamethasone. *Int. J. Biol. Macromol.* **2016**, *89*, 127–136. [CrossRef] [PubMed]
27. Panith, N.; Wichaphon, J.; Lertsiri, S.; Niamsiri, N. Effect of physical and physicochemical characteristics of chitosan on fat-binding capacities under in vitro gastrointestinal conditions. *LWT Food Sci. Technol.* **2016**, *71*, 25–32. [CrossRef]
28. Li, J.; Xie, B.; Xia, K.; Li, Y.; Han, J.; Zhao, C. Enhanced Antibacterial Activity of Silver Doped Titanium Dioxide-Chitosan Composites under Visible Light. *Materials* **2018**, *11*, 1403. [CrossRef]
29. Benito-Pena, E.; González-Vallejo, V.; Rico-Yuste, A.; Barbosa-Pereira, L.; Cruz, J.M.; Bilbao, A.; Alvarez-Lorenzo, C.; Moreno-Bondi, M.C. Molecularly imprinted hydrogels as functional active packaging materials. *Food Chem.* **2016**, *190*, 487–494. [CrossRef]

30. Sung, S.Y.; Sin, L.T.; Tee, T.T.; Bee, S.T.; Rahmat, A.R.; Rahman, W.A.W.A.; Tan, A.C.; Vikhraman, M. Antimicrobial agents for food packaging applications. *Trends Food Sci. Technol.* **2013**, *33*, 110–123. [CrossRef]
31. Limbo, S.; Mousavi Khaneghah, A. Active packaging of foods and its combination with electron beam processing. In *Electron Beam Pasteurization and Complementary Food Processing Technologies*; Pillai, S., Ed.; Woodhead Publishing: Cambridge, UK, 2015; pp. 195–217.
32. Mousavi Khaneghah, A.; Hashemi, S.M.B.; Limbo, S. Antimicrobial agents and packaging systems in antimicrobial active food packaging: An overview of approaches and interactions. *Food Bioprod. Process.* **2018**, *111*, 1–19. [CrossRef]
33. Diblan, S.; Kaya, S. Antimicrobials used in active packaging films. *Food Health* **2018**, *4*, 63–79. [CrossRef]
34. Malhotra, B.; Keshwani, A.; Kharkwal, H. Antimicrobial food packaging: Potential and pitfalls. *Front. Microbiol.* **2015**, *6*, 611. [CrossRef] [PubMed]
35. Tripathi, S.; Mehrotra, G.K.; Dutta, P.K. Chitosan based antimicrobial films for food packaging applications. *E-Polymers* **2008**, *8*, 093. [CrossRef]
36. Mousavi Khaneghah, A.; Hashemi, S.M.B.; Eş, I.; Fracassetti, D.; Limbo, S. Efficacy of Antimicrobial Agents for Food Contact Applications: Biological Activity, Incorporation into Packaging, and Assessment Methods: A Review. *J. Food Prot.* **2018**, *81*, 1142–1156. [CrossRef] [PubMed]
37. Xu, J.D.; Niu, Y.S.; Yue, P.P.; Hu, Y.; Bian, J.; Li, M.F.; Peng, F.; Sun, R.C. Composite Film Based on Pulping Industry Waste and Chitosan for Food Packaging. *Materials* **2018**, *11*, 2264. [CrossRef] [PubMed]
38. Constantin, S.; Lupascu, F.G.; Apotrosoaei, M.; Vasincu, I.M.; Lupascu, D.; Buron, F.; Routier, S.; Profire, L. Synthesis and biological evaluation of the new 1,3-dimethylxanthine derivatives with thiazolidine-4-one scaffold. *Chem. Cent. J.* **2017**, *11*, 12–25. [CrossRef]
39. Li, L.; Li, J.; Si, S.; Wang, L.; Shi, C.; Sun, Y.; Liang, Z.; Mao, S. Effect of formulation variables on *in vitro* release of a water-soluble drug from chitosan-sodium alginate matrix tablets. *Asian J. Pharm. Sci.* **2014**, *10*, 314–321. [CrossRef]
40. Venkatesan, B.; Tumala, A.; Subramanian, V.; Vellaichamy, E. Data on synthesis and characterization of chitosan nanoparticles for in vivo delivery of siRNA-Npr3: Targeting NPR-C expression in the heart. *Data Brief* **2016**, *8*, 441–447. [CrossRef]
41. Papadimitriou, S.; Bikiaris, D. Novel self-assembled core-shell nanoparticles based on crystalline amorphous moieties of aliphatic copolyesters for efficient controlled drug release. *J. Control Release* **2009**, *138*, 177–184. [CrossRef]
42. Anirudhan, T.S.; Divya, P.L.; Nima, J. Synthesis and characterization of novel drug delivery system using modified chitosan based hydrogel grafted with cyclodextrin. *Chem. Eng. J.* **2016**, *284*, 1259–1269. [CrossRef]
43. Dash, M.; Chiellinia, F.; Ottenbriteb, R.M.; Chiellinia, E. Chitosan—A versatile semi-synthetic polymer in biomedical applications. *Prog. Polym. Sci.* **2011**, *36*, 981–1014. [CrossRef]
44. Oprea, A.M.; Nistor, M.T.; Popa, M.I.; Lupusoru, C.E.; Vasile, C. In vitro and in vivo theophylline release from cellulose/chondroitin sulfate hydrogels. *Carbohydr. Polym.* **2012**, *90*, 127–133. [CrossRef] [PubMed]
45. Deora, P.S.; Mishra, C.K.; Mavani, P.; Asha, R.; Shrivastava, B.; Rajesh, K.N. Effective alternative methods of LD50 help to save number of experimental animals. *J. Chem. Pharm. Res.* **2010**, *2*, 450–453.
46. Yadav, R.; Bansal, R.; Rohilla, S.; Kachler, S.; Klotz, K.N. Synthesis and pharmacological characterization of novel xanthine carboxylatae amides as A2A adenosine receptor ligands exhibiting bronchospasmolytic activity. *Bioorg. Chem.* **2016**, *65*, 26–37. [CrossRef] [PubMed]
47. Koeth, L.M.; DiFranco-Fisher, J.M.; Scangarella-Oman, N.E.; Miller, L.A. Analysis of MIC and disk diffusion testing variables for gepotidacin and comparator agents against select bacterial pathogens. *J. Clin. Microbiol.* **2017**, *55*, 1767–1777. [CrossRef] [PubMed]
48. Ahmadi, F.; Oveisi, Z.; Samani, S.M.; Amoozgar, Z. Chitosan based hydrogels: Characteristics and pharmaceutical applications. *Res. Pharm. Sci.* **2015**, *10*, 1–16. [PubMed]
49. Bhattarai, N.; Gunn, J.; Zhang, M. Chitosan-based hydrogels for controlled, localized drug delivery. *Adv. Drug Deliv. Rev.* **2010**, *62*, 83–99. [CrossRef] [PubMed]
50. Berger, J.; Reist, M.; Mayer, J.M.; Felt, O.; Peppas, N.A.; Gurny, R. Structure andinteractions in covalently and ionically crosslinked chitosan hydrogels forbiomedical applications. *Eur. J. Pharm. Biopharm.* **2004**, *57*, 19–34. [CrossRef]
51. Grenha, A. Chitosan nanoparticles: A survey of preparation methods. *J. Drug Target.* **2012**, *20*, 291–300. [CrossRef]

52. Onishi, H. Chitosan microparticles. *J. Drug Deliv. Sci. Tech.* **2010**, *20*, 15–22. [CrossRef]
53. Ko, J.A.; Park, H.J.; Hwang, S.J.; Park, J.B.; Lee, J.S. Preparation and characterization of chitosan microparticles intended for controlled drug delivery. *Int. J. Pharm.* **2002**, *249*, 165–174. [CrossRef]
54. Dinu-Pîrvu, C.; Simona Ivan, S. A Study ofthe Influence of Crosslinking Degree Onthe Physicochemical Properties of Gelatin Microparticles. *Cellulose Chem. Technol.* **2013**, *47*, 721–726.
55. Heydari, M.; Moheb, A.; Ghiaci, M.; Masoomi, M. Effect of Cross-Linking Time on the Thermal and Mechanical Properties and PervaporationPerformance of Poly(vinyl alcohol) Membrane Cross-Linked with Fumaric Acid Used for Dehydration of Isopropanol. *J. Appl. Polym. Sci.* **2013**, *128*, 1640–1651.
56. Priimagi, A.; Cavallo, G.; Metrangolo, P.; Resnati, G. The Halogen Bond in the Design of Functional Supramolecular Materials: Recent Advances. *Acc. Chem. Res.* **2013**, *46*, 2686–2695. [CrossRef] [PubMed]
57. Deshpande, S.S. *Handbook of Food Toxicology*; Marcel Dekker: New York, NY, USA, 2002; ISBN 0-8247-0760-5.

10

Characterization of Active Edible Films based on Citral Essential Oil, Alginate and Pectin

Valentina Siracusa [1,*], Santina Romani [2], Matteo Gigli [3,5], Cinzia Mannozzi [2], Juan Pablo Cecchini [4], Urszula Tylewicz [2] and Nadia Lotti [5]

1 Department of Chemical Science, University of Catania, Viale A. Doria 6, 95125 Catania (CT), Italy
2 Department of Agricultural and Food Sciences-DISTAL, Campus of Food Science, University of Bologna, P.zza Goidanich 60, 47521 Cesena, Italy; santina.romani2@unibo.it (S.R.); cinzia.mannozzi2@unibo.it (C.M.); urszula.tylewicz@unibo.it (U.T.)
3 Department of Chemical Science and Technologies, University of Rome Tor Vergata, Via della Ricerca Scientifica 1, 00133 Rome, Italy; matteo.gigli@unibo.it
4 Universidad Nacional de Rafaela, Bv. Roca 989, Rafaela, 3000 Santa Fe, Argentina; jpcecchini@gmail.com
5 Department of Civil, Chemical, Environmental and Materials Engineering, University of Bologna, via Terracini 28, 40131 Bologna, Italy; nadia.lotti@unibo.it
* Correspondence: vsiracus@dmfci.unict.it

Abstract: Thermal, structural and physico-chemical properties of different composite edible films based on alginate and pectin with the addition of citral essential oil (citral EO) as an agent to improve barrier properties, were investigated. The obtained films were clear and transparent, with a yellow hue that increased with citral EO addition. All the films displayed good thermal stability up to 160 °C, with a slight improvement observed by increasing the amount of citral EO in the composites. Gas transmission rate (GTR) strongly depended on the polymer structure, gas type and temperature, with improvement in barrier performance for composite samples. Also, citral EO did not exert any weakening action on the tensile behavior. On the contrary, an increase of the elastic modulus and of the tensile strength was observed. Lastly, water contact angle measurements demonstrated the dependence of the film wettability on the content of citral EO.

Keywords: edible film; alginate film; pectin film; essential oil; barrier properties; mechanical properties

---

1. Introduction

Edible coatings and films belong to an environmental-friendly technology aimed at enhancing food safety, quality and handling properties by creating a biodegradable semi-permeable barrier protection to water vapor, oxygen and carbon dioxide coming from the external environment. Edible coating is a thin edible substance layer, applied in a liquid form directly on the surface of different food products, mainly fruits and vegetables [1–5]. Edible films are a thin layer obtained as a solid sheet, which can be applied as a wrapping on the food product. For fresh fruit and vegetables, the creation of a wrong moisture and gas atmosphere may lead to weight loss and respiration rate reductions, with a consequent acceleration of the senescence process and a worsening of the visual gloss of coated commodities. Several studies have identified the necessity to evaluate mechanical (flexibility, tension), thermal, optical (brightness, opacity), wettability and morphological properties of edible films, as it creates a modified atmosphere that influences the gas transfer and further becomes a barrier for aromatic compound transferring [6,7]. The above-mentioned characteristics depend on several parameters related to the coating and film composition, such as preparation conditions (solvent, pH, components concentration, temperature) and type of added additives (cross-linking agents, antimicrobials, plasticizers, emulsifiers).

Recently, further potential applications were studied in order to formulate edible coatings and films with special functionalities [8]. They could be used as carriers of antioxidants, flavoring and/or coloring agents and antimicrobials to improve food safety and quality [9,10].

Edible coating and film formulations require several components. At least one of them needs to have structural properties. Between the different compounds that could be used in the formulation, alginate and pectin have been widely investigated [1]. They are hydrocolloid compounds coming from respectively seaweed extracts (brown algae) and plant tissue. Both belong to the polyuranoates group and are natural ionic polysaccharides giving rise to a chain–chain association and forming hydrogels. In order to design new films with improved and selective mechanical and barrier properties, it is important to understand their compatibility [1,11,12]. Further, essential oils represent a very interesting ingredient for edible film formulation with the purpose of realizing active and natural packaging materials [13,14]. Due to their natural origin and special functionality (antioxidant/antimicrobial) they could be selected to extend food shelf-life and add value to the product. However, the introduction of edible films for food packaging application may affect several properties such as optical, tensile, permeability and so on [15]. This, in turn, could influence the consumer acceptability. Citral EO was selected since it has been proved to have a positive effect against food spoilage microorganisms, as described by Guerreiro et al. [16].

The objective of this study was to investigate the final properties of edible films based on sodium alginate (SA) and pectin (Pe), with the addition of citral essential oil (citral EO).

The optimization of an edible film composition is one of the most important steps [17,18], as it must be formulated in order to maintain or increase the quality of fruit and vegetables. In this framework, physico-chemical and structural analyses have been performed on the prepared formulations to evaluate the properties of interest for the in-use application.

2. Materials and Methods

2.1. Materials

Citral essential oil, sodium alginate, glycerol (≥99.5%), Tween®20 and Pectin from citrus peel (Galacturonic acid ≥74.0%) were purchased all from Sigma-Aldrich (Darmstadt, Germany, Europe) and used as received.

2.2. Film Preparation

Edible films were prepared using the solution casting method. Three different compositions were tested: 2% (w/w) of SA, 2% (w/w) of Pe and a combination of 1% (w/w) sodium alginate and 1% (w/w) pectin (SA + Pe). 1.5% (w/w) of glycerol and 0.2% (w/w) of Tween® 20 were added to each solution, which was subsequently dissolved in distilled water. Moreover, different amounts (0, 0.15, 0.3% w/w) of citral essential oil (citral EO) as antimicrobial agent were introduced in the formulation and homogeneously dispersed. In total, nine different films were obtained, as reported in Table 1.

Table 1. Concentration (w/w %) of the components in the film forming solutions.

Sample	Sodium Alginate	Pectin	Glycerol	Tween® 20	Citral EO
SA	2	-	1.5	0.2	-
$SA_{0.15}$	2	-	1.5	0.2	0.15
$SA_{0.3}$	2	-	1.5	0.2	0.3
Pe	-	2	1.5	0.2	-
$Pe_{0.15}$	-	2	1.5	0.2	0.15
$Pe_{0.3}$	-	2	1.5	0.2	0.3
SA + Pe	1	1	1.5	0.2	-
SA + $Pe_{0.15}$	1	1	1.5	0.2	0.15
SA + $Pe_{0.3}$	1	1	1.5	0.2	0.3

All solutions were heated at 70 °C for 30 min under conditions of stirring (until complete dissolution) and homogenized at 5000 rpm for 2 min. Afterwards, about 27 g of solution was used to obtain 1 g of soluble solids in petri dishes (ø 90 mm). Air bubbles were removed by placing the film forming solution under vacuum (60 mbar, 15 min). Films were dried at 40 °C for 5 h to ensure a complete solvent removal, and then stored at room temperature in a desiccator before characterization.

2.3. Film Thickness and Gas Barrier Properties

Film thickness was measured by using a Sample Thickness Tester DM-G, consisting of a digital indicator (MarCator 1086 type, Mahr GmbH, Esslingen, Germany) connected to a PC. The reading was made twice per second measuring a minimum, a maximum and an average value, with a resolution of 0.001 mm. The reported results represent the mean thickness value of three experimental tests run in 10 different positions of each film at room temperature.

The determination of the gas barrier properties was performed by a manometric method using a Permeance Testing Device, type GDP-C (Brugger Feinmechanik GmbH, Munchen, Germany), in accordance with the ASTM 1434, DIN 53536, ISO 15105-1 and with the Gas Permeability Testing Manual-2008. The equipment consisted of an upper and a lower chamber, as previously described [19]. The film sample, of approximately 3 cm × 3 cm in size, was placed between the two chambers. A film mask was used to cover the remaining section of the permeation chamber. The top chamber was filled with the dry test gases at ambient pressure. The gas permeation was determined by evaluating the pressure increase in the bottom chamber, which had been previously evacuated. Ambient temperature fluctuations during the test were controlled by an automatic temperature compensation software, which minimizes gas transmission rate (GTR) temperature deviations. The following conditions were adopted: gas stream of 100 cm^3/min, pure food grade gases with 0% RH, active sample area of 0.785 cm^2. *Method A* (with evacuation of top and bottom permeation chamber) was employed in the analysis, as reported in the Gas Permeability Testing Manual (2008).

In order to determine the activation energies of the permeation process, the films were tested at three different temperatures (8, 15 and 23 °C), in the same operating conditions as reported above. Chamber and sample temperature were kept constant by an external thermostat HAAKE-Circulator DC10-K15 type (Thermo Fischer Scientific, Waltham, MA, USA). All experiments were done in triplicate and the results are presented as mean values.

2.4. Mechanical Properties

Tensile testing was performed by using a Zwick Roell Texture machine mod. Z2.5 (Ulm, Germany), equipped with a rubber grip and a 500 N load cell. A pre-load of 1MPa with a pre-load speed of 5 mm/min was applied. Stress-strain measurements were performed on rectangular films (5 mm width and 50 mm length) with an initial grip separation of 23 mm and a crosshead speed of 5 mm/min. Five different specimens was analyzed for each film and the results are provided as the average value ± standard deviation. All measurements have been carried out in accordance with ASTM D882-09.

2.5. Thermo-Gravimetric Analysis (TGA)

Thermo-Gravimetric Analysis (TGA) was carried out under nitrogen atmosphere by means of a Perkin Elmer TGA7 apparatus (Perkin Elmer, Waltham, MA, USA). Gas flow of 30 mL/min and heating scan of 10 °C/min, over a temperature range 40–800 °C, were used for the analyses. Samples mass of 10 mg were used for the experiments.

2.6. Attenuated Total Reflectance Infrared Spectroscopy (ATR-IR) Analysis

Experimental ATR modified absorbance spectra were collected on a Perkin Elmer FTIR (Perkin Elmer 1725× Spectrophotometer, Labexchange Gropu, Burlandingen, Germany) over the range 650–4000 cm^{-1}, with a resolution of 4.0 cm^{-1}. The results are presented as an average of 10 experimental tests, run on 10 different samples point. 64 scans were recorded on each sample. The experiments were performed at room temperature, without any preliminary samples' treatments.

2.7. *Water Contact Angle Determination (WCA)*

Static contact angle measurements were performed on edible films using a KSV CAM101 (KSV Instrument, Elsinki, Finland) instrument at ambient conditions, by recording the side profiles of deionized water drops (4 μL) deposited on the film surface. The measurement was taken 2 s after the drop deposition to ensure its stabilization, yet to minimize water absorption and evaporation. At least five drops were observed on different areas for each film (both on the top and on the bottom side), and contact angles were reported as the average value ± standard deviation.

2.8. *Color*

The color of film samples was measured using a HunterLab ColorFlex EZ 45/0° color spectro-photometer (Hunterlab, Reston, VA, USA), with D65 illuminant, 10° observer, according to ASTM E308. The measurements were made using CIE Lab scale. The instrument was calibrated with a black and white tile (L* 93.47, a* 0.83, b* 1.33) before the measurements. Results are expressed as L* (luminosity), a* (red/green) and b*(yellow/blue) parameters.

The total color difference (ΔE) was calculated using the following equation:

$$\Delta E = [(L^* - L')^2 + (a^* - a')^2 + (b^* - b')^2]^{1/2} \quad (1)$$

where L*, a* and b* are the color parameter values of the sample and L′, a′ and b′ are the color parameter values of the standard white plate used as the film background (L′ = 66.39, a′ = −0.74, b′ = 1.25). The analyses were conducted in five repetitions. A mean value recorded for the top side and bottom side is reported.

2.9. *Morphology Evaluation*

The films' morphology was determined by using a Nikon upright microscope (Eclipse Ti-U, Nikon Co, Shanghai, Japan) with a standard light. Samples were observed in black and white and the images were recorded at a 20× magnification.

2.10. *Statistical Analysis*

Significance of the different edible films effects was evaluated by means of one-way analysis of variance (ANOVA, 95% significance level) using the software STATISTICA 6.0 (Statsoft Inc., Tulsa, OK, USA).

3. Results and Discussions

In general, all films appeared homogeneous and transparent, without evident pores and crakes. It can be noticed that SA-based films were visually more transparent than Pe-based ones. Similar observations were made by Galus and Lenart [11] in their study on edible films based on sodium alginate and pectin. Visually, the addition of citral essential oil added a bright yellow color to the prepared films.

3.1. *Thickness and Gas Barrier Properties*

3.1.1. Thickness

Films thickness ranged from 94 to 127 μm as reported in Table 2, showing a slight variation of thickness within the different formulations, associated to the film casting technique. This is a crucial parameter for the calculation of mechanical and barrier properties. It depends on the film preparation method, on the flatness of the dish surface and on the film formation during the drying process. Despite the same volume of film forming solutions were used for the film casting, the drying time for complete solvent evaporation increased with the citral EO concentration (anyway remaining in all cases within 5 h). The homogeneity of the films increased with the solvent evaporation time. Consequently, as reported in literature, the thickness tent to be lower in a well-organized and dense network [20]. Differences in the film structure may be attributed to the effect on the drying kinetics of the liquid film-forming dispersion thickness [21]. Several authors reported the effects of thickness on the permeability of edible matrix [22,23].

The thickness values have been used to calculate gas permeability and tensile properties.

3.1.2. Gas Barrier Properties

As it is well known, oxygen might cause food oxidation, which in turn influences various food properties such as odour, colour, flavour and nutrient content [24,25]. The film ability to retard oxidation or degradation of the product is an important characteristic that affects the final product quality and food shelf-life. Therefore, gas barrier properties should be taken into due account. Furthermore, the incorporation of other compounds, such as essential oils, into the polymer matrix could contribute to the modification of their barrier behavior.

Gas transmission rate (GTR) strongly depends on the polymer structure, gas type and temperature. Edible films barrier properties were examined at three different temperatures [26,27]: 8 °C (home and supermarket refrigerated storage temperature), 15 °C (abusing temperature) and 23 °C (room temperature). As reported by Marklinder and Eriksson [27], a typical recommended storage temperature for perishable food items should not exceed 8 °C, with the exception for minced meat that should be stored at 4 °C maximum. Further, in almost all the refrigerators the temperature ranges between 5–8 °C, but it is not uncommon that in certain zones of the refrigerators the temperature increases to values higher than 12 °C.

Data recorded with 100% pure food grade O_2 and CO_2 are reported in Table 2.

The matrix composition influenced the diffusion of the gas molecules through the polymer, resulting in a substantial variation of the gas transmission. Citral EO caused a modification of the barrier performances, correlated to the compatibility between alginate and pectin, ultimately resulting in effective gas molecule permeation through the films. The impact of lipid addition on the microstructure of the coating films is a determining factor in barrier efficiency. The microstructure of the films is affected by the physical state of the essential oil and its distribution in the polymer matrix. The liquid state of citral EO could favor molecular mobility, promoting the transport of gas molecules through the films. In turn, its distribution is another important factor. The more homogeneous is the distribution, the lower is the GTR. Antarés and Chiralt [28] reported that the addition of essential oil causes an increase of the oxygen permeability due to its hydrophobic character. However, a different trend can be described for the materials object of the present study. Indeed, a GTR decrease, particularly evident at lower temperatures, has been observed. The data here presented are well in accordance with the results reported by Rojas-Graü and coworkers [8]. They observed that the introduction of antimicrobials agents (oregano, carvacrol and lemongrass oil) did not affect the oxygen permeability of the films, while a slight decrease was observed by adding citral EO to alginate-based edible films.

It must be considered that the film barrier behavior at 8 °C is the most crucial for food packaging applications, as it corresponds to home and commercial (supermarket) storage conditions. At this temperature the gas transmission rate of O_2 and CO_2 is very similar: CO_2 permeates maximum 1.4 times faster than O_2, as it can be seen from the perm-selectivity ratio reported in Table 2.

Table 2. Film thickness, O_2 and CO_2 gas transmission rate (GTR) at 8, 15 and 23 °C, with the corresponding perm-selectivity ratio and GTR Activation Energy.

Sample	Thickness (μm)	O_2-GTR (cm^3/m^2 d bar)			CO_2-GTR (cm^3/m^2 d bar)			CO_2/O_2	CO_2/O_2	CO_2/O_2	CO_2 E_{GTR} (J/mol K)	O_2 E_{GTR} (J/mol K)
		8 °C	15 °C	23 °C	8 °C	15 °C	23 °C	8 °C	15 °C	23 °C		
SA	127 ± 13^{a}	102 ± 1^{a}	118 ± 1^{b}	148 ± 2^{c}	110 ± 0^{c}	111 ± 1^{d}	222 ± 1^{a}	1.1	0.9	1.5	33 (0.8)	17 (1)
$SA_{0.15}$	111 ± 6^{b}	94 ± 1^{c}	112 ± 2^{de}	134 ± 1^{d}	99 ± 2^{d}	96 ± 1^{f}	105 ± 1^{i}	1.1	0.7	0.8	3 (0.4)	16 (1)
$SA_{0.30}$	110 ± 4^{b}	70 ± 1^{f}	92 ± 1^{f}	110 ± 0^{g}	99 ± 1^{d}	129 ± 1^{b}	124 ± 2^{h}	1.4	1.4	1.1	10 (0.6)	21 (1)
Pe	110 ± 5^{b}	102 ± 1^{a}	123 ± 1^{a}	159 ± 1^{b}	134 ± 2^{a}	144 ± 2^{a}	195 ± 1^{b}	1.3	1.2	1.2	−16 (0.6)	20 (1)
$Pe_{0.15}$	107 ± 1^{bc}	99 ± 2^{ab}	109 ± 1^{e}	119 ± 1^{f}	97 ± 1^{d}	101 ± 1^{e}	176 ± 1^{d}	1.0	0.9	1.6	28 (0.8)	8 (1)
$Pe_{0.30}$	94 ± 9^{c}	84 ± 2^{d}	117 ± 1^{bc}	127 ± 0^{e}	66 ± 2^{g}	110 ± 1^{d}	180 ± 1^{c}	0.8	0.9	1.4	46 (1)	19 (1)
SA + Pe	101 ± 3^{bc}	98 ± 1^{b}	117 ± 1^{bc}	166 ± 1^{a}	131 ± 1^{b}	117 ± 1^{c}	173 ± 1^{e}	1.3	1.0	1.0	13 (0.5)	24 (1)
SA + $Pe_{0.15}$	104 ± 1^{bc}	75 ± 1^{e}	114 ± 4^{cd}	136 ± 1^{d}	94 ± 1^{e}	116 ± 2^{c}	165 ± 1^{f}	1.3	1.0	1.2	26 (1)	27 (0.9)
SA + $Pe_{0.30}$	104 ± 5^{bc}	84 ± 1^{d}	94 ± 3^{f}	117 ± 1^{e}	92 ± 1^{f}	116 ± 1^{c}	150 ± 1^{g}	1.1	1.2	1.3	23 (1)	44 (0.9)

Values with different letters within the same column differ significantly at $p < 0.05$ levels. Values in brackets indicate the R^2 parameter of the fitting curve. Samples with *a* present the highest value, while samples with *d* the lowest.

As reported in the literature [29], for non-condensable gases the perm-selectivity ratio is relatively constant and is not correlated to the polymer chemical structure. Moreover, CO_2 transmission across the matrix is in general about six times faster than O_2 ones.

Furthermore, the perm-selectivity ratio changes with the temperature, i.e., it usually increases as the temperature decreases. On the contrary, our study evidenced a different trend, as the permselectivity remained almost constant in all cases.

The Arrhenius model was used to describe the permeation dependence on the temperature and to calculate the activation energy of the gas transmission (E_{GTR}) process. In Figure 1, natural logarithm (ln) of GTR has been plotted as a function of the reciprocal of the absolute temperature (1/T).

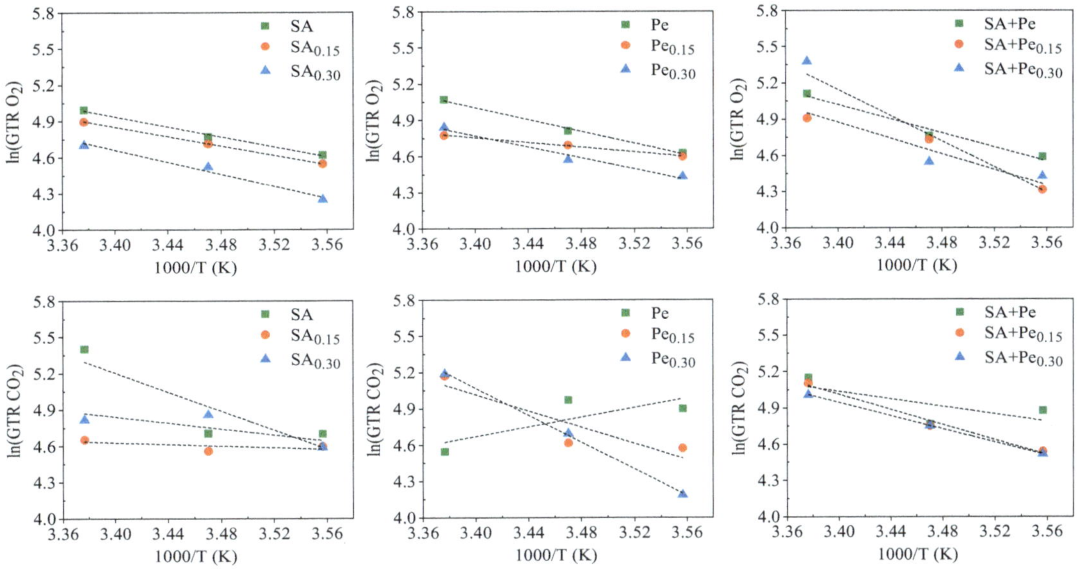

Figure 1. Arrhenius plot of O_2 and CO_2 GTR coefficient for SA, Pe and SA + Pe films.

Data reported in Table 2 well fit the theoretical relation (high R^2 value) in most cases, indicating a good correlation between permeability and temperature for both CO_2 and O_2. As expected, the higher the temperature, the higher the permeability. In addition, E_{GTR} reflected the dependence of the barrier behavior from the temperature for each polymer matrix. The low activation energy demonstrated a low dependence of the gas transmission process on the temperature.

In conclusion, the GTR data here reported confirm the antioxidant action of citral EO, which is due to the promotion of oxygen barrier capacity, as already reported by Rojas-Graü et al. and Shahbazi [8–30]. The samples containing citral EO showed an interesting improvement of the gas barrier behavior, making them suitable as edible films for food more susceptible to oxidative degradation. As can be evicted from the data reported, the GTR value decreased when citral EO was added. The barrier properties improvement is responsible of the feasibility of formed films to meet the desired functionality for the product to be packed, as was reported also by Avila-Sosa et al. [31] in their study of chitosan and oregano essential oil, as well as their combination, for antimicrobial food packaging application.

3.2. Mechanical Properties

The tensile behavior of the films under study is reported in Table 3. Films were analyzed in terms of their elastic modulus (E), stress at yield (σ_y) and at break (σ_b), and elongation at yield (ε_y) and at break (ε_b). Generally speaking, the tensile properties of edible films and coatings are dependent on film constituents, relative proportion and preparation conditions [28].

Neat SA and Pe films showed very comparable tensile properties, although the elongation at break of SA was higher, while SA + Pe films displayed the lowest elastic modulus and stress at break and the highest elongation at break among the studied samples. The addition of citral EO caused a significant change of the mechanical properties of all the formulations. A general increase of E, σ_y and σ_b can be observed, while no significant differences can be noticed for ε_b (Table 3). Also, the introduction of different amounts of citral EO did not influence the tensile behavior of the samples. In all the investigated samples, the yield phenomenon is visible.

Table 3. Mechanical data of SA, Pe and SA + Pe films.

Sample	E (MPa)	σ_y (MPa)	ε_y (%)	σ_b (MPa)	ε_b (%)
SA	526 ± 52 [c]	14 ± 1 [d]	4 ± 1 [b]	25 ± 3 [cd]	30 ± 1 [b]
$SA_{0.15}$	1102 ± 68 [ab]	24 ± 2 [b]	3 ± 1[b]	44 ± 7 [ab]	24 ± 5 [cd]
$SA_{0.30}$	955 ± 74 [b]	17 ± 5 [c]	5 ± 2 [b]	31 ± 11 [c]	30 ± 3 [b]
Pe	492 ± 51 [c]	13 ± 1 [de]	4 ± 1 [b]	21 ± 2 [cd]	18 ± 2 [e]
$Pe_{0.15}$	1120 ± 112 [ab]	21 ± 2 [bc]	3 ± 1 [b]	42 ± 5 [b]	20 ± 1 [de]
$Pe_{0.30}$	1338 ± 115 [a]	32 ± 5 [a]	4 ± 1 [b]	54 ± 7 [a]	20 ± 2 [de]
Pe + SA	173 ± 17 [d]	11 ± 1 [e]	12 ± 1 [a]	18 ± 1 [d]	50 ± 1 [a]
Pe + $SA_{0.15}$	915 ± 109 [b]	19 ± 3 [c]	4 ± 1 [b]	31 ± 4 [c]	18 ± 2 [e]
PE + $SA_{0.30}$	1171 ± 98 [ab]	28 ± 3 [a]	4 ± 1 [b]	50 ± 6 [ab]	26 ± 2 [bc]

Values with different letters within the same column differ significantly at $p < 0.05$ levels. Samples with *a* present the highest value, while samples with *d* the lowest.

The effect of the essential oil addition on the mechanical properties of edible films is quite complex and conflicting results have been reported in the literature, since both enhancement and weakening of the tensile characteristics have been observed [28]. Specific interactions between the film constituents and the oil, such as crosslinks, different structural arrangements of the components or the formation of heterogeneous biphasic structures, must be taken into consideration [28].

Our study demonstrates that citral EO does not have any detrimental effect on the tensile properties of SA and Pe based formulations. On the contrary, an enhancement of the tensile strength and of the film rigidity has been obtained. The results indicate that not only the film microstructure was preserved, yet citral EO played a reinforcement role. Also, the introduction of a hydrophobic compound may have caused a reduction of the film water absorption, this latter acting as plasticizer and causing a decrease of E and σ_b in the neat films as compared to the composites.

In conclusion, the composite films, in particular those containing a higher amount of citral EO, displayed improved mechanical resistance with respect to the neat films, thus making them more suitable for food wrapping applications.

3.3. Thermogravimetric Analysis

Table 4 reports the thermogravimetric data of the materials under study, i.e., T_{onset} (the temperature at which the degradation starts), T_{max} (the temperature of the maximum degradation rate), and $m_{res,600\,°C}$ (the residual mass at 600 °C).

Table 4. Thermogravimetric and wettability data of Sa, Pe and SA + Pe films.

Sample	T_{onset} (°C)	T_{max} (°C)	$m_{res,600\,°C}$ (%)	WCA (°)
SA	163 ± 1 ^{e}	220 ± 1 ^{b}	20 ± 1 ^{de}	44 ± 3 ^{b}
$SA_{0.15}$	172 ± 1 ^{c}	221 ± 1 ^{ab}	23 ± 0 ^{b}	45 ± 3 ^{ab}
$SA_{0.30}$	178 ± 1 ^{ab}	221 ± 0 ^{ab}	24 ± 0 ^{a}	49 ± 4 ^{a}
Pe	171 ± 1 ^{c}	218 ± 1 ^{c}	17 ± 1 ^{f}	35 ± 4 ^{d}
$Pe_{0.15}$	176 ± 3 ^{b}	222 ± 2 ^{ab}	21 ± 1 ^{cd}	38 ± 4 ^{cd}
$Pe_{0.30}$	179 ± 2 ^{a}	222 ± 1 ^{a}	20 ± 1 ^{e}	42 ± 2 ^{bc}
SA + Pe	163 ± 1 ^{e}	210 ± 1 ^{d}	18 ± 1 ^{f}	43 ± 3 ^{bc}
SA + $Pe_{0.15}$	166 ± 2 ^{d}	207 ± 1 ^{e}	22 ± 0 ^{bc}	46 ± 2 ^{ab}
SA + $Pe_{0.30}$	173 ± 2 ^{c}	213 ± 2 ^{c}	20 ± 1 ^{de}	48 ± 1 ^{ab}

Values with different letters within the same column differ significantly at $p < 0.05$ levels. Samples with *a* present the highest value, while samples with *d* the lowest.

The degradation patterns of the different films are very similar, although a slight improvement of the stability can be seen by increasing the amount of citral EO in the composites (Table 4).

Figure 2 reports the thermogravimetric curves of $SA_{0.15}$, $Pe_{0.15}$ and SA + $Pe_{0.15}$, together with their derivative.

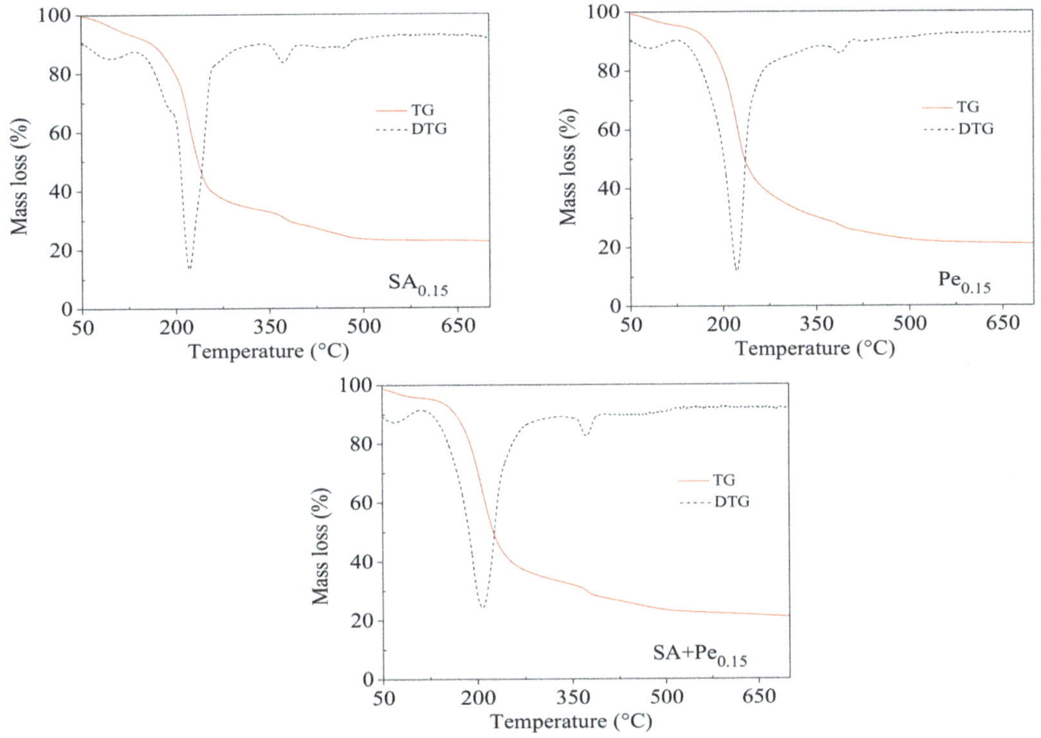

Figure 2. Thermogravimetric curves of $SA_{0.15}$, $Pe_{0.15}$ and SA + $Pe_{0.15}$ under N_2 flow.

Two main degradation steps can be outlined (Figure 2). The first one, which takes place below 160 °C, is due to film dehydration. The second step, which comprises two distinct peaks (as evidenced by the DTG curves), is related to the decomposition of the film components. In particular, the most evident mass loss (which reached the maximum in the range 210–220 °C), can be ascribed to the degradation of the polymer backbone.

3.4. Water Contact Angle (WCA) Measurements

Surface hydrophobicity is an important parameter to control the sensitivity of the films to water or moisture and it is usually evaluated by measuring the contact angle that forms between the film surface and a water droplet [20,32]. Generally, higher contact angle values correspond to higher surface hydrophobicity, with higher surface water repellent ability. Figure 3 collects, as an example, the images of the water drops deposited on the film surface of $SA_{0.15}$, $Pe_{0.15}$ and SA + $Pe_{0.15}$. Water contact angle measurements (Table 4) demonstrated the high wettability of the prepared films, due to the presence of a high amount of ether bonds and free hydroxyl-groups. In fact, WCAs were in all cases below 50°. The addition of citral EO caused a rise of the contact angle. In particular, the higher the content of citral EO, the higher the WCA, because of the hydrophobic nature of this latter Therefore, the presence of citral EO plays a beneficial role in decreasing the wettability of the edible film, thus improving its protective action towards the wrapped food.

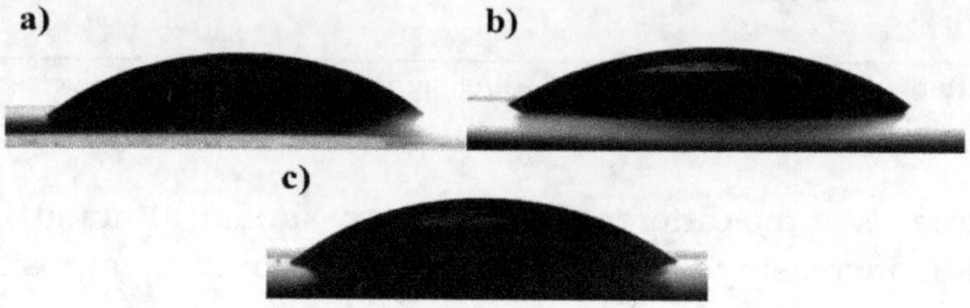

Figure 3. Water drops deposited on the film surface of: (**a**) $SA_{0.15}$; (**b**) $Pe_{0.15}$; (**c**) SA + $Pe_{0.15}$.

3.5. ATR-IR Analysis

The IR pattern of SA and Pe, as well as those of the composites, displayed some common and characteristic peaks (Figure S1). In particular, a broad and intense peak in the range 3270–3300 cm^{-1} due to the stretching of the OH groups, a smaller peak of the CH stretching at 2920–2930 cm^{-1}, and an intense and sharp peak at 1015–1025 cm^{-1} due to the C-O-C stretching of the glycosidic bonds linking two galacturonic sugar units were observed [33–35].

Some differences between the SA and Pe profile can also be highlighted. Indeed, the SA spectrum reveals the presence of two intense peaks at 1602 cm^{-1} and 1408 cm^{-1}, respectively due to the COO- asymmetric and symmetric stretching vibration, and with a weak shoulder at 1298 due to the CO stretching. On the contrary, Pe showed its characteristic band at 1743 cm^{-1}, assigned to the stretching of the methyl esterified carboxyl groups [36]. The peak at 1607 cm^{-1} (due to the COO- asymmetric stretching) was also clearly visible. The IR pattern of the composites presented the peaks of both SA and Pe.

The identification of the characteristics peaks allowed to confirm for all films the expected structure.

3.6. Color

In Table 5, L*, a*, b*, and total color difference (ΔE) values were reported for each sample.

Table 5. L*, a*, b* and total color difference (ΔE) of SA, Pe and SA + Pe films.

Sample	L*	a*	b*	ΔE
SA	86 ± 2^{b}	-1.56 ± 0.06^{de}	5.1 ± 0.4^{e}	9 ± 2^{c}
$SA_{0.15}$	87.4 ± 0.7^{a}	-1.69 ± 0.02^{efg}	6 ± 1^{de}	8 ± 1^{c}
$SA_{0.30}$	87.1 ± 0.3^{a}	-1.75 ± 0.02^{fg}	6.9 ± 0.6^{d}	8.5 ± 0.6^{c}
Pe	83.7 ± 0.7^{c}	-0.17 ± 0.03^{a}	10 ± 1^{b}	13 ± 1^{ab}
$Pe_{0.15}$	84.3 ± 1.3^{c}	-1.2 ± 0.2^{b}	13 ± 1^{a}	15 ± 2^{a}
$Pe_{0.30}$	87.9 ± 0.4^{a}	-1.62 ± 0.08^{def}	8 ± 2^{cd}	8 ± 2^{c}
SA + Pe	83.7 ± 0.8^{c}	-1.40 ± 0.04^{c}	7.9 ± 0.6^{cd}	11.8 ± 0.7^{b}
SA + $Pe_{0.15}$	85 ± 2^{bc}	-1.5 ± 0.2^{cd}	11 ± 2^{a}	13 ± 3^{ab}
SA + $Pe_{0.30}$	87.5 ± 0.8^{a}	-1.80 ± 0.03^{g}	8.9 ± 0.9^{bc}	10 ± 1^{c}

Values with different letters within the same column differ significantly at $p < 0.05$ levels. Samples with *a* present the highest value, while samples with *d* the lowest. L* (luminosity), a* (red/green index) and b* (yellow/blue index).

All the films were transparent. Since the color of edible films may affect the consumer acceptance, it is of primary importance that its transparency is preserved or that at least they display a color as close as possible to the natural pigment of foods on which the film is going to be applied [37]. SA based films showed slightly higher values of L* (luminosity) as compared with those composed of Pe and SA + Pe, in agreement with Galus and Lenart [11]. The addition of citral EO led to a general enhancement of L*. In particular, the increment was of the same entity in the SA samples containing different amounts of citral EO, while for Pe and SA + Pe films the L* values gradually increased with the increment of citral EO content. In terms of a* (red/green index) and b* (yellow/blue index) parameters, the samples containing Pe (both pure or in combination with SA) tended to a more yellowish color. In this case, 0.15 w/w of citral EO significantly enhanced ($p < 0.05$) the b* values. The obtained ΔE data ranged from 8 to 15. Pe-based samples showed the highest value, indicating a greater color variation, in agreement with the results reported by Galus and Lenart [11]. Samples displaying higher ΔE values (Pe, $Pe_{0.15}$, SA + Pe and SA + $Pe_{0.15}$) appeared also less transparent.

By citral EO addition, only a slight yellowish hue was recorded, making all those films suitable for food application.

3.7. Microstructure

As reported by Antarés and Chiralt [28], while conventional plastics are non-polar materials, edible films are usually hydrophilic. The incorporation of essential oil in the film forming dispersion is usually carried out by emulsification or homogenization of the aqueous solution containing the polymer. When the film is dried, droplets of lipid remain embedded into the polymer matrix, as observed by microscopy. The drying time plays an important role in determining the arrangement of the components during the film-forming step, thus the final microstructure of the edible films.

Morphological analysis of the films under study was carried out by optical microscopy (20× magnification), to evaluate the homogeneity and the structure of the prepared films. Figure 4 shows the surface micrographs of SA (a), Pe (c) and SA + Pe (e) films and of $SA_{0.30}$, $Pe_{0.30}$ and SA + $Pe_{0.30}$ (respectively (b), (d) and (f)).

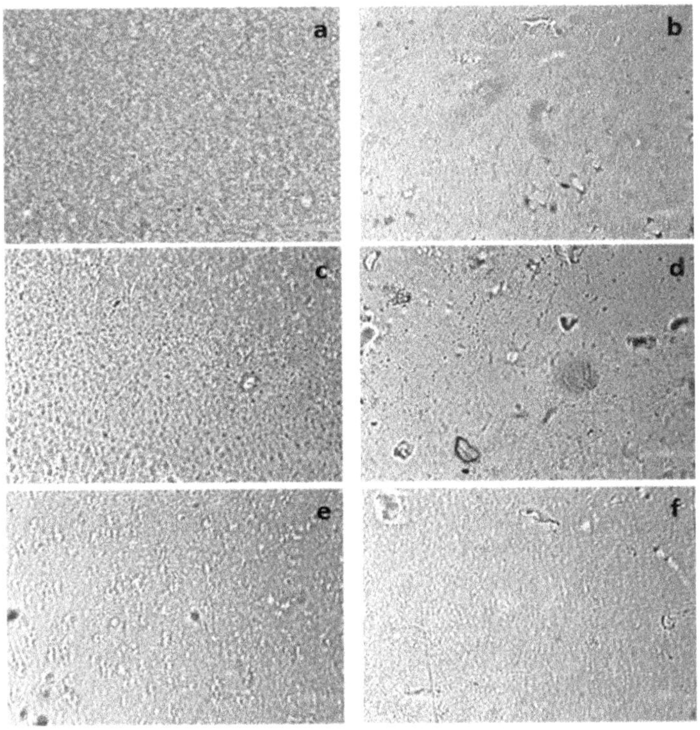

Figure 4. Surface micrographs of: (**a**) SA; (**b**) $SA_{0.30}$; (**c**) Pe; (**d**) $Pe_{0.30}$; (**e**) SA + Pe and (**f**) SA + $Pe_{0.30}$ films.

SA films presented a more homogenous and uniform structure than Pe ones, in accordance with Bierhalz et al. [38], who performed the microstructural analysis on alginate and pectin based films with the addition of natamycin as antimicrobial agent. The addition of citral EO caused a mild change in the structure of all films. Furthermore, some agglomerates can be noticed, possibly indicating a non-uniform distribution of citral EO. The most irregular structure was observed in Pe-based films. The morphology of the films supported the tensile results, evidencing that a different structural arrangement of the components in the film forming dispersion significantly influences both mechanical and gas barrier properties.

4. Conclusions

Citral EO has been incorporated into edible films through emulsification. Film structural, physical, mechanical and barrier properties were evaluated. The combination of citral EO with SA and/or Pe gives continuous and transparent edible films. Samples with citral EO addition showed improved performances with respect to the neat ones. As compared to neat films, no substantial differences were observed in the visual appearance of the composite samples, with the exception of a slight yellowish hue. On the other hand, citral EO addition caused a different microstructure arrangement and some agglomerates. However, the film structure was not weakened as testified by the gas barrier and mechanical tests, rather a reinforcement role was exerted. Furthermore, by increasing the amount of citral EO, a slight increase of the thermal stability as well as of the surface hydrophobicity was recorded.

With the aim to extend the shelf-life of food, the addition of essential oils can provide edible films with antimicrobial properties, depending both on their nature and interaction with the polymer matrix. Further studies are necessary to optimize and verify the suitability of these natural films on food products. The oil type and its interaction with the matrix determine the effectiveness of the edible films as food packaging material. In order to understand the feasibility of edible films to improve the shelf-life of food products susceptible to oxidation or to microbial spoilage, a real food systems needs to be investigated, together with specific tests on antioxidant activity [39,40].

Author Contributions: V.S., S.R. and N.L. conceived of and designed the experiments; C.M., J.P.C. and U.T. prepared the edible films and performed microscopic examination and statistical analysis; V.S. performed the thickness determination, permeation analyses, mechanical analyses and color evaluation; M.G. and N.L. performed thermogravimetric analyses, infrared spectroscopy and water contact angle determination; all authors analyzed the data; V.S., S.R. and N.L. contributed to reagents/materials/analysis tools; V.S. wrote the paper with the fundamental contribution of all authors for the final revision, especially M.G.

Acknowledgments: V.S. wish to thanks the "Piano della Ricerca di Ateneo 2016–2018", University of Catania, Italy, for financial support for gas barrier measurements.

References

1. Falguera, V.; Quintero, J.P.; Jiménez, A.; Munoz, J.A.; Ibarz, A. Edible films and coatings: Structures, active functions and trends in their use. *Trends Food Sci. Technol.* **2011**, *22*, 292–303. [CrossRef]
2. Mannozzi, C.; Cecchini, J.P.; Tylewicz, U.; Siroli, L.; Patrignani, F.; Lanciotti, R.; Romani, S. Study on the efficacy of edible coatings on quality of blueberry fruits during shelf-life. *LWT-Food Sci. Technol.* **2017**, *85*, 440–444. [CrossRef]
3. Mannozzi, C.; Tylewicz, U.; Chinnici, F.; Siroli, L.; Rocculi, P.; Dalla Rosa, M.; Romani, S. Effects of chitosan based coatings enriched with procyanidin by-product on quality of fresh blueberries during storage. *Food Chem.* **2018**, *251*, 18–24. [CrossRef] [PubMed]
4. Pasha, I.; Saeed, F.; Sultan, M.T.; Khan, M.R.; Rohi, M. Recent Developments in Minimal Processing: A Tool to Retain Nutritional Quality of Food. *Crit. Rev. Food Sci. Nutr.* **2014**, *54*, 340–351. [CrossRef] [PubMed]
5. Gomes, M.S.; Cardoso, M.G.; Guimarães, A.C.; Guerreiro, A.C.; Gago, C.M.; Vilas Boas, E.V.; Dias, C.M.; Manhita, A.C.; Faleiro, M.L.; et al. Effect of edible coatings with essential oils on the quality of red raspberries over shelf-life. *J. Sci. Food Agric.* **2017**, *97*, 929–938. [CrossRef] [PubMed]
6. Benbettaïeb, N.; Kurek, M.; Bornaz, S.; Debeaufort, F. Barrier, structural and mechanical properties of bovine gelatin-chitosan blend films related to biopolymer interactions. *J. Sci. Food Agric.* **2014**, *94*, 2409–2419.

[CrossRef] [PubMed]

7. Yoo, S.; Krochta, J.M. Whey protein-polysaccharide blended edible film formation and barrier, tensile, thermal and transparency properties. *J. Sci. Food Agric.* **2011**, *91*, 2628–2636. [CrossRef] [PubMed]
8. Rojas-Graü, M.A.; Avena-Bustillos, R.J.; Olsen, C.; Friedman, M.; Henika, P.R.; Martín-Belloso, O.; Pan, Z.; McHugh, T.H. Effects of plant essential oils and oil compounds on mechanical, barrier and antimicrobial properties of alginate-apple puree edible films. *J. Food Eng.* **2007**, *81*, 634–641. [CrossRef]
9. Valencia-Chamorro, S.A.; Palou, L.; Del Río, M.A.; Pérez-Gago, M.B. Antimicrobial Edible Films and Coatings for Fresh and Minimally Processed Fruits and Vegetables: A Review. *Crit. Rev. Food Sci. Nutr.* **2011**, *51*, 872–900. [CrossRef] [PubMed]
10. Vargas, M.; Pastor, C.; Chiralt, A.; McClements, D.J.; González-Martínez, C. Recent Advances in Edible Coatings for Fresh and Minimally Processed Fruits. *Crit. Rev. Food Sci. Nutr.* **2008**, *48*, 496–511. [CrossRef] [PubMed]
11. Galus, S.; Lenart, A. Development and characterization of composite edible films based on sodium alginate and pectin. *J. Food Eng.* **2013**, *115*, 459–465. [CrossRef]
12. Morillon, V.; Debeaufort, F.; Blond, G.; Capelle, M.; Voilley, A. Factors affecting the moisture Permeability of Lipid-Based Edible films: A review. *Crit. Rev. Food Sci. Nutr.* **2002**, *42*, 67–89. [CrossRef] [PubMed]
13. Sánchez-González, L.; Vargas, M.; González-Martínez, C.; Chiralt, A.; Cháfer, M. Use of Essential Oils in Bioactive Edible Coatings: A Review. *Food Eng. Rev.* **2011**, *3*, 1–16. [CrossRef]
14. Silva-Weiss, A.; Ihl, M.; Sobral, P.J.A.; Gómez-Guillén, M.C.; Bifani, V. Natural Additives in Bioactive Edible Films and Coatings: Functionality and Applications in Foods. *Food Eng. Rev.* **2013**, *5*, 200–216. [CrossRef]
15. Prommakool, A.; Sajjaanantakul, T.; Janjarasskul, T.; Krochta, J.M. Whey protein-okra polysaccharide fraction blend edible films: Tensile properties, water vapor permeability and oxygen permeability. *J. Sci. Food Agric.* **2011**, *91*, 362–369. [CrossRef] [PubMed]
16. Guerreiro, A.C.; Gago, C.M.L.; Faleiro, M.L.; Miguel, M.G.C.; Antunes, M.D.C. The effect of alginate-based edible coatings enriched with essential oils constituents on Arbutus unedo L. fresh fruit storage. *Postharvest. Biol. Technol.* **2015**, *100*, 226–233. [CrossRef]
17. Bonilla, J.; Antarés, L.; Vargas, M.; Chiralt, A. Edible films and coatings to prevent the detrimental effect of oxygen on food quality: Possibilities and limitations. *J. Food Eng.* **2012**, *110*, 208–213. [CrossRef]
18. Emmambux, N.M.; Minnaar, A. The effect of edible coatings and polymeric packaging films on the quality of minimally processed carrots. *J. Sci. Food Agric.* **2003**, *83*, 1065–1071. [CrossRef]
19. Siracusa, V.; Ingrao, C. Correlation among gas barrier behavior, temperature and thickness in BOPP films for food packaging usage: A lab-scale testing experience. *Polym. Test.* **2017**, *59*, 277–289. [CrossRef]
20. Kokoszka, S.; Debeaufort, F.; Lenart, A.; Voilley, A. Water vapor permeability, thermal and wetting properties of whey protein isolate based edible films. *Int. Dairy J.* **2010**, *20*, 53–60. [CrossRef]
21. Debeaufort, F.; Quezada-Gallo, J.-A.; Voilley, A. Edible films and Coatings: Tomorrow's Packagings: A review. *Crit. Rev. Food Sci. Nutr.* **1998**, *38*, 299–313. [CrossRef] [PubMed]
22. Debeaufort, F.; Voilley, A. Methyl cellulose-based films and coatings I. Effect of plasticizer content on water and 1-octan-e-ol sorption and transport. *Cellulose* **1995**, *2*, 205–513. [CrossRef]
23. Kurek, M.; Descours, E.; Galic, K.; Voilley, A.; Debeaufrot, F. How composition and process parameters affect volatile active compounds in biopolymer films. *Carbohydr. Polym.* **2012**, *88*, 646–656. [CrossRef]
24. Wang, X.; Sun, X.; Liu, H.; Li, M.; Ma, Z. Barrier and Mechanical properties of puree films. *Food Bioprod. Proc.* **2011**, *89*, 149–156. [CrossRef]
25. Bonilla, J.; Atarés, L.; Vargas, M.; Chiralt, A. Effect of essential oils and homogenization conditions on properties of chitosan-based films. *Food Hydrocol.* **2012**, *26*, 9–16. [CrossRef]
26. Pao, S.; Brown, G.E.; Schneider, K.R. Challenge studies with selected pathogenic bacteria on freshly peeled Hamlin orange. *J. Food Sci.* **1998**, *63*, 359–362. [CrossRef]
27. Marklinder, I.; Eriksson, M.K. Best-before date—Food storage temperatures recorded by Swedish students. *Br. Food J.* **2015**, *117*, 1764–1776. [CrossRef]
28. Antares, L.; Chiralt, A. Essential oils as additives in biodegradable films and coatings for active food packaging. *Trends Food Sci. Technol.* **2016**, *48*, 51–62. [CrossRef]
29. Robertson, G.L. *Food Packaging: Principles and Practice*, 2nd ed.; CRC Press for Taylor & Francis Group: Boca Raton, FL, USA, 2006; pp. 63–64.
30. Shahbazi, Y. Characterization of nanocomposite films based on chitosan and carboxymethylcellulose containing Ziziphora clinopodioides essential oil and methanolic Ficus carica extract. *J. Food Proc. Pres.* **2018**,

42, e13444. [CrossRef]
31. Avila-Sosa, R.; Ochoa-Velasco, C.E.; Navarro-Cruz, A.R.; Palou, E.; López-Malo, A. *Combinational Approcahes for Antimicrobial Packaging: Chitosan and Oregano Oil in Antimicrobial Food Packaging*; Elsevier Inc.: Amsterdam, The Netherlands, 2016; pp. 581–588.
32. Ramírez, C.; Gallegos, I.; Ihl, M.; Bifani, V. Study of contact angle, wettability and water vapor permeability in carboxymethylcellulose (CMC) based film with murta leaves (Ugni molinae Turcz) extract. *J. Food Eng.* **2012**, *109*, 424–429. [CrossRef]
33. Guillen, M.D.; Cabo, N. Some of the most significant changes in the Fourier transform infrared spectra of edible oils under oxidative conditions. *J. Sci. Food Agric.* **2000**, *80*, 2028–2036. [CrossRef]
34. Mishra, R.K.; Majeed, A.B.A.; Banthia, A.K. Development and characterization of pectin/gelatin hydrogel membranes for wound dressing. *Int. J. Plast. Technol.* **2011**, *15*, 82–95. [CrossRef]
35. Rezvanian, M.; Mohd Amin, M.C.I.; Shiow-Fern, N. Development and physicochemical characterization of alginate composite film loaded with simvastatin as a potential wound dressing. *Carbohydr. Polym.* **2016**, *137*, 295–304. [CrossRef] [PubMed]
36. Kurita, O.; Fujiwara, T.; Yamazaki, E. Characterization of the pectin extracted from citrus peel in the presence of citric acid. *Carbohydr. Polym.* **2008**, *74*, 725–730. [CrossRef]
37. Rhim, J.W.; Gennadios, A.; Weller, C.L.; Hanna, M.A. Sodium dodecyl sulfate treatment improves properties of cast films from soy protein isolate. *Ind. Crops Prod.* **2002**, *15*, 199–205. [CrossRef]
38. Bierhalz, A.C.K.; Da Silva, M.A.; Kieckbusch, T.G. Natamycin release from alginate/pectin films for food packaging applications. *J. Food Eng.* **2012**, *110*, 18–25. [CrossRef]
39. Sacchetti, G.; Maietti, S.; Muzzoli, M.; Scaglianti, M.; Manfredini, S.; Radice, M.; Bruni, R. Comparative evaluation of 11 essential oils of different origin as functional antioxidants, antiradicals and antimicrobials in foods. *Food Chem.* **2005**, *91*, 621–632. [CrossRef]
40. Baschieri, A.; Ajvazi, M.D.; Tonfack, J.L.F.; Valgimigli, L.; Amorati, R. Explaining the antioxidant activity of some common non-phenolic components of essential oils. *Food Chem.* **2017**, *232*, 656–663. [CrossRef] [PubMed]

11

Evaluation of Reliefs' Properties on Design of Thermoformed Packaging Using Fused Deposition Modelling Moulds

Lucía Rodríguez-Parada *, Pedro F. Mayuet and Antonio J. Gámez

Department of Mechanical Engineering & Industrial Design, Faculty of Engineering, University of Cadiz, Av. Universidad de Cádiz 10, E-11519 Puerto Real-Cadiz, Spain; pedro.mayuet@uca.es (P.F.M.); antoniojuan.gamez@uca.es (A.J.G.)

* Correspondence: lucia.rodriguez@uca.es

Abstract: The increased consumption of food requiring thermoformed packaging implies that the packaging industry demands customized solutions in terms of shapes and sizes to make each packaging unique. In particular, food industry increasingly requires more transparent packaging, with greater clarity and a better presentation of the product they contain. However, in turn, the differentiation of packaging is sought through its geometry and quality, as well as the arrangement of food inside the packaging. In addition, these types of packaging usually include ribs in the walls to improve their physical properties. However, these ribs also affect the final aesthetics of the product. In accordance with this, this research study analyses the mechanical properties of different relief geometries that can affect not only their aesthetics but also their strength. For this purpose, tensile and compression tests were carried out using thermoformed PET sheets. The results provide comparative data on the reliefs studied and show that there are differences in the mechanical properties according to shape, size and disposition in the package.

Keywords: packaging design; product design; mechanical properties; thermoforming; tensile test; 3D printing; simulation; technology

---

1. Introduction

Thermoformed food packaging, usually called rigid or semi-rigid containers, have as main functions protection, containment, preservation and distribution [1,2]. Different types of thermoformed packaging can be observed in the market according to the specific needs of the food. These include heat-sealable containers for processed or semi-processed products [3,4]. Packaging usually applied to fresh products, such as fruit or vegetables [5], all of which are therefore often used for products with a short life cycle and with the aim of protecting and making food more functional [6]. For this reason, during the development of these containers, the aim is to reduce the amount of material used to increase sustainability by reducing the large amount of waste generated [7]; and to optimise production costs while maintaining their functional properties [5].

Although distribution is one of their main functions [8,9], most of these packages reach the consumer, especially because of the new trend of food on the move [10]. For this reason, the design and finish of these packages must also be taken into consideration. Thus, the ergonomic and functional aspects of the packaging must be aimed at adapting the product to the consumer's needs for use and protection of the food it contains [2].

Throughout the short life cycle of freshly packaged food there are very different situations to which the packaging is exposed to external aggressions. This is linked to the sustainability of both the product inside and the cost of material involved. Thus, food packaging can delay and protect food from physical, chemical and biological spoilage, that is, packaging can extend shelf life and ensure product quality [7] and therefore can be understood as a sustainability system for food. Light, humidity, microorganisms, shocks and other mechanical forces are examples of some of the external agents that can adversely affect food [2]. In general, both the handling of the package, its transport or its stacking on the supermarket shelf are considered external aggressions to the food contained inside before consumption [5]. Therefore, elastic deformation of the packaging may be required to preserve its content.

Specifically, the weight of the food generates tensile stresses in the walls. One of the situations where this occurs is when the container is suspended during transport or handling. Also, during transport the containers are normally stacked and this means that the container must also withstand the stacking weight. These mechanical forces cause a compressive stress on the container walls. Also, in situations where a user, whether a consumer or not, handles the packaging, a functional situation occurs where the packaging must protect the food [11].

According to these approaches, thermoformed containers usually incorporate patterns on their walls in the form of reliefs, which are used to give them some mechanical properties, without the need to increase the thickness of the plastic sheet used during the manufacturing of the packaging. In addition, reliefs can attribute a different geometric aspect by creating a semantic value on the product. However, the containers currently used in the market usually incorporate reliefs in the form of vertical ribs or columns, generally with square geometry.

A related point to consider is the package visual appearance, which influences the decision of consumption due to its symbiotic or simply aesthetic qualities [12,13]. In these circumstances, the container may be a key factor in the consumer decision-making, allowing inferences to be drawn about the product, like attributes or taste. Therefore, the package can influence subsequent experiences in the product so much so that researchers study its influence concerning materials, shapes and sizes [3,13–17], although only a few studies focus on the tactile sensation that a container causes. This is a new trend in packaging, such as in cosmetics or drinks [18]. As a designed texture or relief using differentiating factors can attract consumers as much as shape or colour, designing custom geometries is of utmost interest, although it may alter the functional properties of the package.

Given the above conditions, the influence of relief geometries on the mechanical properties of thermoformed sheet is studied and analysed in this work. By means of this method, the properties of the reliefs of the packaging can be evaluated in order to guarantee sustainability and also to reduce the expense of plastic material. Furthermore, the improvement of the mechanical properties of the containers also contributes to increase the useful life of the food. It should be also stressed that, conventionally, reliefs are not specifically evaluated until the physical package is tested. This means that design and development times can be increased by design modifications.

For this purpose, Fused Deposition Modelling (FDM), an Additive Manufacturing (AM) technology, is proposed for making economic moulds that allow the thermoforming of test specimens. AM refers as a general term to manufacturing technologies that build up a product layer by layer [19]. Since the 1980´s, AM first started unsteadily and then moved from laboratory to industrial practice. New applications are constantly announced and they have been developed over time [20]. Thus, the applications of this technology have been increasing. Despite the fact that AM processes, such as Stereolithography (SLA), FDM or Selective Laser Sintering (SLS) were initially created for the purpose of generating rapid prototypes, it is currently sought to use them to create final products as well [21,22].

In the context of thermoformed products, moulds from additive manufacturing are being created nowadays [23]. Specifically, the applications focus on the study of this technology in the field of design and development processes, in order to create moulds for short series [24]. To do this, additive manufacturing is used to validate proposals as well as for research support, as it allows to generate moulds in a fast and economic way [25,26].

In the field of thermoforming, Laser Sintering is widely used. In this research, in contrast, the application of the FDM technology is considered since it is an affordable technology and accessible to any design team [27]. In effect, the use of FDM offers the possibility of creating moulds of high quality, though it may also be slow and costly [28]. Thus, printers provide the possibility of generating moulds in order to create thermoforming prototypes in a rapid way, resulting in shortened deadlines and reduction of errors.

This work presents an experimental procedure that provides a novel scheme to evaluate the relief patterns designed on food packages with a double objective: to improve the mechanical resistance without increasing the quantity of raw material and to build more efficient products. Moreover, this evaluation technique, can be used with thermoplastic materials like Polyethylenne Terephtalate (PET), due to their predominance in disposable containers but can be extended to new biodegradable materials such as Bio Polyethylenne Terephtalate (Bio-PET) and Polylactic Acid (PLA). In addition, its application could be extended to other materials such as thin cardboard sheets.

In short, additive manufacturing allows new applications and procedures to emerge in the field of engineering, as is the case with the work proposed here.

2. Construction of Test Specimens

In this work, a study of the mechanical behaviour of reliefs has been carried out. Initially, the reliefs were evaluated as a unit, performing tensile tests with the aim of comparatively analyse the existing differences between geometries and relief sizes. After that, the influence of the position and number of the reliefs on the faces of a container, used as a test tube, was studied by performing compression tests.

The overall procedure for the manufacture and preparation of the specimens consists of several phases, as seen in Figure 1. Specifically, the generation of the specimen and mould in Computer Aided Design (CAD) was done by using the Solidworks® software (2016 version, Dassault Systèmes SE, Velizy-Villacoublay, France) and the 3D printing of the specimen was done with the Simplify3D® software (Cincinnati, OH, USA) for the generation of the G-code. For the manufacture of the mould of the specimen printed in FDM, a 3D printer machine Witbox® and PLA filament with a diameter of 1.75 mm have been used. The machine used for thermoforming of the sheet was the table-top thermoforming machine Formech 450DT. Finally, Minitab® software (18 version, Minitab Inc., State College, PA, USA) was used for the treatment of the statistical data.

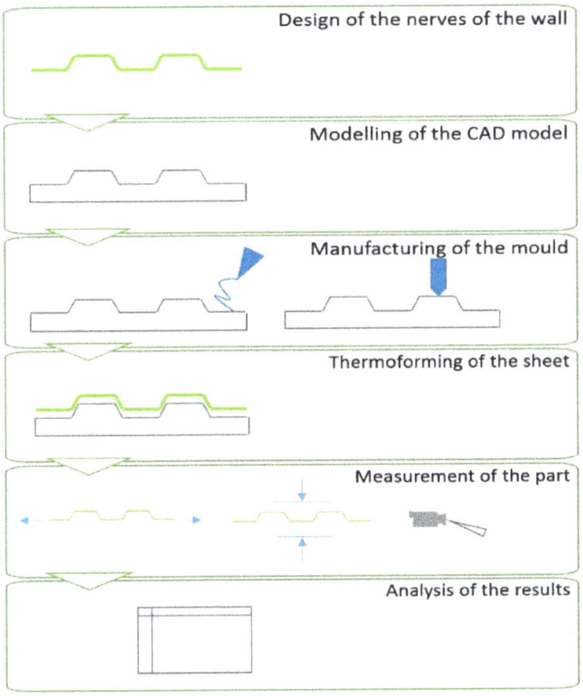

Figure 1. Methodological procedure.

2.1. *Procedure for the Construction and Preparation of Tensile Specimens*

In the case of the generation of tensile specimens, to ensure the flatness on the surface, a two parts hybrid mould was designed, Figure 2a. One part was common to all the specimens (labelled 1 in Figure 2a), consisting of a wooden profile with perforations along the perimeter to ensure that the sheet was fixed along the entire surface during the thermoforming. This profile also served as a cutting template to remove the excess of material to ensure repeatability during specimen creation. The second part corresponded to the mould of the specific specimen obtained by FDM (labelled 2 in Figure 2a) and it was fitted into the wooden profile to generate a mould with a flat surface, Figure 2b. This ensured that there were no roundings along the perimeter of the base of the specimen.

Then, Figure 2c shows the procedure for obtaining the final specimen. Once the specimen was thermoformed, the PET sheet was cut. The wooden profile was used to mark the cutting perimeter over the thermoformed sheet. To make the cut, a sheet shear has been used, according to [29].

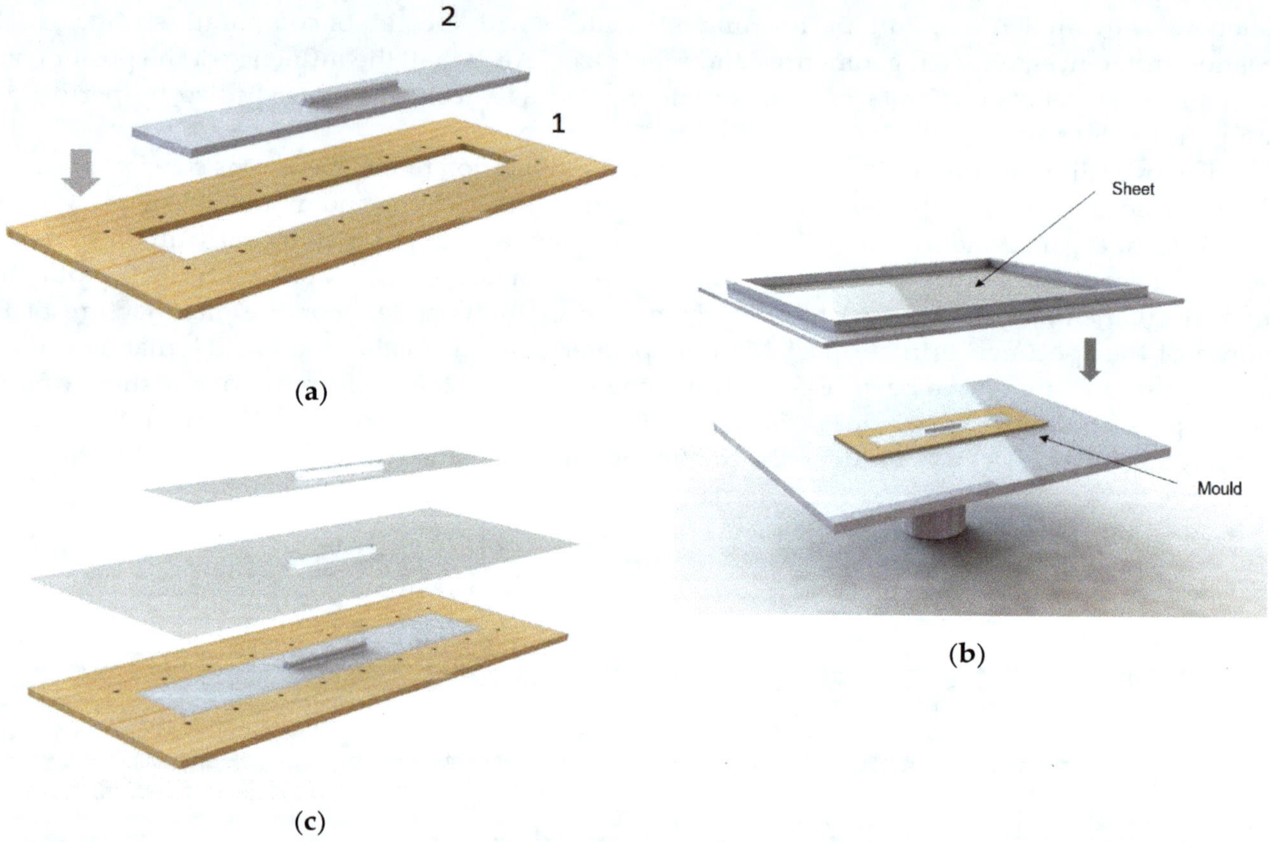

Figure 2. Thermoforming of specimens for tensile tests: (**a**) Hybrid mould; (**b**) Thermoforming of the sheet with hybrid mould to obtain the thermoformed sheet specimen; (**c**) Procedure for obtaining the thermoformed specimen.

2.2. *Procedure for the Construction and Preparation of Specimens for Compression Tests*

The specimens used for compression tests had a square shape. A mould made entirely of FDM, Figure 3, was used for their construction, as it had flat sides and this way it could be compared with the traction experiments.

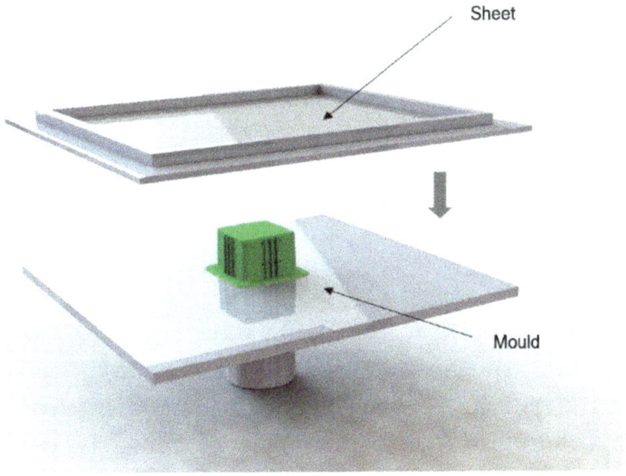

Figure 3. Thermoforming process of specimens for compression tests.

The generated mould, Figure 4a, contained a draft angle and a base with a 45° chamfer to avoid the appearance of defects on the thermoformed sheet, in accordance with [30]. It also contained vacuum channels to ensure proper attachment of the PET sheet to the mould. The channels have been made at three heights along the walls of the mould with an angle of 45°, Figure 4b.

(**a**)

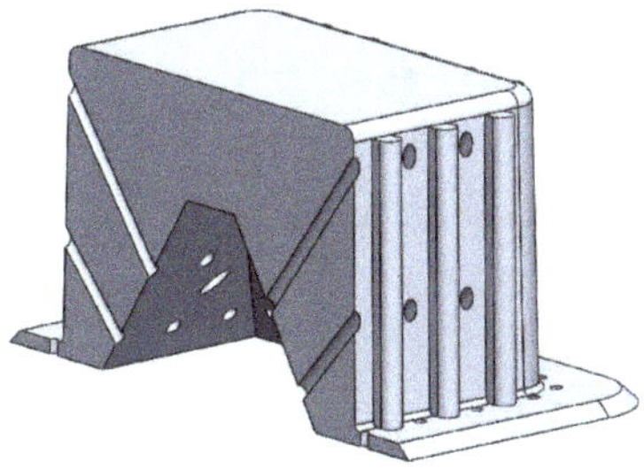

(**b**)

Figure 4. Example images of the mould used for thermoforming compression specimens: (**a**) image of the mould, (**b**) vacuum channels included inside the mould.

Once the thermoforming process had been carried out, Figure 3, the excess of material has been eliminated using the Iberolaser IL-1390 laser cutting machine. This cutting procedure was precise and left no relevant defects on the cutting surface.

3. Materials and Methods

3.1. Tested Material

The material used in this test was a three-layer PET laminate roll with a recycled PET sheet inside and the thickness used was 180 micrometres. This material is commonly used for food packaging in industrial applications. The material had undergone an extrusion manufacturing process. The properties provided by the supplier are detailed in Table 1, according to UNE-EN-ISO 527-3 [31] and DIN 53479-B [32].

Table 1. Properties of PET laminate.

Thickness	Surface Treatment	Tensile Strength	Ductility	Impact Resistance	Density
0.180 mm ± 3%	Silicone on both outer sides	>50 N/mm^2	220%	>3 KJ/m^2	1.33 gr/cm^2

3.2. Tensile Test Method

The main objective of these tensile tests was to deepen the knowledge of the behaviour of the different geometries in relation to the dimensions of the sheet protrusion. In addition, for the correct evaluation of the reliefs, a morphological study of the thermoformed geometries was performed. For this reason, the reliefs were included in a unitary way on each specimen.

3.2.1. Morphological Study

A total of 9 different types of test tubes, based on three different reliefs, were studied. On one hand, three basic geometries were analysed: semi-circular (A), square (B) and triangular (C). These geometries were transferred to the PET sheet in a straight line. On the other hand, three sizes or scale relationships were studied: a, a/2 and 2a where a was 3 mm in the designs, Figure 5. That particular value was chosen after analysing the reliefs of different commercial packages.

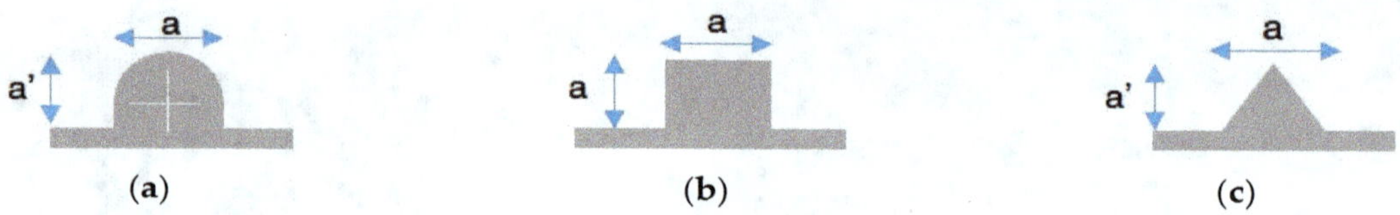

Figure 5. Geometries of the studied reliefs: (**a**) semi-circular, A; (**b**) square, B.; (**c**) triangular, C.

Obviously, the section, S, of these three geometries is different, directly affecting the mechanical properties, Equation (1).

$$\sigma = F_y/S, \tag{1}$$

where F_y corresponds to the axial force applied to the specimen, either in tensile or compression tests.

Sections were easily calculated as the chosen reliefs were simple geometrical shapes. Relief type

A corresponded to a semicircle attached to a rectangle, Figure 6a. Relief B was chosen to be a square, Figure 6b, while relief C was designed as a triangle, Figure 6c. Their respective sections are:

$$S_A = \left(\pi * \frac{a^2}{8}\right) + a * \left(a' - \frac{a}{2}\right), \tag{2}$$

$$S_B = a^2, \tag{3}$$

$$S_C = (a * a')/2 \tag{4}$$

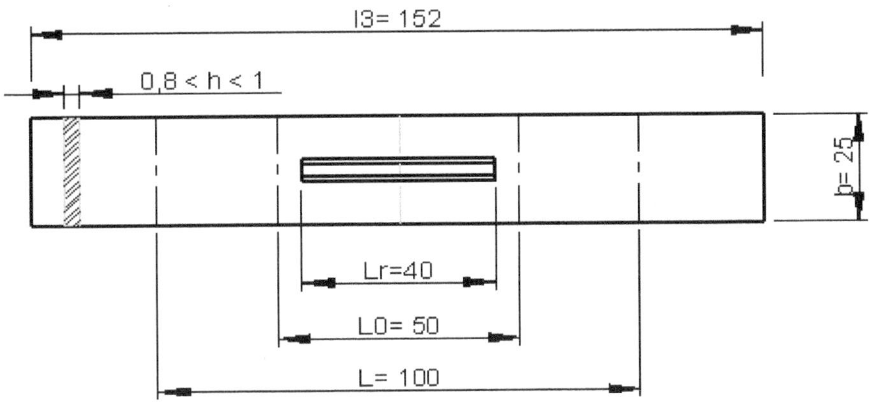

Figure 6. Dimensions of specimen according to standard [31].

3.2.2. Mechanical Characterization Study

In accordance with standard UNE-527-3 [31], Figure 6, the designed dimensions used for a′ and l3 in the test specimen corresponded to 25 mm and 152 mm, respectively. All the other relevant dimensions are depicted in Figure 7.

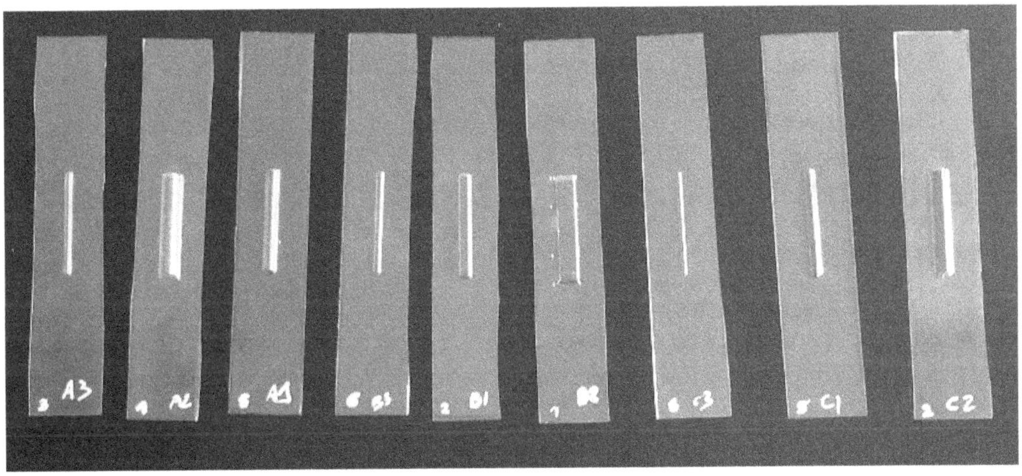

Figure 7. Thermoformed specimens.

The guidelines established in References [31] and [33] had been taken as a reference for carrying out the tensile tests. Thus, a total of 5 specimens per geometry were studied.

The machine used for mechanical testing was the equipment Shimadzu, model AG-X, with a load cell of 50 KN, Figure 8a. All test parameters were managed with the help of the universal test software Trapezium® for Windows®. For the development of the tests, plastic-specific jaws were used, Figure 8b.

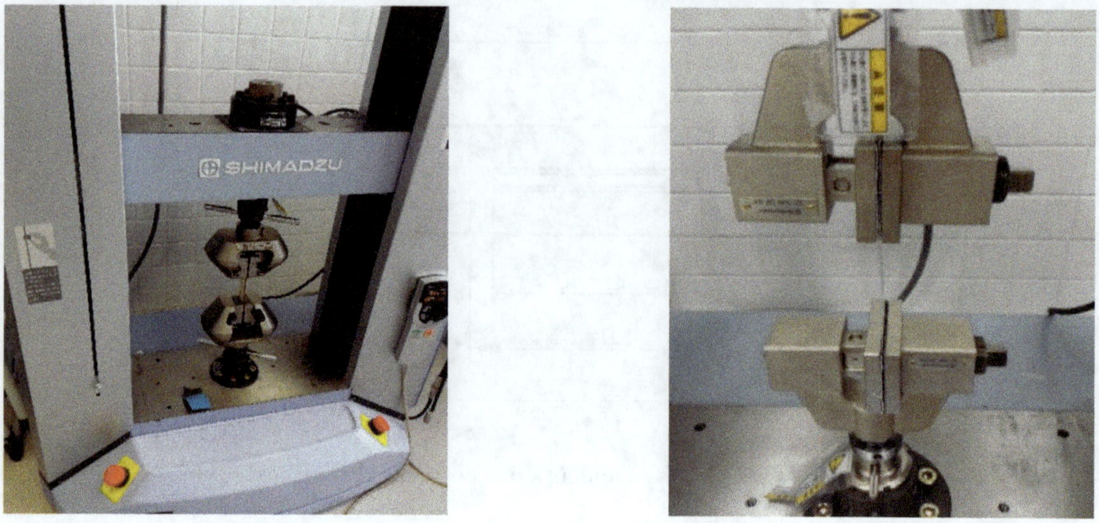

(a) (b)

Figure 8. Testing machine with clamps for tensile tests: (**a**) testing machine for the experimental (**b**) clamps with specimen in the middle.

After the laboratory tests, from which experimental data were extracted, simulations of the tensile tests using the Finite Elements Method (FEM) were carried out with the software Solidworks® and Hyperworks® Radios for geometric modelling and dynamic simulation, respectively.

Due to the fact that the range of displacement increases with the test time, the study time was defined at 6 s.

3.3. *Method for Compression Tests*

The objective of the compression tests, with a structure similar to that used for tensile tests, is to deepen the knowledge of the behaviour of the reliefs included on the walls of the packaging. The main idea of this study is to evaluate the influence of the number of reliefs on the packaging and the distance between them.

1, 3 and 5 reliefs had been inserted on each face in order to study their influence on the mechanical properties. In this case, the study was restricted to the study of type A geometry, Figure 5. All the reliefs were placed symmetrically from the centre of each face.

3.3.1. Macrogeometric Analysis

The moulds and the thermoformed packaging had been visually inspected to analyse the final result, Figure 9. Likewise, all the tests carried out had been recorded in order to study the behaviour of the specimens subjected to compression.

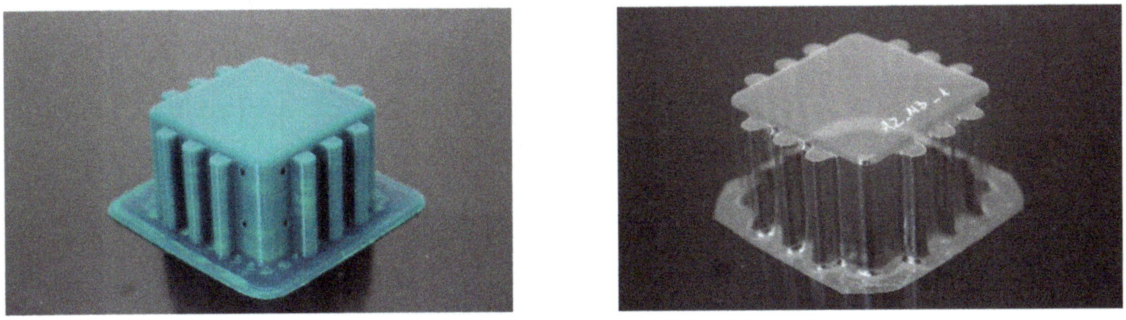

(a) (b)

Figure 9. Example images of the results obtained: (**a**) mould A2_M3, (**b**) packaging created with mould A2_M3.

Two distances between reliefs had been studied: 3 mm and 6 mm. M1 being the test piece that included 1 relief, M2 and M3 corresponded to 3 reliefs, with a distance of 3 and 6 mm, respectively and M4 and M5 corresponded to the test pieces that had 5 reliefs, 3 and 6 mm, respectively. A2 geometry with 5 reliefs and 6 mm spacing was not evaluated because the size of the specimen face was too small to include these reliefs. Table 2 details the full nomenclature of the designed test specimens.

Table 2. Nomenclature of reliefs according to the type of relief, distance and dimensional correspondence.

Relief	Specimen	N° Reliefs	Relief Dimensional Proportion	Relief Size (mm)	Distance between Reliefs (mm)
-	O2	0	-	-	-
A0	M1	1	a/2 × a/2	1.5 × 1.5	-
A0	M2	3	a/2 × a/2	1.5 ×1.5	3
A0	M3	3	a/2 × a/2	1.5 × 1.5	6
A0	M4	5	a/2 × a/2	1.5 × 1.5	3
A0	M5	5	a/2 × a/2	1.5 × 1.5	6
A1	M1	1	a × a	3 × 3	-
A1	M2	3	a × a	3 × 3	3
A1	M3	3	a × a	3 × 3	6
A1	M4	5	a × a	3 × 3	3
A1	M5	5	a × a	3 × 3	6
A2	M1	1	2 a × 2a	6 × 6	-
A2	M2	3	2 a × 2a	6 × 6	3
A2	M3	3	2 a × 2a	6 × 6	6
A2	M4	5	2 a × 2a	6 × 6	3

3.3.2. Mechanical Characterization

A total of 15 tests were performed, including the specimen test without reliefs. The specimens had a square base of dimensions 50 × 50 mm, according to [34]. Figure 10 shows an image with the dimensions of the specimen O2.

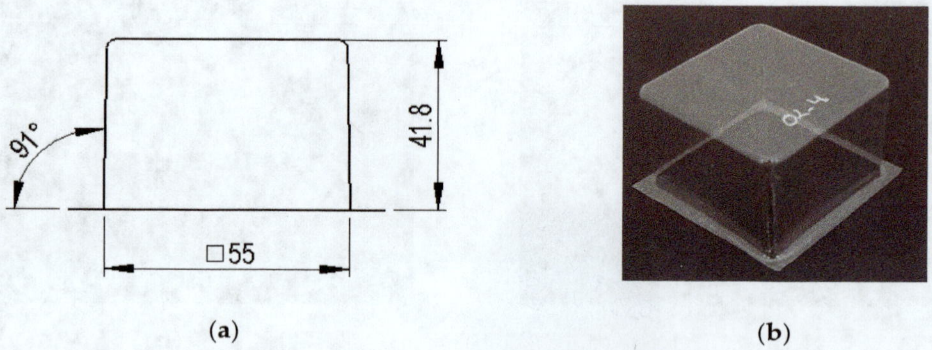

Figure 10. Images and dimensions of the plain test specimen, O2: (**a**) general dimensions; (**b**) test specimen.

The compression tests were carried out according to the guidelines established in Reference [34] and [35]. 4 tests were realized per test specimen typology.

The machine used to carry out the mechanical compression tests was the same used for the tensile ones, Figure 11. The load cell was set to 50 KN. As for the tensile tests, the test parameters were handled with Trapezium®. The compression speed of the tests was 10 mm/min, according to [35].

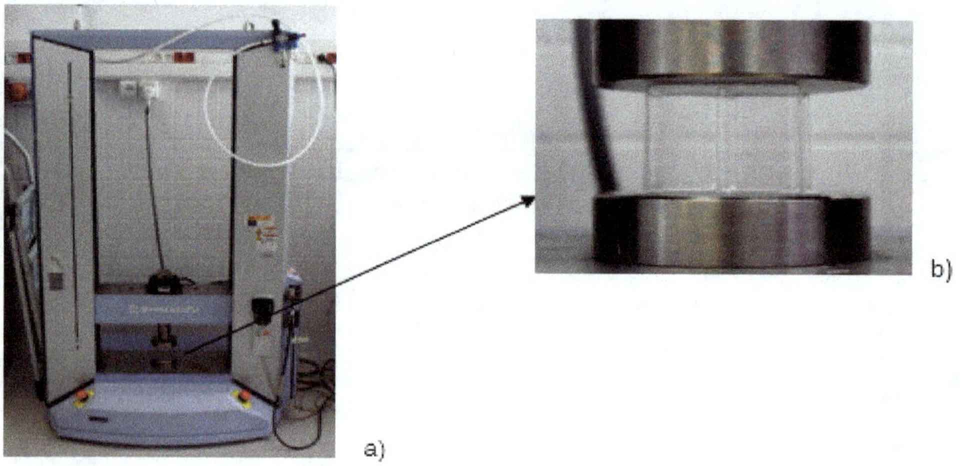

Figure 11. Compression tests: (**a**) general machine view; (**b**) specimen positioning.

The compression tests were recorded using a high-precision digital camera Canon EOS 650D. The recorded images of the compressed specimens were used to relate the real deformation to the measured stress-strain data. This is important, because the recorded axial force, F_y, could be overestimated if folding of the packaging occurs. Also, plastic deformation could be admitted if the packaged product was not affected by it.

The test procedure started with the placement of the specimens on the cylindrical platform of the testing machine, Figure 11b. The load cell approached the specimen and, once the machine had been calibrated, the compression test began. The control parameters set in the program were force in N, time in s and displacement in mm. After carrying out the laboratory tests, a case study was carried out using the same methodology as in the tensile tests, with one of the test pieces used to compare the simulation with the data obtained in the real tests. The simulation, as in the tensile tests, was carried out with Hyperworks Radioss® and Solidworks® where the base was established as fixed and constant speed was included in the upper side.

4. Results and Discussion

4.1. Tensile Strength

4.1.1. Mechanical Evaluation

Table 3 shows the measured section of geometries studied. A study of the thicknesses on the physical specimens was carried out to obtain the real section, bearing in mind that the material was stretched where the thermoforming process took place. According to this, the specimens presented a different tensile strength given by the type of area and volume generated.

Table 3. Results of the area of specimens and tensile strength.

Relief	Measured S (mm^2)	F_y (N)	Standard Deviation (N)
A0	4.60	187.24	15.87
A1	4.82	163.49	22.04
A2	5.31	109.75	15.02
B0	4.68	182.78	22.56
B1	4.73	128.53	17.74
B2	5.63	112.71	16.07
C0	4.56	184.06	30.90
C1	4.66	180.38	16.14
C2	4.93	112.84	10.09

As can be seen in Figure 12, geometries A0, B0 and C0 provided higher force values compared to A2, B2 and C2. This is related to the fact that a smaller section provides a greater wall thickness after thermoforming due to the lower stretching suffered by the specimen and generates greater rigidity to the relief.

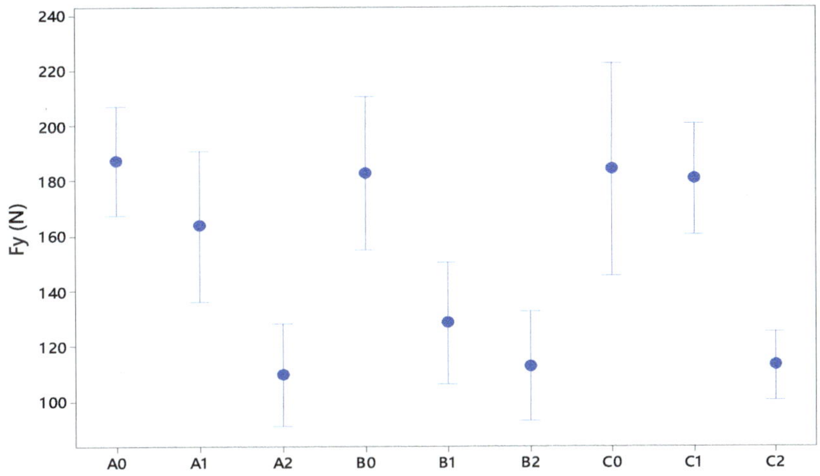

Figure 12. Dispersion of F_y in tensile tests.

Indeed, for the smallest packaging, the values reached forces around 185 N, although it is true that it is the A geometry that seems to maintain a lower dispersion in the repetitiveness of the tests. This phenomenon is especially visible in the results for the B0 and C0 reliefs, although it is in the latter geometry where the dispersion of the results is greater, with values of up to 35% lower than the established maximum. This could be due to the fact that the vertices of the square and the triangular geometry act as stress concentration points, which could lead to defects and micro-cracks that lead to premature failure.

Size 1 presented intermediate force values to those compared for size 2 and size 0 with greater variability in the averages obtained. Thus, C geometry presented values close to those reached for the smallest relief with an average of approximately 180 N. However, B geometry obtained values closer to those found in higher relief tests with an average of about 130 N. This seems to reinforce what was previously discussed with respect to the size of the section.

Size 2 presented mean values of about 110 N, about 40% lower than the other sizes. In a deeper analysis of the results, the geometry 2 tests showed values between 90 N and approximately 130 N with smaller dispersions than in previous cases.

Summarizing, the nominal stress analysis shows the trend followed by each A, B and C geometry as a function of their relief. In all three cases there is a tendency for the stress to decrease as the relief increases in size. However, as discussed, the C1 case shows a different trend because its section is proportionally smaller than in cases A1 and B1 and similar to the values obtained for size 0.

4.1.2. Dimensional Evaluation

Due to the fact that the base of the specimen influences the result of the tests, dimensional analysis has been carried out isolating the stress of the relief by means of the dimensional relation given by Equation (5):

$$\sigma y_{relief} = (\sigma y * S_{rel})/S, \quad (5)$$

where S is the real section of area of the specimen and S_{rel} is the real section area of the relief.

Figure 13 details the results obtained according to the middle section, by calculating the stress from the force and area of the relief. It is observed that the geometries A1 and C1, which contain the smaller reliefs, supported a greater tension, especially C1, considering its size. This reinforces the results obtained previously for circular and triangular reliefs with less size. On the other hand, it can also be seen that the three largest geometries bore less stress, as seen in previous sections, thus highlighting the importance of the relationship between the area of the specimen and the relief.

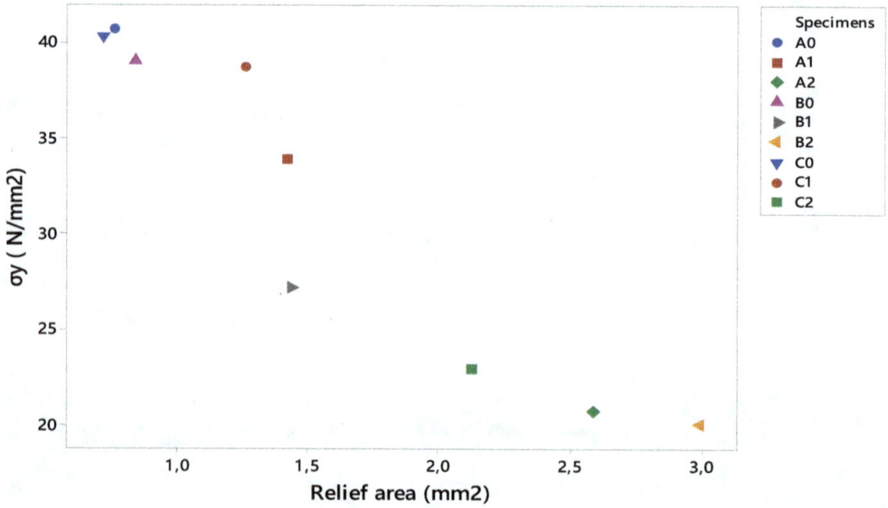

Figure 13. Data of the yield stress with respect to the section of the relief.

To try to have a detailed view of the results by taking into account only the geometry, the dispersion of the results obtained in Figure 14 is shown.

Regarding the results for size 0, Figure 14a, it is observed that the differences of F_y are minimal according to relief and size. This may be due to the fact that the area of the relief with respect to the specimen is minimal and therefore has less influence on the base geometry.

The results of the tests with the specimens with relief size 1, Figure 14b, evidence that there are differences with respect to the dimension of the relief. They also show that the C1 geometry withstands the highest stress and that B1 has the lowest tensile strength.

Finally, Figure 14c shows the results for size 2. The A2 and B2 reliefs present similar creep stress despite the dimensional differences between them, although the triangular geometry, C2, presents greater strength.

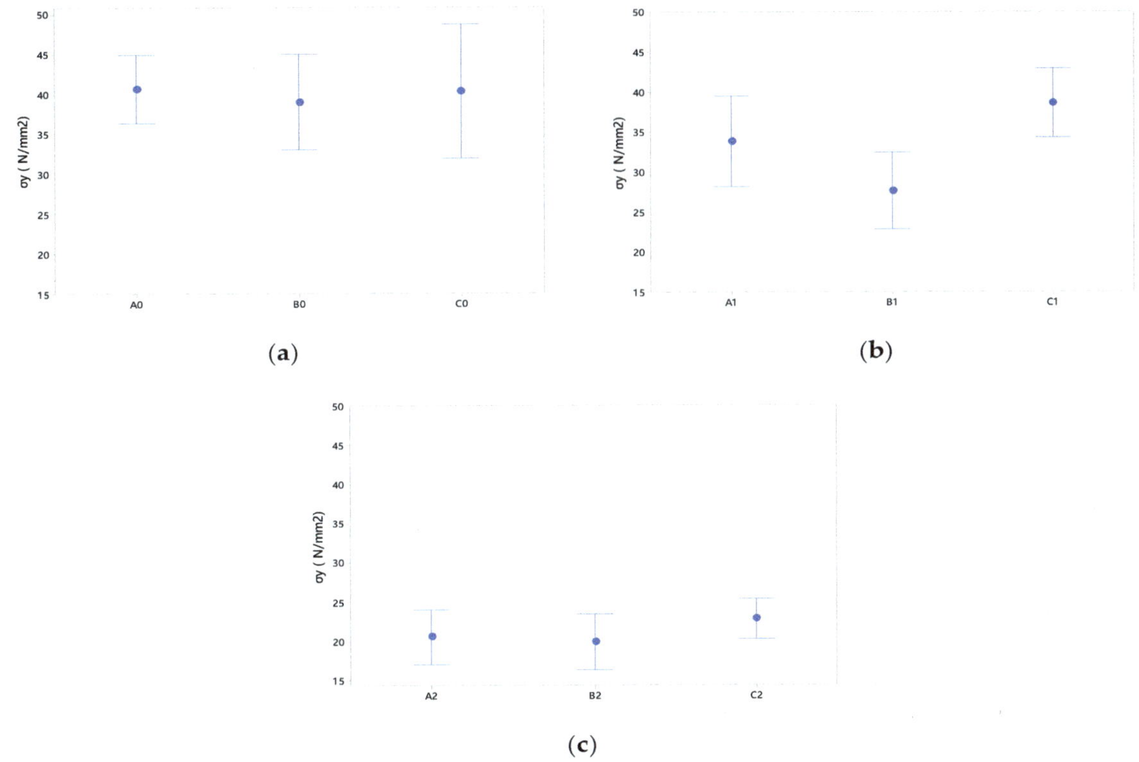

Figure 14. Comparative graph of the average results obtained for the yield stress of the three types of relief (A, B and C): (**a**) for the lower relief measure 0, 1.5 × 1.5 mm; (**b**) for the lower relief measure 1 of 3 × 3 mm; (**c**) for the relief measure 2 of 6 × 6 mm.

4.1.3. Simulation Validation

In general, the curve obtained in the different geometries is partially shifted to the left with respect to the experimental ones, probably due to the placement of the specimen in the testing machine, Figure 15. Another explanation can be given by the flexible characteristic of the specimens where the initial tension varies at the beginning of the test. In this way, it is observed that the specimen began to stress when the jaws move less than 1 mm, that is, there was a delay reflected by the described displacement.

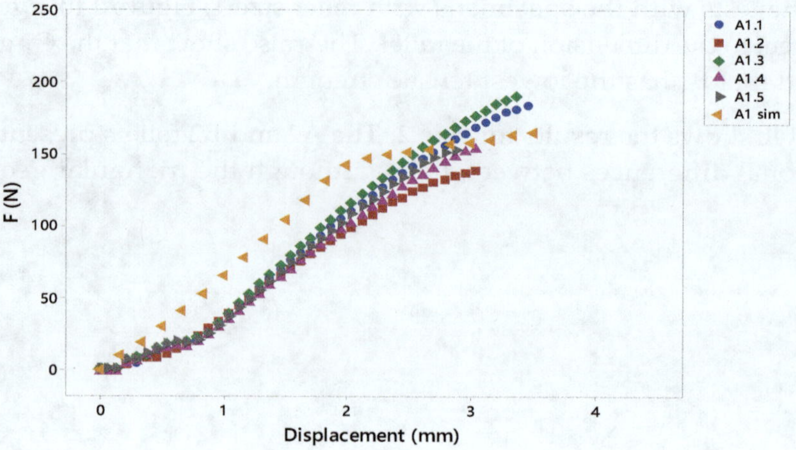

Figure 15. Example comparing the results obtained in the simulation test and the experimental studies.

On the other hand, the average values of σy obtained experimentally from the five tests and the values obtained in the simulation are detailed in Figure 16. Following [36], it is possible to the values obtained in the simulation with respect to the tests using the following equation:

$$Er = \frac{\text{Test value} - \text{simulation value}}{\text{test value}} \times 100, \quad (6)$$

where Er is the relative error.

In view of the results obtained, it can be considered that the simulation model adequately reproduces the mean specimen rupture with a relative error below 2% for circular and square geometries and slightly higher for triangular geometry specimens. This may be due to the concentration of stresses in the main corner of the triangular specimen as a consequence of the scale of the meshing and the simplification of the geometries, this being a singular point of study that shows a distorted behaviour in comparison to the rest of the reliefs.

The graph in Figure 16 shows that the C0 and C2 specimens, both with triangular geometry, are the ones with the greatest relative error for the reasons discussed above, although it is the C2 type specimen that lies outside the range of dispersion obtained experimentally, being the only set of tests under these conditions.

It should be noted that for general purposes the simulation results comply to a large extent with the experimental results validating the simulation methodology applied.

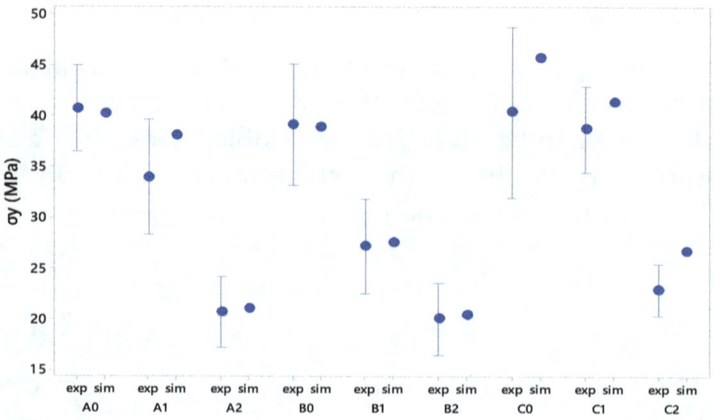

Figure 16. Comparative results of the experimental study (exp) and the simulation carried out (sim) for the yield stress.

4.2. Compression Tests

4.2.1. Visual Characterization

As shown in Table 4, in general, for the same size of relief, the film adapts better to the contour of the plastic mould for bigger separation of the reliefs. This is due to the stretching of the film. Thus, the lack of definition that occurs when the reliefs are located at a shorter distance can cause the compressive strength of the container to be reduced.

Similarly, the relief size also affects the type of adaptation of the film on the reliefs. Thus, the larger the relief size, the poorer the definition of the geometry obtained after thermoforming. According to [37] and [38], this is due to the fact that the machine has to exert a greater force to adapt the sheet to the geometry, also producing greater stretching. Thus, in the smaller geometries, A0 and A1, the sheet adapts to the relief, while in the A2 geometry, with a height of 6 mm, there is less definition in the intermediate spaces between reliefs.

It is observed that in the lateral walls the thickness of the sheet was reduced considerably being the part of the sheet where the greatest stretching took place. Therefore, the smaller the distance between reliefs, the less defined the relief geometry.

Table 4. Kinds of specimens made according to the number of reliefs and disposition.

	M1	M2	M3	M4	M5
A0					
A1					
A2					

4.2.2. Mechanical Evaluation

The force in the tests made on the thermoformed specimen without reliefs, O2, presented an average F_y of 5.13 N, Figure 17a. The standard deviation presented an interval amplitude range of 1.18 N. According to the images collected in the test video, this deviation in the results was due to the fact that the container began to deform on the walls and folds were produced that could result in a different F_y. Thus, it was observed that, when there was a displacement of 0.5 mm, the container was deformed but no folds were observed. At that moment, the average force collected was 3.10 N and its deviation was 0.43 N, Figure 17b. From the data collected, it can be seen that the 3 sizes, A0, A1 and A2, improved the F-ε properties with respect to the O2 container, although it is true that A2 shows higher values than the rest. Comparing the results obtained by separation and reliefs, the graph shows that slightly higher results are obtained when the reliefs were arranged with a separation of 6 mm (M3 and M5) conserving the same number of reliefs per face: M3 with respect to M2 and M5 with respect to M4. This phenomenon was in good agreement with the videos studied since the greater the relief, the less the deformation of the walls was, due to the fact that a greater distance between reliefs favoured the sheet adaptation to the mould during the thermoforming process.

On the other hand, it should be pointed out that, although A2 has the highest values of force with respect to A0 and A1, it stands out for reflecting higher deviations, especially in A2-M2 and A2-M3, indicating that the behaviour of these containers was more irregular when tested as a consequence of having a higher relief height.

In Figure 17b, it is observed that the recorded force data, for a 0.5 mm of displacement, have greater homogeneity when compared at the same point and not at the yield point, where each test shows different values. Thus, in this case, the results show a similar trend in terms of behaviour based on geometry, relief and separation.

In addition, the dispersion of the data is considerably reduced, which indicates that the tests carried out had a greater homogeneity in the first stages of the test, making possible to clearly distinguish the package that presented better properties. Also, it should be noted that, for deformities greater than 0.5 mm, the packaging could lose its functional properties, thus deteriorating the product contained inside.

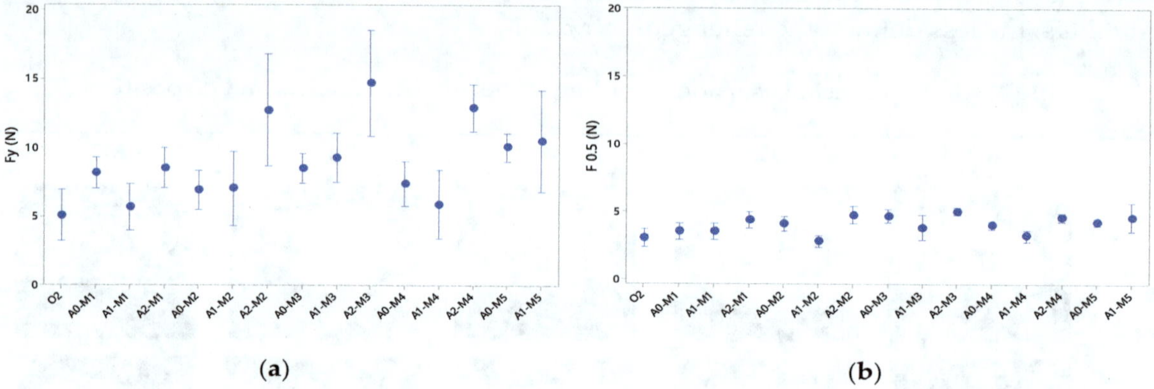

Figure 17. (**a**) Compression forces at stress yield; (**b**) Compression forces for a displacement of 0.5 mm.

4.2.3. Case Study of Validation by FEM

Figure 18 shows that the compression curve obtained by simulation for the A1-M1 specimen is within the range of the experimentally measured values. When a displacement of 0.5 mm occurred, the obtained force data was also homogeneous with respect to laboratory tests.

Thus, the relative error, according to Equation (6), is 0.14% with respect to the mean data. It is therefore within the dispersion value collected in the performance of the 4 experimental tests.

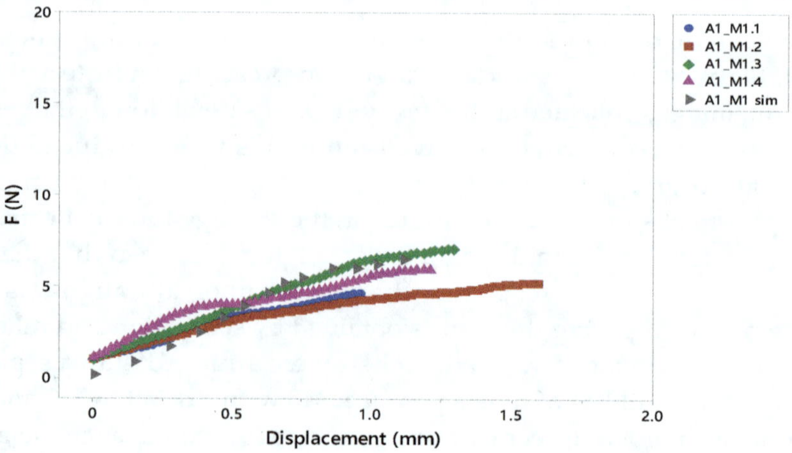

Figure 18. Comparative graph of the results obtained in the compression test by FE.

4.3. *Relationship between Tensile and Compressive Tests*

Figure 19 shows the area graphs that relate the force data obtained in the tensile tests, F_y-T, where the reliefs have been studied in a unitary way, with respect to the results of the compressive strength in two stages, F0.5-C and F_y-C, according to the number of reliefs included per face. Figure 19a relates the results of the F_y-T with respect to the force at the moment when the displacement has reached 0.5 mm. On the other hand, in Figure 19b the performance is made with respect to the force to compression at the creep point, F_y-C.

It is observed that the most favourable data of traction and compression, for the force at the moment of 0.5 mm of displacement, were collected with 3 and 5 reliefs per face, Figure 19a. In the case of the F_y-C, Figure 19b, data reflects that the results that obtain an equilibrium of compressive force between 8 and 10 N in addition to good traction results are the containers presenting 5 reliefs per face.

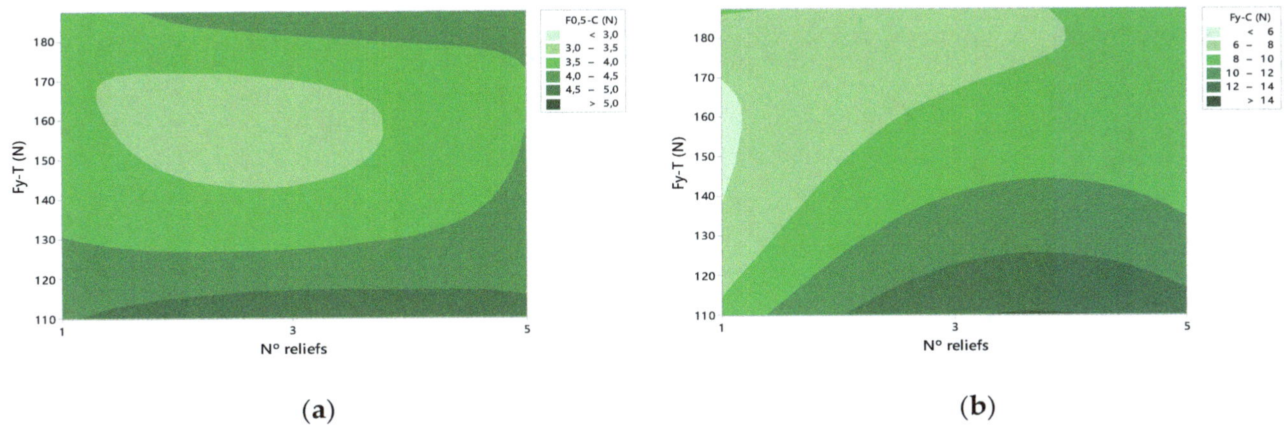

Figure 19. Graph of the compression force, F-C, in relation to the number of reliefs per face and the tensile force at the yield point, F_y-T, (**a**) at the moment of displacement of 0.5 mm, F0.5-C (**b**) at the yield point, F_y-C.

5. Conclusions

The physical tests carried out show that, the higher the relief, the lower the tensile strength compared to smaller sizes. This may be due to the stretching of the sheet on the side walls. However, it provides greater rigidity to the container by favouring compressive strength by introducing relief patterns on the surface of the thermoforming container.

From the geometries studied it can be observed that the semi-circular geometry, A and triangular, C, are the ones that present the best results. Also, the semi-circular geometry, A, adapts better to the shape of the mould since the shape of the relief has no edges.

The analysis carried out shows that the main structural factors are the relationship between the width of the relief and its height. In addition, the type of geometry performed affects the mechanical properties. Likewise, from the comparison of the physical tests and the simulation it is concluded that the correct positioning and tension of the specimen on the testing machine influences the displacement registered in the laboratory tests.

Thus, it can be stated that, by simulation tests, a virtual evaluation method can be established as a basis for the optimization of the design applied to thermoforming packaging. In addition, validation by means of FEM tests gives rise to the possibility of testing new reliefs with more complex geometries. Therefore, it is proposed that, by means of the individualized study of the behaviour of the reliefs, it is possible to compare their mechanical properties.

Moreover, it was also possible to develop a method that has allowed satisfactory results after the compression tests carried out. The results obtained show that the correct design of the reliefs, according to the number of reliefs used and their disposition, can favour the improvement of the mechanical properties of the containers. Specifically, the reliefs with greater distance increase the resistance of the packaging.

In comparison with the conventional design process of thermoforming packaging, this new procedure, aimed at the study of reliefs using low-cost moulds of AM, facilitates the realization of more innovative and efficient designs through the study of geometries, number of reliefs and their positioning. In this sense, it is possible to design containers with lower material costs by improving the mechanical properties using materials of reduced thickness and increasing the protection of the food.

Along these lines, a field of study was opened on the inclusion of complex relief patterns on thermoformed surfaces. The realization of complex patterns can be possible by means of the application of FDM moulds. Also, these moulds could be used to make cut series and thus improve personalization and brand identity. In short, this work presents a tool for designers that facilitates customization, eco-design and allows them to optimize their mechanical properties.

Author Contributions: Conceptualization, L.R.-P.; methodology, L.R.-P., P.F.M. and A.J.G.; software, L.R.-P.; validation, L.R.-P., P.F.M. and A.J.G.; formal analysis, L.R.-P., P.F.M. and A.J.G.; investigation, L.R.-P. and P.F.M.; resources, A.J.G.; data curation, L.R.-P., P.F.M. and A.J.G.; writing—original draft preparation, L.R.-P.; writing—review and editing, P.F.M. and A.J.G.; supervision, P.F.M. and A.J.G. All authors read and approved the final manuscript.

References

1. Paine, F.A. *Fundamentals of Packaging*; Institute of Packaging: Leicestershire, UK, 1981; ISBN 0950756709.
2. Trinetta, V. Definition and Function of Food Packaging. In *Reference Module in Food Science*; Elsevier: Amsterdam, The Netherlands, 2016. [CrossRef]
3. Hannay, F. *Rigid Plastics Packaging: Materials, Processes and Applications*; Rapra: Shawbury, UK, 2002; ISBN 1859573584.
4. Selke, S.; Culter, J. *Plastics Packaging*, 3rd ed.; Hanser: Munich, Germany, 2016; ISBN 9781569904435.
5. Hellström, D.; Olsson, A. *Managing Packaging Design for Sustainable Development: A Compass for Strategic Directions*; Wiley: Chichester, UK, 2017; ISBN 1119150930.
6. Rosato, D.V. Plastics End Use Application Fundamentals. In *Plastics End Use Applications. Springer Briefs in Materials*; Springer: Berlin, Germany, 2011; pp. 11–18.
7. Siracusa, V.; Rosa, M.D. Sustainable Packaging. In *Sustainable Food Systems from Agriculture to Industry*; Academic Press: Cambridge, MA, USA, 2018; pp. 275–307.
8. Baselice, A.; Colantuoni, F.; Lass, D.A.; Nardone, G.; Stasi, A. Trends in EU consumers' attitude towards fresh-cut fruit and vegetables. *Food Qual. Prefer.* **2017**, *59*, 87–96. [CrossRef]
9. Cagnon, T.; Méry, A.; Chalier, P.; Guillaume, C.; Gontard, N. Fresh food packaging design: A requirement driven approach applied to strawberries and agro-based materials. *Innov. Food Sci. Emerg. Technol.* **2013**, *20*, 288–298. [CrossRef]
10. Santeramo, F.; Carlucci, D.; De Devitiis, B.; Seccia, A.; Stasi, A.; Viscecchia, R.; Nardone, G. Emerging trends in European food, diets and food industry. *Food Res. Int.* **2018**, *104*, 39–47. [CrossRef] [PubMed]
11. Simmonds, G.; Spence, C. Thinking inside the box: How seeing products on, or through, the packaging influences consumer perceptions and purchase behaviour. *Food Qual. Prefer.* **2017**, *62*, 340–351. [CrossRef]
12. Becker, L.; van Rompay, T.J.L.; Schifferstein, H.N.J.; Galetzka, M. Tough package, strong taste: The influence of packaging design on taste impressions and product evaluations. *Food Qual. Prefer.* **2011**, *22*, 17–23. [CrossRef]
13. Westerman, S.; Sutherland, E.; Gardner, P.; Baig, N.; Critchley, C.; Hickey, C.; Mehigan, S.; Solway, A.; Zervos, Z. The design of consumer packaging: Effects of manipulations of shape, orientation, and alignment of graphical forms on consumers' assessments. *Food Qual. Prefer.* **2013**, *27*, 8–17. [CrossRef]
14. Celhay, F.; Boysselle, J.; Cohen, J. Food packages and communication through typeface design: The exoticism of exotypes. *Food Qual. Prefer.* **2015**, *39*, 167–175. [CrossRef]
15. Rundh, B. Packaging design: Creating competitive advantage with product packaging. *Br. Food J.* **2009**, *111*, 988–1002. [CrossRef]
16. Klimchuk, M.R.; Krasovec, S.A. *Packaging Design: Successful Product Branding from Concept to Shelf*; John Wiley & Sons: Hoboken, NJ, USA, 2012; ISBN 1118358546.

17. Marcos, B.; Bueno-Ferrer, C.; Fernández, A. Innovations in Packaging of Fermented Food Products. In *Novel Food Fermentation Technologies*; Springer: Cham, Switzerland, 2016; pp. 311–333.
18. Ritnamkam, S.; Chavalkul, Y. The Design Factors of Cosmetic Packaging Textures for Conveying Feelings. *Asian Soc. Sci.* **2017**, *13*, 86. [CrossRef]
19. Falck, R.; Goushegir, S.M.; dos Santos, J.F.; Amancio-Filho, S.T. AddJoining: A novel additive manufacturing approach for layered metal-polymer hybrid structures. *Mater. Lett.* **2018**, *217*, 211–214. [CrossRef]
20. Quinlan, H.E.; Hasan, T.; Jaddou, J.; Hart, A.J. Industrial and Consumer Uses of Additive Manufacturing: A Discussion of Capabilities, Trajectories, and Challenges. *J. Ind. Ecol.* **2017**, *21*, S15–S20. [CrossRef]
21. Collins, P.K.; Leen, R.; Gibson, I. Industry case study: Rapid prototype of mountain bike frame section. *Virtual Phys. Prototyp.* **2016**, *11*, 295–303. [CrossRef]
22. Ko, H.; Moon, S.K.; Hwang, J. Design for additive manufacturing in customized products. *Int. J. Precis. Eng. Manuf.* **2015**, *16*, 2369–2375. [CrossRef]
23. Jiménez, M.; Romero, L.; Domínguez, M.; Espinosa, M.M. Rapid prototyping model for the manufacturing by thermoforming of occlusal splints. *Rapid Prototyp. J.* **2015**, *21*, 56–69. [CrossRef]
24. Serrano-Mira, J.; Gual-Ortí, J.; Bruscas-Bellido, G.; Abellán-Nebot, V. Use of additive manufacturing to obtain moulds to thermoform tactile graphics for people with visual impairment. *Procedia Manuf.* **2017**, *13*, 810–817. [CrossRef]
25. Boisse, P.; Wang, P.; Hamila, N. Thermoforming simulation of thermoplastic textile composites. In Proceedings of the 16th European Conference on Composite Materials, Seville, Spain, 22–26 June 2014; pp. 1–7, ISBN 9780000000002.
26. Van Mieghem, B.; Desplentere, F.; Van Bael, A.; Ivens, J. Improvements in thermoforming simulation by use of 3D digital image correlation. *Express Polym. Lett.* **2015**, *9*, 119–128. [CrossRef]
27. Zhang, Y.; Tong, Y.; Zhou, K. Coloring 3D Printed Surfaces by Thermoforming. *IEEE Trans. Vis. Comput. Graph.* **2017**, *23*, 1924–1935. [CrossRef] [PubMed]
28. Haldane, D.W.; Casarez, C.S.; Karras, J.T.; Lee, J.; Li, C.; Pullin, A.O.; Schaler, E.W.; Yun, D.; Ota, H.; Javey, A.; et al. Integrated Manufacture of Exoskeletons and Sensing Structures for Folded Millirobots. *J. Mech. Robot.* **2015**, *7*, 021011. [CrossRef]
29. D 6287-98. *Standard Practice for Cutting Film and Sheeting Test Specimens*; ASTM: West Conshohocjen, PA, USA, 1998.
30. Throne, J.L. *Understanding Thermoforming*, 2nd ed.; Hanser: Cincinnati, OH, USA, 2008; ISBN 9783446407961. [CrossRef]
31. UNE-EN ISO 527-3. *Determination of Tensile Properties. Part 3: Test Conditions for Films and Sheets*; AENOR: Madrid, Spain, 1995.
32. DIN 53479. *Testing of Plastics and Elastomers; Determination of Density*; DIN: Berlin, Germany, 1976.
33. D 882-18. *Standard Test Method for Tensile Properties of Thin Plastic Sheeting*; ASTM: West Conshohocjen, PA, USA, 2018.
34. UNE-EN ISO 12048:2001. *Pakaging: Comprete, Filled transport Packages. Compression and Stacking Test Using a Compression Tester*; AENOR: Madrid, Spain, 2001.
35. UNE-EN ISO 604. *Determination of Compressive Properties*; AENOR: Madrid, Spain, 2003.
36. Carvalho, C.; Baltar, J. Simulation of uniaxial tensile test through of finite element method. *INOVAE J. Eng. Archit. Technol. Innov.* **2017**, *5*, 3–13.
37. Loepp, D. Emerging Trends in Thermoforming. *Plasticnews*, 2015. Available online: http://www.plasticsnews.com/article/20150507/BLOG01/150509942/emerging-trends-in-thermoforming (accessed on 21 March 2018).
38. Sreedhara, V.S.M.; Mocko, G. Control of thermoforming process parameters to increase quality of surfaces using pin-based tooling. In Proceedings of the 20th Design for Manufacturing and the Life Cycle Conference, Boston, MA, USA, 2–5 August 2015; Volume 4, p. V004T05A016. [CrossRef]

12

Evaluation of the Rosemary Extract Effect on the Properties of Polylactic Acid-Based Materials

Raluca Nicoleta Darie-Niţă [1], Cornelia Vasile [1,*], Elena Stoleru [1,*], Daniela Pamfil [1], Traian Zaharescu [2], Liliana Tarţău [3], Niţă Tudorachi [1], Mihai Adrian Brebu [1], Gina Mihaela Pricope [4], Raluca Petronela Dumitriu [1] and Karol Leluk [5]

1 Department of Physical Chemistry of Polymers, "Petru Poni" Institute of Macromolecular Chemistry, 41A Gr. Ghica Voda Alley, 700487 Iasi, Romania; darier@icmpp.ro (R.N.D.-N.); pamfil.daniela@icmpp.ro (D.P.); ntudor@icmpp.ro (N.T.); bmihai@icmpp.ro (M.A.B.); rdumi@icmpp.ro (R.P.D.)

2 National Institute for Electrical Engineering (INCDIE ICPE CA), 313 Splaiul Unirii, P.O. Box 149, 030138 Bucharest, Romania; traian_zaharescu@yahoo.com

3 Grigore T. Popa University of Medicine and Pharmacy Iasi, 16 University Street, 700115 Iasi, Romania; lylytartau@yahoo.com

4 Veterinary and Food Safety Laboratory, Department of Food Safety, 700115 Iasi, Romania; ginacornelia@yahoo.com

5 Institute of Environmental Protection Engineering, Wroclaw University of Technology, Plac Grunwaldzki 9, 50-377 Wroclaw, Poland; kleluk@yahoo.com

* Correspondence: cvasile@icmpp.ro (C.V.); elena.paslaru@icmpp.ro (E.S.)

Abstract: New multifunctional materials containing additives derived from natural resources as powdered rosemary ethanolic extract were obtained by melt mixing and processed in good conditions without degradation and loss of additives. Incorporation of powdered rosemary ethanolic extract (R) into poly(lactic acid) (PLA) improved elongation at break, rheological properties, antibacterial and antioxidant activities, in addition to the biocompatibility. The good accordance between results of the chemiluminescence method and radical scavenging activity determination by chemical method evidenced the increased thermoxidative stability of the PLA biocomposites with respect to neat PLA, with R acting as an antioxidant. PLA/R biocomposites also showed low permeability to gases and migration rates of the bioactive compounds and could be considered as high-performance materials for food packaging. In vitro biocompatibility based on the determination of surface properties demonstrated a good hydrophilicity, better spreading and division of fibroblasts, and increased platelet cohesion. The implantation of PLA/R pellets, was proven to possess a good in vivo biocompatibility, and resulted in similar changes in blood parameters and biochemical responses with the control group, suggesting that these PLA-based materials demonstrate very desirable properties as potential biomaterials, useful in human medicine for tissue engineering, wound management, orthopedic devices, scaffolds, drug delivery systems, etc. Therefore, PLA/R-based materials show promising properties for applications both in food packaging and as bioactive biomaterials.

Keywords: powdered rosemary ethanolic extract; poly(lactic acid); bioactive food packaging; biomaterials

---

1. Introduction

The use of natural additives is gaining increasing interest in the development of new multifunctional materials as a key for new active materials strategies. Different natural essential oils have been proposed for incorporation into the polymer matrices to improve the functionality as well as the products (food and pharmaceuticals) quality and safety.

The action of natural additives is essential in reducing or even eliminating some of the main spoilage causes, such as rancidity, color loss/change, active compounds losses, dehydration, microbial proliferation, senescence, gas build-up, and off-odors, etc. [1].

Nature offers several solutions to obtain appropriate antioxidants with higher activity than those which are synthesized. Ethanolic extraction is a method by which the highest quantity of phenolic compounds can be extracted from plant leaves. Isolated compounds, as well as crude ethanol extract, showed antibacterial activities against four Gram-negative bacteria strains, *Escherichia coli*, *Pseudomonas aeruginosa*, *Klebsiella pneumoniae*, and *Enterobacter aerogenes* [2], in addition to antioxidant activity because the content in phenolic compounds is preserved [3].

The use of extracts from rosemary (*Rosmarinus officinalis*) as food preservatives is well established [4,5]. A broad range of beneficial health effects can be attributed to rosemary, such as antidepressant, antihypertensive, antiproliferative, antibacterial, antiatherogenic, hypocholesterolemic, hepatoprotective, and anti-obesity properties. The biological properties of rosemary are attributed to the contribution of its different bioactive compounds belonging mainly to the classes of phenolic acids, flavonoids, diterpenoids, and triterpenes. Rosemary samples have the highest levels of flavonoids and other compounds such as carnosol, rosmaridiphenol, rosmadial, rosmarinic acid, and carnosic acid [6]. It was found to be very efficient to protect food against lipid oxidation [7] and is known as rosemary active packaging [8]. Antioxidative efficiency is imparted by at least 20 specific phenols, the most effective compounds are carnosol, rosmarinic acid, and carnosic acid, followed by caffeic acid, rosmanol, rosmadial, genkwanin, and cirsimaritin. Rosemary extracts derived from *Rosmarinus officinalis* L. contain several compounds which have been shown to possess antimicrobial, antioxidative, anti-inflammatory, antiviral, and anti-tumor functions. Literature reports either rosmarinic acid, an ester of caffeic acid and 3,4-dihydroxyphenyllactic acid or/and the phenolic diterpenes carnosol and carnosic acid as the principal antioxidative components of the rosemary extract (Figure 1) [9,10].

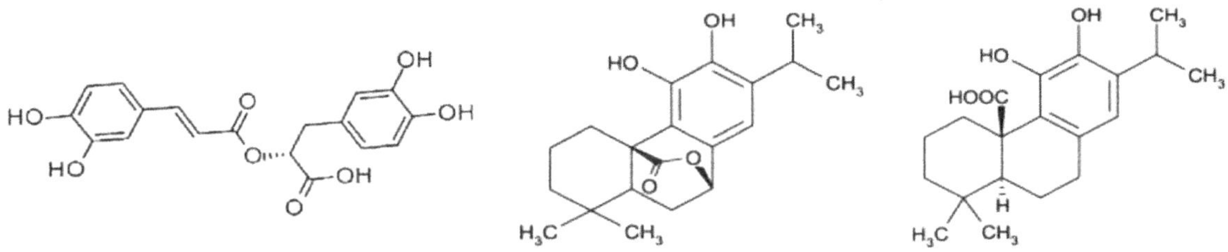

Rosmarinic Acid Carnosol Carnosic acid

Figure 1. Chemical structure of the major antioxidative compounds in rosemary extracts.

Almost 90% of the rosemary leaf extract's antioxidant activity can be attributed to carnosol and carnosic acid [11].

Knowing that oxidative stress plays a determinant role in the pathogenesis of liver diseases, at this moment the antioxidant products from natural sources are being increasingly used to treat various pathological liver conditions. Due to its antioxidant and antimicrobial properties, rosemary essential oil (REO) is already largely used in the food industry as a preservative, while additionally possessing other health benefits. In addition to free radical scavenging activity, the REO mediates the hepatoprotective effects by activating the physiological defense mechanisms. Rašković et al. [12] have found 29 chemical compounds of a selected REO, and the main ones identified were 1,8-cineole (43.77%), camphor (12.53%), and α-pinene (11.51%). The essential oil they tested has exerted hepatoprotective effects on rats with carbon tetrachloride-induced acute liver damage, at doses of 5 mg/kg and 10 mg/kg, by diminishing aspartate transaminase (AST) and alanine aminotransferase (ALT) activities by up to two-fold and preventing lipid peroxidation in liver homogenates. Moreover, pre-treatment with the tested essential oil for seven days has significantly reversed the activities of antioxidant enzymes (such as catalase, peroxidase, glutathione peroxidase, and glutathione reductase) in liver homogenates, mainly using a dose of 10 mg/kg.

Polylactic acid (PLA) is one of the well-known biodegradable, biocompatible, non-toxic, and eco-friendly polyesters. The low cost and attractive materials properties of PLA would open many applications. Despite the great advantages of PLA, it presents poor toughness, thus its use in obtaining materials that require high deformation capabilities is limited. There are several ways to improve PLA's processability, flexibility, and ductility, such as: blending with polymers, copolymerization with other monomers, plasticization using biocompatible plasticizer, or incorporation of filler materials [13]. The most promising method to increase PLA flexibility for film manufacturing as well as to improve the compatibility between PLA and other additives in the blend has been proven to be blending with low-molecular-weight polymers, which act as plasticizers [14]. An efficient plasticizer for PLA must reduce the glass transition temperature and the crystallinity, not migrate and show miscibility with the matrix, and possess a low volatility and a lack of toxicity [15]. Depending on the targeted application, there are various types of plasticizers for PLA, but in view of obtaining sustainable materials, mostly eco-friendly plasticizers are used today [16]. Literature shows possible plasticizers used for PLA processing, such as, oligomeric lactic acid, L-lactide, poly(ethylene glycol) (PEG), epoxidized soybean oil (USE), citrate esters, glycerol, and glucose monoesters [14,17]. Among these, the most common and suitable plasticizer for PLA has been reported to be low-molecular-weight PEG due to its miscibility, biodegradability, glass transition temperature (T_g) reduction, and approval for food-contactable materials [13,17]. The major characteristics of PEG are its biocompatibility, water solubility, and low cost that recommend its use in biomaterials composition [18].

PLA and its composites are currently being extensively studied to obtain degradable food packaging and are also widely used in human medicine for tissue engineering, wound management, orthopedic devices, biomedical and clinical applications in scaffolds, bone fixation devices such as screws and plates, surgical suture and meshes, medical implants, drug delivery systems, etc. [19–22].

As natural additives are too volatile to be directly used in the processing technologies, a viable alternative is to incorporate these compounds into materials as natural active additives with the possibility to be released from the material of interest and to reduce their volatility. The release rate of active substances such as antioxidants from a packaging material can be evaluated through migration studies, usually performed using food simulants and conditions specified in the European food packaging regulations [23,24]. As it is desirable to place the samples in contact with the food simulant in the worst foreseeable conditions of time and temperature pertaining to actual use, and as increasing the temperature accelerates the migration, the conditions for the release studies were 40 °C for a minimum of ten days with a 50% aqueous ethanol solution used as a food simulant. This is known as a modified D1 food simulant, which is assigned as a conventional simulant selected for foods with lipophilic character, oil-in-water emulsion character, hydrophilic foods containing relevant amounts of organic ingredients, and for hydrophilic alcohol-containing foods with over 20% alcohol content. Generally, food simulant D1 can be considered to be the most severe aqueous food simulant for hydrophilic non-acidic foods [23]. Since ethanol is commonly used as a food simulant, the 50% aqueous ethanol solution was chosen to perform these tests in order to comply with the latest approved directives on migration testing (EU EC. No. 10/2011) [23]. Through ethanol sorption in the PLA matrix, voids or swelling of the film can occur, facilitating penetration within PLA chains and promoting the migration of the active substances [25,26].

In this paper antimicrobial and antioxidant activity was offered to the PLA by mixing with ethanolic rosemary extract as a powder. This was obtained by alcoholic extraction of the rosemary leaves followed by precipitation [27].

2. Experimental

2.1. Materials

Polylactic acid (PLA) obtained from renewable resources from NatureWorks LLC (trade name: PLA 2002D, Minnetonka, MN, USA) is a transparent material with a melt flow index of 5–7 g/10 min (conditions 210 °C/2.16 kg) and a content of 96% L-lactide and 4% isomer D. Average molecular weight determined by GPC was 4475 kDa. It has a density of 1.25 g/cm^3, melting point of 152 °C, glass transition temperature of 58 °C. The crystallinity depends on the isomer content and thermal history. Water permeability at 25 °C was 172 g/m^2*day and percentage of biodegradation/mineralization is 100% [16].

Poly(ethylene glycol) (PEG) BioUltra 4000 purchased from Sigma-Aldrich (St. Louis, MO, USA) was used as a plasticizer.

Powdered rosemary ethanolic extract (R) was obtained following a previously reported procedure by the solvent extraction method in a Soxhlet unit [27] involving the alcoholic extraction of the rosemary leaves followed by precipitation. Fresh rosemary leaves were collected from local farms, dried at ambient temperature and subsequently milled, cleaned, and powdered in a grinder (Laboratory of Radiation Chemistry, INCDIE-ICPE CA. Bucharest, Romania). The vegetal material was then refluxed in 2 L of ethanol at 65–75 °C for 10 h during continuous extraction in the Soxhlet apparatus (Laboratory of Radiation Chemistry, INCDIE-ICPE CA. Bucharest, Romania). The extract in ethanol solution was subjected to precipitation induced by the addition of water as non-solvent and a solid phase was finally filtered, washed with acetone, dried in air, and further dried under vacuum at ambient temperature. A greenish-yellow fine powder was obtained and stored in desiccators to avoid the absorption of moisture.

All other reagents were of analytical grade purity.

The composition of the rosemary ethanolic extracts was detailed and studied because of the importance of its biological activity in various applications. As previously mentioned in the introduction, the rosemary extracts, especially the ones derived from the leaves, are usually herbal products used as natural antioxidant and flavoring agents in food processing and cosmetics. In the present study, the phenolic and flavonoids contents have been determined according to the following methods.

2.1.1. Determination of Total Phenolic Content

Total phenols content was determined using Folin-Ciocalteu's reagent (FC) method as presented in the work of Scalbert et al. [28]. Powdered rosemary extract (0.01 g) was added to 10 mL methanol and then 0.1 mL of methanolic solution was diluted with 0.4 mL of double distilled H_2O. FC/H_2O solution (1:10 *v*/*v*) (1 mL) was added to 0.5 mL diluted methanol/R extract and left for 10 min at 25 °C after a strong mixing. A 2 mL solution of 15% sodium carbonate ($Na_2CO_3 \cdot 10H_2O$) was added, and after 1 h of incubation, the absorbance was read at 740 nm with a Cary 60 UV-Vis spectrophotometer (Agilent Technologies, Santa Clara, CA, USA) against a freshly prepared blank sample. The blank sample contained 0.1 mL methanol without powdered rosemary ethanolic extract, 0.4 mL of water, 1 mL FC reagent solution, and 2 mL of $Na_2CO_3 \cdot 10H_2O$ solution. Solutions of gallic acid (used as standard) in methanol of different concentrations (0.01–1 mg/mL) were prepared and based on the measured absorbance at 740 nm, the calibration curve was drawn. The obtained calibration curve was then used for the determination of the total phenolics content expressed as mgGAE/g dw (mg of gallic acid equivalent per g of dry weight). The determination was made in triplicate and an average value was obtained. The resulted total phenolic content was of 112.5 mg GAE/g dw.

2.1.2. Determination of Total Flavonoids Content

Total flavonoid content was determined by the aluminum chloride colorimetric assay [29]. A quantity of 1 mL of extract (1 mg/mL powdered rosemary ethanolic extract) or standard Quercetin (of different concentrations) was added to 4 mL of H_2O. Then, at different time intervals the following solutions were added consecutively: 0.30 mL of 5% $NaNO_2$; 0.3 mL of 10% $AlCl_3$ after 5 min; and 2 mL of 1 M NaOH after another 5 min. At the end, the volume was made up to 10 mL with H_2O, mixed, and absorbance was measured against the blank sample at 510 nm. The total content of flavonoids was expressed as mg Quercetin Equivalents (QE)/g dw. The measurement was made in triplicate and an average value was reported. Thus, a total flavonoids content of 261.5 (mg QE/g dw) was found.

The obtained values both for phenolic and flavonoids content of the powdered ethanolic rosemary extract are close to those found in the literature, which explain, in part, its good antioxidant properties [30].

2.2. *PLA-Based Blends Processing*

PLA-based blends were prepared using different amounts of powdered ethanolic rosemary extract (R) by incorporation into a PLA matrix in the melt state. Before mixing, drying of the additives and PLA was done in a vacuum oven (Binder, Tuttingen, Germany) for 6 h at a temperature of 80 °C. The compounding was performed for 10 min, at 175 °C, and 60 rpm, by means of a Brabender mixer (30EHT, Duisburg, Germany). A Carver press (Wabash MPI, IN, USA) and special parameters for compression molding (175 °C, or 165 °C for PLA/R blends, pre-pressing for 3 min at 50 atm and a pressing for 2 min at 150 atm) were used to obtain specimens for different analyses. Both films (thickness of ~0.15 mm), as well as sheets (thickness of ~1 mm), were prepared for different analyses that required specific thickness. Sheets were used for mechanical (tensile) and rheological tests (1 mm distance between parallel plates) while films were used for spectral characterization, permeability tests, etc. The compositions of the prepared systems are shown in Table 1.

Table 1. Compositions of the prepared poly(lactic acid) (PLA)-based systems.

No.	Sample	PLA (wt %)	Powdered Rosemary Ethanolic Extract (R) (wt %)	PEG (wt %)
1	PLA	100	-	-
2	PLA/0.25R	99.75	0.25	-
3	PLA/0.5R	99.5	0.5	-
4	PLA/0.75R	99.25	0.75	-
5	PLA/PEG	80	-	20
6	PLA/PEG/0.5R	79.5	0.5	20

2.3. *Investigation Methods*

2.3.1. Processing Behavior

Processing behavior was evaluated by analysis of processing characteristics following the torque-time curves registered during blending on a Brabender mixer.

2.3.2. Scanning Electron Microscopy (SEM)

The samples were fixed on copper supports for the SEM investigation and the surface of the films was examined as such, without metal coating. An Environmental Scanning Electron Microscope (ESEM) type Quanta 200 Instrument (FEI Company, Hillsboro, TX, USA) was used, operating at 25 kV with secondary electrons in low vacuum mode (LFD detector).

2.3.3. ATR–FTIR Spectroscopy

A Bruker VERTEX 70 spectrometer (Ettlingen, Germany) was used to record the Attenuated Total Reflectance-Fourier Transformed Infrared ATR-FTIR spectra by a 4 cm^{-1} resolution. Both the spectra of samples as well as background were recorded in the wavenumber interval of 4000–600 cm^{-1}, and the OPUS program was used for the processing of spectra.

2.3.4. Stress-Strain Measurements

Tensile measurements were performed at room temperature on dumbbell-shaped samples, by means of an Instron Single Column machine (3345, Norwood, MA, USA), according to EN ISO 527-2/2012. The load cell was of 1 kN, the loading speed was of 10 mm/min, and the gauge length was 40 mm. Young's modulus, tensile strength at break, and strain at break have been evaluated.

2.3.5. Dynamic Rheology

An Anton Paar rheometer (MCR301, Graz, Austria) was used to study the rheological properties of the PLA/R composites. Oscillatory frequency sweeps were performed using a geometry with a parallel plate of 25 mm in diameter, at 175 °C, and a strain of 10% (linear viscoelasticity region) in the range between 0.05 to 500 rad/s.

2.3.6. Differential Scanning Calorimetry (DSC)

Thermal analysis of PLA/R composites was performed under a nitrogen atmosphere by means of a TA Instruments Q20 Dynamic Scanning Calorimeter (New Castle, DE, USA). Each sample of ~10 mg was cooled down to 0 °C and heated up to +250 °C, which was below the glass transition and above the melting temperature of the materials. After the first heating run, each sample was kept for 2 min and then cooled down to 0 °C with a cooling rate of 5 °C/min and reheated up to 250 °C with a heating rate of 10 °C/min. As PLA is a semicrystalline polymer with a low crystallinity, the cooling rate must be slower than the heating rate in order to increase the crystallinity. An empty crucible was used as a reference. The parameters achieved through the DSC experiments were the glass transition temperature, cold crystallization, as well as the melting temperature and crystallinity. The degree of crystallinity of the investigated samples was obtained by dividing the melting enthalpy of the sample by ΔH_m = 93.7 J/g [31], which is equilibrium enthalpy of PLA sample with 100% crystallinity (PLA 100%).

2.3.7. Thermogravimetry–Infrared Spectroscopy/Mass Spectrometry (TG-FTIR/MS) Coupled Analysis (TG-FTIR/MS)

The thermal behavior of the samples and evolved gases analyses were performed using a TG-FTIR/MS system. The system is equipped with a thermobalance STA 449F1 Jupiter (Netzsch, Germany), spectrophotometer FTIR model Vertex-70 Bruker, (Ettlingen, Germany) and mass spectrometer QMS 403C Aeolos (Netzsch, Germany). The thermogravimetric analyzer was calibrated for temperature and sensitivity using the melting point of some standard metals (Hg, In, Sn, Bi, Al) from −38.5 °C to 800 °C. The temperature reproducibility was ±2 °C. The heating program was between 30–600 °C, with a 10 °C min^{-1} heating rate and nitrogen as a carrier gas with a flow rate of 50 mL min^{-1}. The samples (7–8 mg) were heated in an open Al_2O_3 crucible, and Al_2O_3 was used as the reference material. The gases released during the thermal degradation were transferred by two isothermal transition lines to the FTIR and mass spectrometer. The transfer line to the FTIR spectrophotometer is heated at 190 °C, the gases are introduced in TGA-IR external modulus (gas cell V = 8.7 mL) and the spectra are recorded on a 600–4000 cm^{-1} domain with a resolution of 4 cm^{-1}. The acquisition of FTIR spectra in 3D was done with OPUS 6.5 software (Ettlingen, Germany) that measures from the start and synchronizes across all the apparatus. On the other hand, the transfer line to the mass spectrometer is heated at 300 °C and the volatile degradation products are directly transferred to an electron impact ion source of 70 eV. The data acquisition was achieved with Aeolos® 7.0 software (Netzsch, Germany), in the scan bar graph mode on the range m/z = 1–300.

2.3.8. Chemiluminescence

Chemiluminescence (CL) spectra were recorded on small-weight samples (max. 5 mg) by means of a LUMIPOL 3 unit (SAS, Bratislava, Slovakia) instrument as non-isothermal dependencies of the recorded intensity on the temperature of samples. The temperature interval used for measurements started from room temperature until 250 °C, and a low error (±0.5 °C) was associated with measured temperatures. CL measurements were performed in air under static conditions using different heating rates, namely 2, 3.5, 5, and 10 °C/min. The intensity values of CL were normalized to sample mass for their reliable comparison.

2.3.9. Antioxidant Activity Evaluation

The ABTS•⁺ (2,2′-azino-bis(3-ethylbenzothiazoline-6-sulfonic acid) diammonium salt radical cation scavenging assay can be used to assess the antioxidant activity of both hydrophilic and hydrophobic compounds. To obtain the ABTS•⁺ radical cation, 2.5 mL of 7 mM ABTS is mixed with 2.5 mL 14.7 mM potassium persulphate (KPS) aqueous solutions and the mixture is left in the dark, at room temperature, for 16 h. By the direct reaction between ABTS and KPS, the ABTS•⁺ radical cation is generated, which is a blue-green chromophore, that can undergo a reduction reaction in the presence of an antioxidant. This reaction determines a loss of absorbance at 750 nm. Using Equation (1), the antioxidant capacity was calculated and the resulting values were expressed as percentage inhibition.

$$Inhibition(\%) = \left[\frac{A_{control} - A_{sample}}{A_{control}} \times 100\right] \quad (1)$$

In the case of polymeric composites, the reaction mixture consisted of adding different volumes (mL) of the reaction mixture and 2 mL of ABTS radical solution. The reaction mixture was obtained by placing 80 mg of sample in 5 mL ethanol and stirred for 24 h at room temperature (20 °C).

2.3.10. Antimicrobial Activity

The antimicrobial activity against *Bacillus cereus, Salmonella typhymurium,* and *Escherichia coli,* included four experimental steps as follows. The first stage was the sterilization of samples in an autoclave at 110 °C and 0.5 bars for 20 min. The second step consisted in the American Type Culture Collection (ATCC) culture bacteria contamination where the preparation of ATCC cultures was done by seeding of the pre-enrichment medium and incubation for 24 h at 37 °C; then, the colonies were counted in 0.1 mL culture by selective culture medium separation; and 0.1 mL bacterial culture ATCC was seeded using a sterile swab samples surface. The third step consisted in the incubation of samples contaminated with the ATCC for 24 h at 25 °C in the dark, in sterilized glass Petri dishes, repeated for another 24 h incubation. The last step involved the identifying of target germs. The following standardized methods of bacteriology procedures were used, according to standards in force: SR ISO 16649—*E. coli*; ISO 7932:2004 ISO 21871:2006(en)—*Bacillus cereus*, SR EN ISO 6579—*Salmonella* spp.

2.3.11. Gas Permeability

The gas transmission rates of CO_2 and O_2 were recorded by means of a Lyssy L100-5000 (Systech Instruments Ltd., Johnsburg, IL, USA) manometric gas permeability tester (~4 h to equilibrium for CO_2 and ~7 h to equilibrium for O_2). Specimens with 108 mm × 108 mm × 0.2 mm dimensions were used. Four determinations were made for each sample and the average value was reported.

2.3.12. Migration Study

The migration of the active components belonging to the powdered rosemary ethanolic extract from the different polymeric PLA-based films was investigated by a total immersion migration test (EC, 1997) [32] using 50% aqueous ethanolic solution as a food simulant assigned for fatty foods which frequently undergo oxidation [33]. The release studies were performed at 40 °C for a minimum of 10 days by keeping the samples in an oven, and the selected testing conditions simulate storage at ambient temperature for an unlimited duration. The migration tests were performed with pieces of films of about 1 cm^2 in 5 mL of simulant. A blank test for the simulant and each type of control sample was carried out previously.

Aliquots of ~1 mL withdrawn from the release medium at predetermined time intervals were analyzed at 285 nm by means of a Cary 60 UV-VIS spectrophotometer (Agilent Technologies, Santa Clara, CA, USA) by scanning from 200 to 600 nm. Samples were run in quartz cuvettes with a 1 mm path length.

The active components' concentrations were calculated based on the calibration curve previously determined at 285 nm for the main components of rosemary extract, sincefor rosmarinic acid the λ_{max} values are reported in literature as 254, 290, and 328 nm, and for carnosol and carnosic acid at 283 and 246 nm [34,35]. The chosen λ_{max} value can be attributed to overlapped maxima corresponding to rosemary extract active ingredients since all of them contain the aromatic double bond.

The corresponding release curves were represented as time-dependent plots of the cumulative percentage of active component released.

In order to evaluate the migration kinetic parameters and to establish the release mechanism involved, the release data were fitted first to the Korsmeyer-Peppas model using the Equation (2) [36,37]:

$$M_t/M_\infty = \mathrm{k}\cdot t^n \tag{2}$$

The migration data were also fitted to the Higuchi kinetic model, using the Equation (3):

$$M_t/M_\infty = \mathrm{k_H}\cdot t^{1/2} \tag{3}$$

where M_t/M_∞ represents the fraction of bioactive compound (s) released at time t, n is the release exponent and k and $\mathrm{k_H}$ are the release rate constants for each model considered.

The power law release exponent n describes the release mechanism from a thin polymer sample: a value of $n = 0.5$ corresponds to a Fickian diffusion mechanism, $0.5 < n < 1$ to non-Fickian/anomalous transport, $n = 1$ to Case II transport, and $n > 1$ to super case II transport [36,37].

2.3.13. Biocompatibility Evaluation

In Vitro Biocompatibility Evaluation—Contact Angle (CA) and Surface Free Energy (SFE).

By the sessile drop method, the contact angle (CA) of the polymeric sample surfaces was measured using a CAM-200 instrument from KSV (Helsinki, Finland) at room temperature and a controlled humidity. A 1 μL drop of water was placed on the film's surface and after 10 s the static contact angle was recorded. To evaluate the wettability at least ten contact angle CA measurements were realized in different locations on the surface and the obtained average values were used further. For more details on the method see references [38,39]. To determine the surface free energy (SFE) components, the CA at equilibrium between the film surface and three pure liquids (in addition to water, methylene iodide, and formamide were used, as-purchased at maximum obtainable purity) were measured by fitting the drop profile using the Young-Laplace equation [40–43]. The acid/base (LW/AB) approach of van Oss and Good, see Equation (4) [44,45], was used to calculate the total SFE and its components, namely the dispersive component, also named the Lifshitz–van der Waals interaction, (γ_{sv}^{LW}) and polar Lewis acid-base interactions (γ_{sv}^{AB}), respectively, see Equation (5). The acid-base interactions are subdivided into electron donor γ_{sv}^{-} (Lewis base) and electron acceptor γ_{sv}^{+} (Lewis acid) parts, see Equations (4)–(6). The subscripts "lv" and "sv" denote the liquid-vapor and surface-vapor interfaces.

$$(1+\cos\theta)\gamma_{lv} = 2\left(\sqrt{\gamma_{sv}^{LW}\gamma_{lv}^{LW}} + \sqrt{\gamma_{sv}^{+}\gamma_{lv}^{-}} + \sqrt{\gamma_{sv}^{-}\gamma_{lv}^{+}}\right) \tag{4}$$

$$\gamma_{sv}^{AB} = 2\sqrt{\gamma_{sv}^{+}\gamma_{sv}^{-}} \tag{5}$$

$$\gamma_{sv}^{TOT} = \gamma_{sv}^{LW} + \gamma_{sv}^{AB} \tag{6}$$

where θ is the contact angle, γ_{sv} is the liquid's total surface tension, and γ_{lv}^{LW} and γ_{sv}^{LW} are the apolar (dispersive) Lifshitz–van der Waals components of the liquid and the solid, respectively, whereas $\gamma_{sv}^{+}\gamma_{lv}^{-}$ and $\gamma_{sv}^{-}\gamma_{lv}^{+}$ are the Lewis acid–base contributions of either the solid or the liquid phase as indicated by the subscripts. To determine the total surface free energy of the solid material (γ_{sv}^{TOT}) a system based on Equation (4) must be used. To solve the resulting systems of equations it is necessary to use at least three test liquids with known γ_{lv}, γ_{lv}^{LW}, γ_{lv}^{-}, and γ_{lv}^{+}, see Table 2 [45–48].

Table 2. Surface tension parameters (mN/m) of the liquids used for contact angle measurements.

Liquid	γ_{lv}^{TOT}	γ_{lv}^{LW}	γ_{lv}^{AB}	γ_{lv}^{+}	γ_{lv}^{-}
Water	72.80	21.80	51.00	25.50	25.50
Formamide	58.00	39.00	19.00	2.28	39.6
Methylene iodide	50.80	50.80	0.00	0.72	0.00
Red blood cells (rbc) [49]	36.56	35.2	1.36	0.01	46.2
Platelets (p) [49]	118.24	99.14	19.1	12.26	7.44

The manner in which the surface of the materials interacts with the constituents of blood, such as platelets and red blood cells, determines if the tested materials are blood compatible. Evaluating the surface/interfacial free energy of a material can represent an in vitro method to determine biocompatibility. Equation (7) was used to establish if the obtained materials present blood compatibility, where $W_{s/rbc}$ and $W_{s/p}$ denote the red blood cells and platelets work spreading [49]. When a biomaterial is exposed to blood it can cause the adhesion of blood cells onto surface, and the extent of this adhesion will determine the life of the implanted biomaterials. By such cells adherence to biomaterial surfaces the coagulation and immunological cascades could be activated [50].

$$W_s = W_a - W_c = 2\left(\sqrt{\gamma_{sv}^{LW}\gamma_{lv}^{LW}} + \sqrt{\gamma_{sv}^{+}\gamma_{lv}^{-}} + \sqrt{\gamma_{sv}^{-}\gamma_{lv}^{+}}\right) - 2\gamma_{lv} \quad (7)$$

where W_s—work of spreading (the negative free energy associated with spreading liquid over the solid surface); W_a—work of adhesion (defined as the work required separating the liquid and solid phases) and W_c—work of cohesion (defined as the work required separating a liquid into two parts) [41].

In Vivo Biological Evaluation

White male Wistar rats (200–250 g) were used in the experiment. The animals were housed under a standard laboratory environment (relative humidity 55–65%, chamber temperature 23.0 ± 2.0 °C and 12 h of light: Dark sequence (lights on at 6:00 a.m.) and fed with a specific diet and water ad libitum, excluding the time of the investigations. Before the assessment, the animals were positioned on a raised wire mesh, under a clear Plexiglass container and allowed 2 h to familiarize themselves to the testing room.

Rats were randomly assigned into eight groups of six animals each according to the following protocol: Group 1: Control (C)—distilled water; Group 2: R—powdered rosemary ethanolic extract; Group 3: PLA; Group 4: PLA/0.25R; Group 5: PLA/0.50R; Group 6: PLA/0.75R; Group 7: PLA/PEG; and Group 8: PLA/PEG/0.5R.

In the first day of the investigations, under general anesthesia (with ketamine 50 mg/kg body weight and xylazine 10 mg/kg body weight), the samples (weighing 62 mg) were placed subcutaneously, in one side in the rat's dorsal region, after a small incision, and subsequently sutured. Cotton pellets, of 62 mg, impregnated with distilled water (0.3 mL) and respectively with R solution (0.3 mL), were subcutaneously inserted into control animals. Pellets and the implanted samples behaved like foreign materials generating a subacute inflammatory response.

The in vivo biocompatibility of PLA-based materials containing rosemary ethanolic extract was evaluated by estimating the influences on the hematological and serum biochemical tests and on certain immune system parameters [51,52].

After certain time intervals (24 h and 7 days) following the implantation of PLA-based materials containing rosemary ethanolic extract, 0.3 mL of blood samples were collected from retro-orbital plexus and the following parameters were determined: Blood count, aspartate transaminase (AST), alanine aminotransferase (ALT), and lactic dehydrogenase (LDH) activity, as well as, serum urea and creatinine levels [53].

At the same time in the experiment, the serum complement activity, and the phagocytic capacity of peripheral neutrophils (Nitro Blue Toluene—NBT test) were also established [54]. These parameters belong to specific tests used to quantify the influence of pharmacologic agents on the immune defense capacity of laboratory animals [55].

The data were expressed as mean +/− standard deviation (S.D.) and processed using SPSS variant 17.0 for Windows 10, to estimate the differences between the control group and the groups receiving the investigated substances. The values of coefficient p (probability) below 0.05 compared with those of the control group were considered to be statistically significant.

Ethics statement: The experimental research protocol was approved by the local Animal Ethics Committee of the "Grigore T. Popa" University of Medicine and Pharmacy, Iași, Romania, in strict observance of the international ethical regulations on laboratory animal work (AVMA Guidelines on Euthanasia, 2007). Biocompatibility tests were performed according to the "Grigore T. Popa" University of Medicine and Pharmacy guidelines for the handling and use of experimental animals and in accordance with the recommendations and policies of the International Association for the Study of Pain [56].

Each animal was used only once and the length of the experiments was kept as short as possible. For ethical considerations, all the animals were euthanized at the end of the experiment [57].

3. Results and Discussion

Films and sheets resulting from the melt mixing process are transparent and homogeneous and at high R concentration show a little yellowing.

3.1. Processing Behavior

The torque-time curves were registered during blending in a Brabender mixer in order to evaluate the processing behavior by analyzing torque values at different mixing times. Melt processing characteristics of PLA and its blends—Table 3—are not changed by the incorporation of the powdered rosemary ethanolic extract.

Table 3. Melt processing characteristics of PLA and its blends with rosemary solid extract.

Sample	TQ_{max1} (Nm)	TQ_{1min} (Nm)	TQ_{max2} (Nm)	TQ_{5min} (Nm)	TQ_{final} (Nm)
PLA	66.4	17.2	-	13.3	10.2
PLA/0.25R	73.8	15.7	-	11.2	10.2
PLA/0.5R	72.0	17	-	11.4	10.5
PLA/0.75R	74.6	17.6	-	11.1	10.1
PLA/PEG	12.9	0.9	-	7.3	6.5
PLA/PEG/0.5R	10.1	2.8	-	7.3	5.7

TQ_{max1}—maximum torque; TQ_{1min}—torque after one minute of mixing; TQ_{max2}—maximum torque after 1.5 min of mixing; TQ_{5min}—torque after 5 min of mixing (half processing time); TQ_{final}—torque at the end of mixing.

A sharp change in the melt mixing behavior was recorded in the presence of a plasticizer. PEG improved the melt flow, the materials being easily processed as all torque characteristics decreased. PEG favors the processing of polymers due to the increased chain mobility, therefore improving macromolecular movement. This effect was previously observed by other authors for various types of blends containing PLA and PEG, for example in PLA/PEG/organoclays nanocomposites [13] or PLA/PHB blends [58]. No oxidation occurred during processing.

3.2. SEM Results

SEM images, presented in Figure 2, show a homogeneous distribution of the rosemary powder in the PEG plasticized PLA samples without agglomeration, which happens in the absence of the plasticizer.

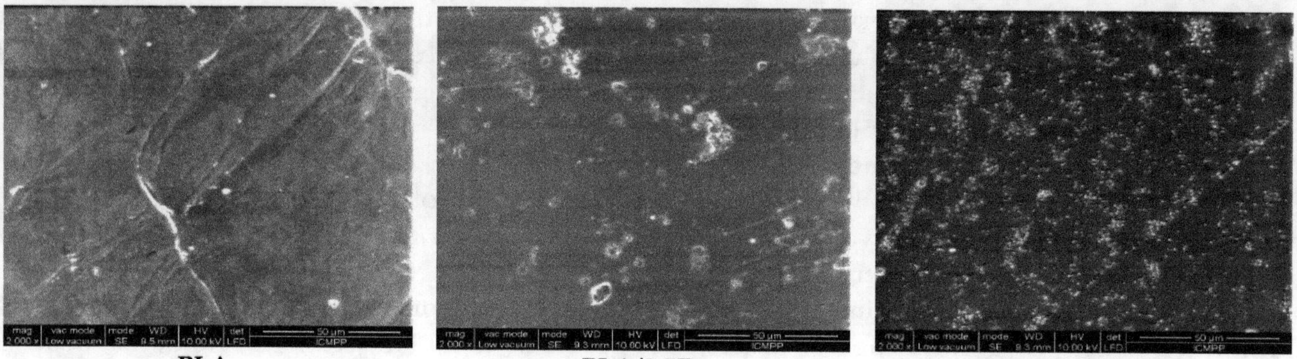

PLA PLA/0.5R PLA/PEG/0.5R

Figure 2. SEM images of PLA, PLA/0.5R and PLA/PEG/0.5R. PEG is defined as poly(ethylene glycol) and R is defined as rosemary ethanolic extract.

From SEM images, the histograms of the distribution of the particle dimensions were evaluated and are shown in Figure 3. For the PLA/0.5R sample, an average dimension of the particles of 14.2 μm was found while for PLA/PEG/0.5R this was approximately 4.1 μm. Therefore, the incorporation of PEG into the PLA-based composites improves the R distribution, with the sample being more homogenous and the particles dimension histogram being narrower.

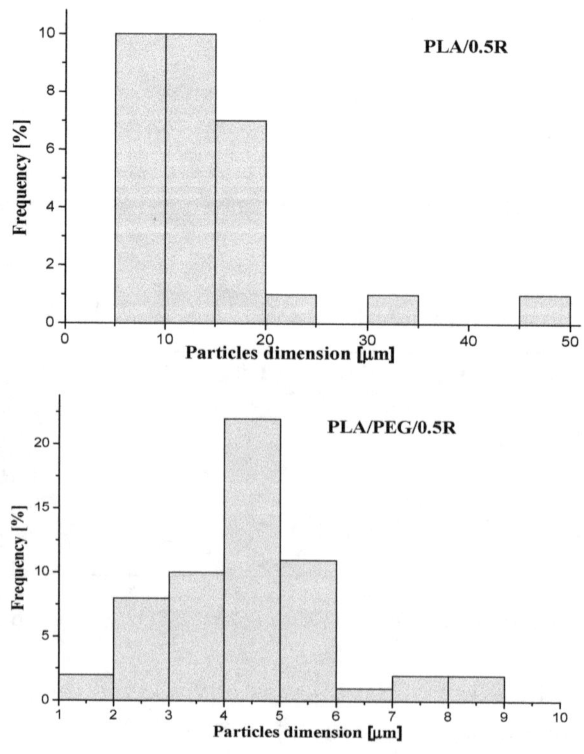

Figure 3. The distribution of particle dimensions determined from SEM images.

3.3. ATR-FTIR Data

The ATR-FTIR spectra of the PLA prior to and after loading with powdered rosemary ethanolic extract are presented in Figure 4. The IR spectrum of the PLA presents a strong band at 1749.3 cm^{-1} due to C=O stretching of the carbonyl group, while the bending vibration of this group appears at 1267.1 cm^{-1}. The bands at 867.9 cm^{-1} and 754.1 cm^{-1} are assigned to the amorphous and respectively crystalline phases of PLA. The IR bands at 2995.2 and 2945.1 cm^{-1} are assigned to the CH stretching as ν_{as} CH_3 and ν_s CH_3 modes.

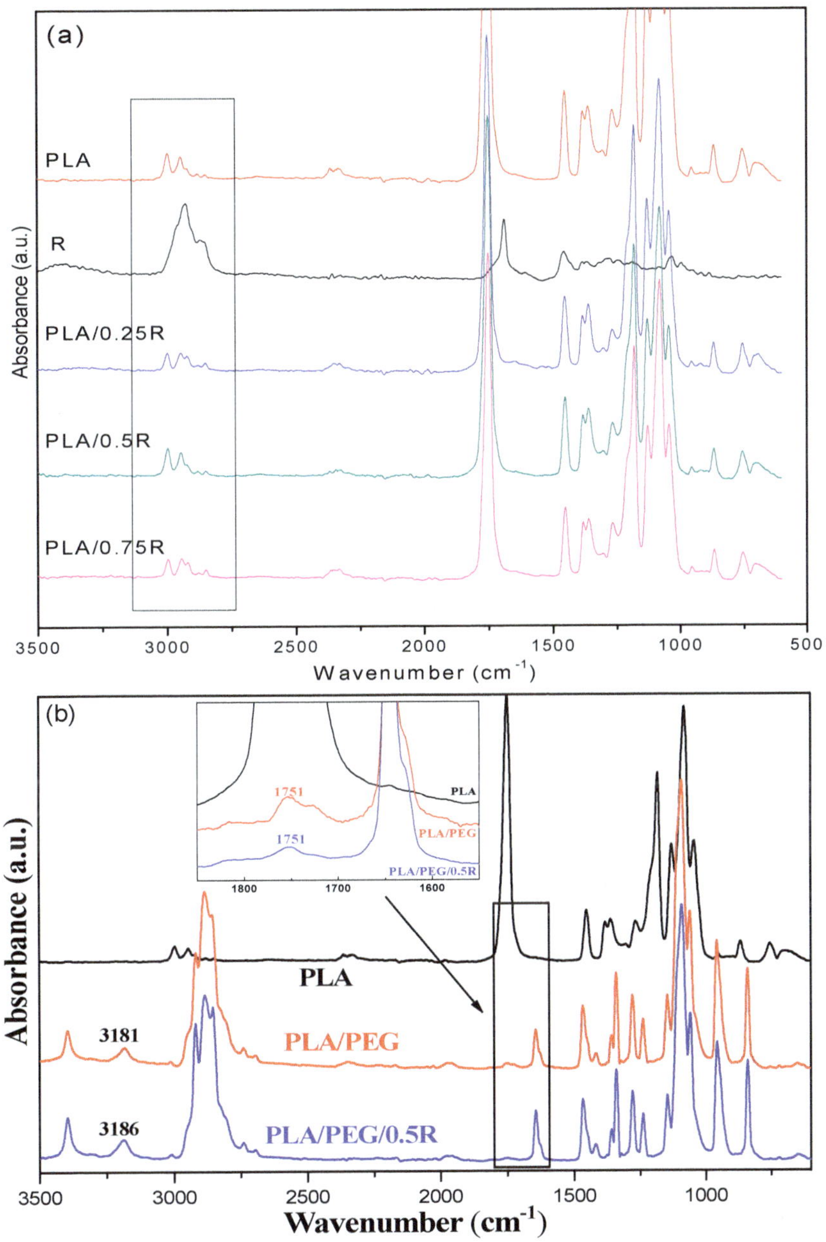

Figure 4. FTIR spectra of the neat PLA and the PLA/rosemary blends (**a**) and PLA/PEG/R (**b**); insert spectrum highlights details of different spectral regions.

The components of powdered rosemary ethanolic extract have different groups in their structures (such as COOH, C–O, phenolic) with corresponding absorption bands in the region of 1750–1600 cm^{-1}. The R sample exhibits a relatively sharp band with a maximum at 1692 cm^{-1} corresponding to C–O vibrations, which are also present in the samples loaded with powdered rosemary ethanolic extract.

The rosemary constituents are mainly aromatic compounds and the band appearing at 1026 cm^{-1} is attributed to the deformation vibration of C–H bonds from aromatic rings [59].

The shape of the bands in the 2500–3500 cm^{-1} region, see the highlighted region of Figure 4a, are specific for each sample, as they are a coupling of the bands of PLA and R with insignificant shifts, while in the case of the systems containing plasticized PLA the bands' shift indicates some interactions between components due to the better distribution of the components. These bands are associated with the stretching modes of C–H overlapped with the –OH stretch from alcohols, carbonyl in aldehydes, and carboxyl groups [60].

3.4. Mechanical Properties

The Young's modulus and tensile strength increase after incorporation of R into PLA and decrease in plasticized PLA, while as is expected, the elongation at break increased and is two times higher for plasticized systems containing R– as revealed in Figure 5. An increase in the elongation at break and a decrease in the tensile strength for PLA/PEG4000 blends were previously reported [61], with ~7% elongation at break for the PLA/PEG4000 blend [62]. The mechanical properties are kept within satisfactory limits with improved elasticity for plasticized samples in the presence of R. Examples in the literature present a lack of systems containing PLA/rosemary, mostly using essential oils; therefore, it is difficult to compare results with different types of materials processed in other conditions. A PLA-based compound comprising rosemary essential oil showed a slightly decreased elongation at break by 2.8% [63].

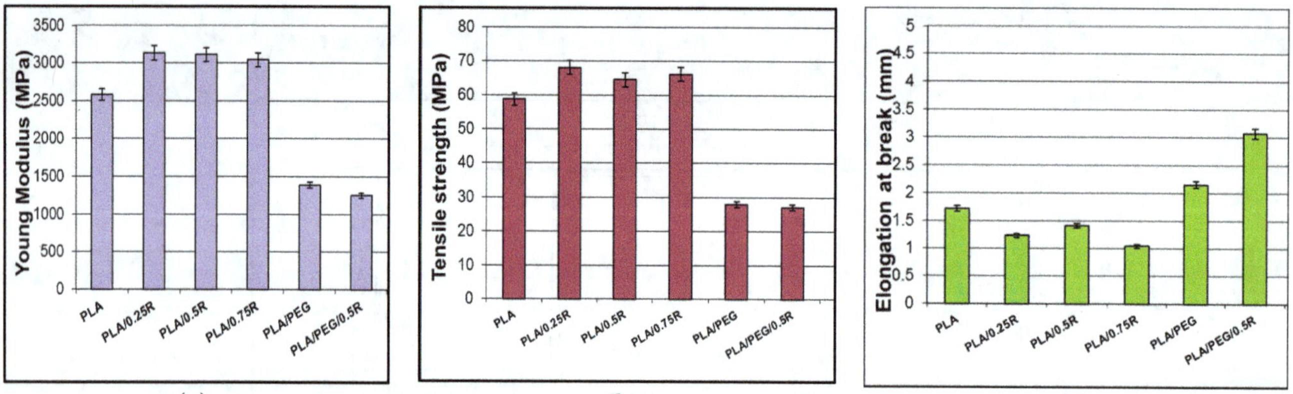

(a) (b) (c)

Figure 5. Mechanical properties of the PLA/R and PEG-plasticized PLA/R: (**a**) Young's modulus; (**b**) tensile strength; (**c**) elongation at beak.

3.5. Rheological Behavior

The incorporation of the powdered rosemary ethanolic extract in the PLA films led to a decrease of all studied rheological parameters: Storage modulus (G'), loss modulus (G''), and complex viscosity, as presented in Figure 6a,b. A predominantly viscous behavior ($G'' > G'$) can be noticed both for PLA and for the PLA-based composites. The loss modulus dependence on the deformation frequency presents the same decreasing trend for the composites investigated with the differences becoming more pronounced at low frequencies. The results obtained are supported by those found by other authors comparing various formulations with that containing rosemary; the flow curve of the rosemary-containing formulation is the lowest one [64].

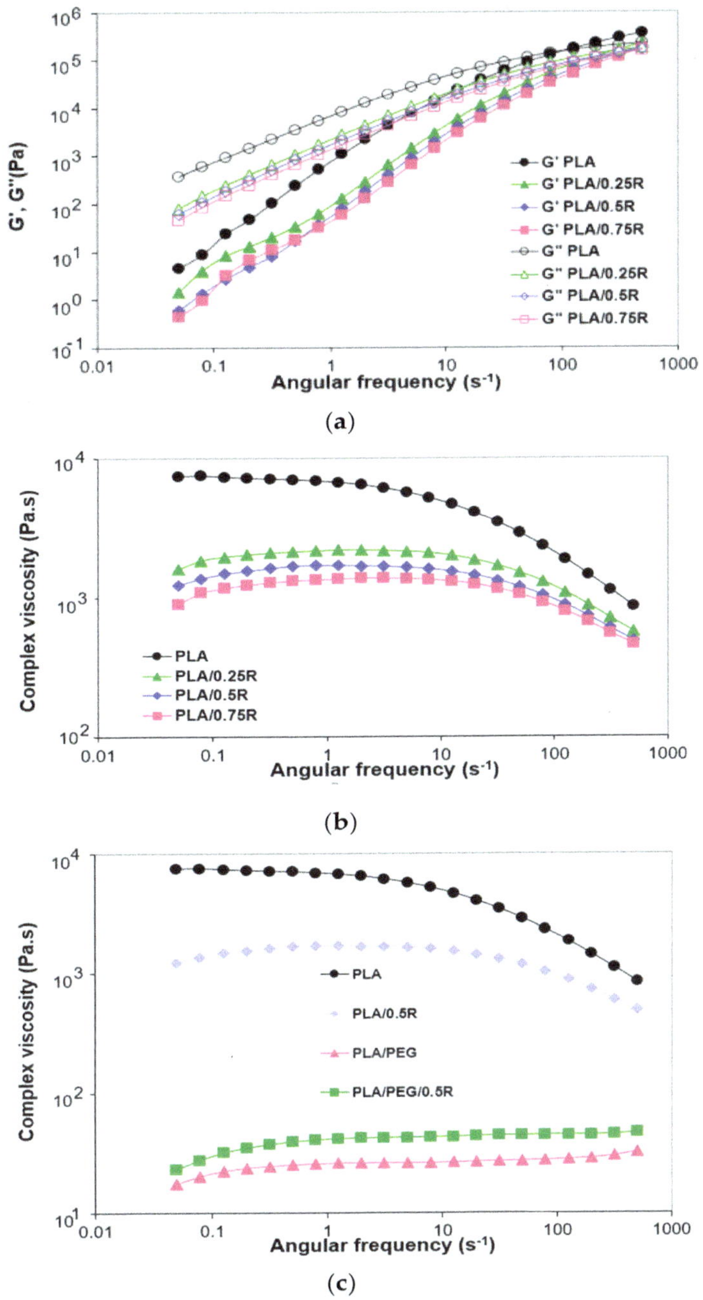

Figure 6. Angular frequency dependence of the storage modulus and loss modulus (**a**) and complex viscosity (**b**,**c**) for PLA, PLA/R and PLA/PEG/R systems.

The incorporated R amount maintains the decreasing order of moduli and viscosity, the highest content (0.75%) causes a drastic decrease of complex viscosity in the melt state as was also observed from the torque-time curves recorded during melt processing.

The lowest value is obtained for plasticized PLA, see Figure 6c, while the addition of R to plasticized PLA resulted in a small increase of the complex viscosity.

As shown in Table 4, the crossover frequency (ωi) and the crossover modulus (corresponding to $G'' = G'$) were obtained at higher oscillation frequencies for the blends with an increasing R content incorporated. The relaxation time (θ)—calculated as $\theta = 1/\omega i$ and the crossover modulus values obtained for PLA and its blends are plotted in Figure 7. For the PLA/0.75R blend, the relaxation time decreased by more than 23% compared with neat PLA, showing that less time was required for reorientation of the entangled chains in these cases, as the rosemary powder induced relaxation and thus flexibility of the chains.

Table 4. Crossover frequency and crossover moduli values for PLA, and PLA/Rosemary blends.

Crossover Characteristic	PLA	PLA/0.25R	PLA/0.5R	PLA/0.75R
ω (s^{-1})	96.37	273.3	363.3	411.4
t (s)	10.38	3.66	2.75	2.43
$G' = G''$ (MPa)	144.5	146.9	145.2	145.4

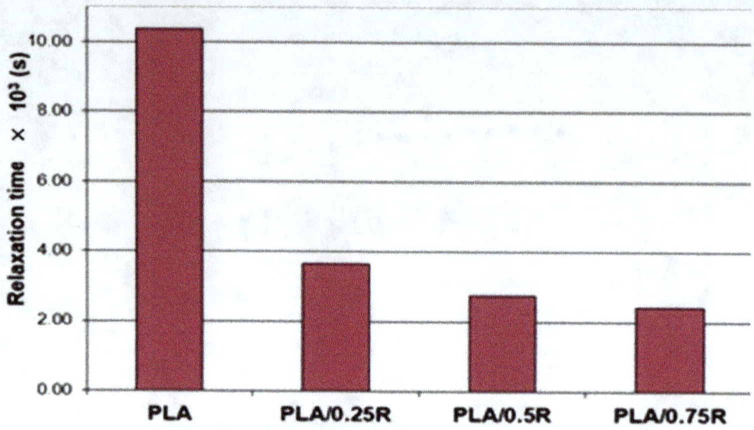

Figure 7. Relaxation time vs. sample composition.

3.6. DSC Results

The DSC curves of the PLA/R and PLA/PEG/R systems recorded both in run I and II are shown in Figure 8 and values of the thermal properties are summarized in Table 5. In the DSC curves, all particular transitions of PLA, namely, glass transition temperature (T_g), cold crystallization temperature (T_{cc}), and melting temperature (T_m) are evidenced, which show some differences between the studied systems.

Table 5. Thermal characteristics of the PLA/R and plasticized PLA/R systems determined by DSC method.

Sample	T_g (°C)	T_{cc} (°C)	ΔH_{cc} (J/g)	T_m (°C)	ΔH_m (J/g)	T_{cr} (°C)	X_{cr}
				Run I			
PLA	60.64	125.89	17.69	151.63	23.12	55.80	24.6
PLA/0.25R	62.95	124.56	18.23	155.96	19.66	55.96	21.1
PLA/0.5R	64.07	124.82	10.40	154.51	10.78	56.26	11.5
PLA/0.75R	64.82	118.28	16.54	156.35	19.05	55.86	20.33
PLA/PEG	49.30	-	-	157.74	23.95	70.03	25.56
PLA/PEG/0.5R	59.59	90.23	2.6	157.07	26.06	66.70	28.39
				Run II			
PLA	61.58	132.03	9.69	152.75	12.05	-	13.34
PLA/0.25R	60.96	128.95	17.14	154.01	16.4	-	17.50
PLA/0.5R	61.15	133.87	6.24	153.90	6.37	-	6.79
PLA/0.75R	60.40	133.48	7.18	154.04	7.9	-	8.43
PLA/PEG	-	91.37	0.36	154.66	25.32	-	27.02
PLA/PEG/0.5R	-	80.93	8.003	153.72	27.28	-	29.11

T_g—glass transition temperature; T_{cc}—cold crystallization temperature; T_m—melting temperature; T_{cr}—crystallization temperature; ΔH_{cc}—cold crystallization enthalpy; ΔH_m—melting enthalpy; X_{cr}—crystallinity index.

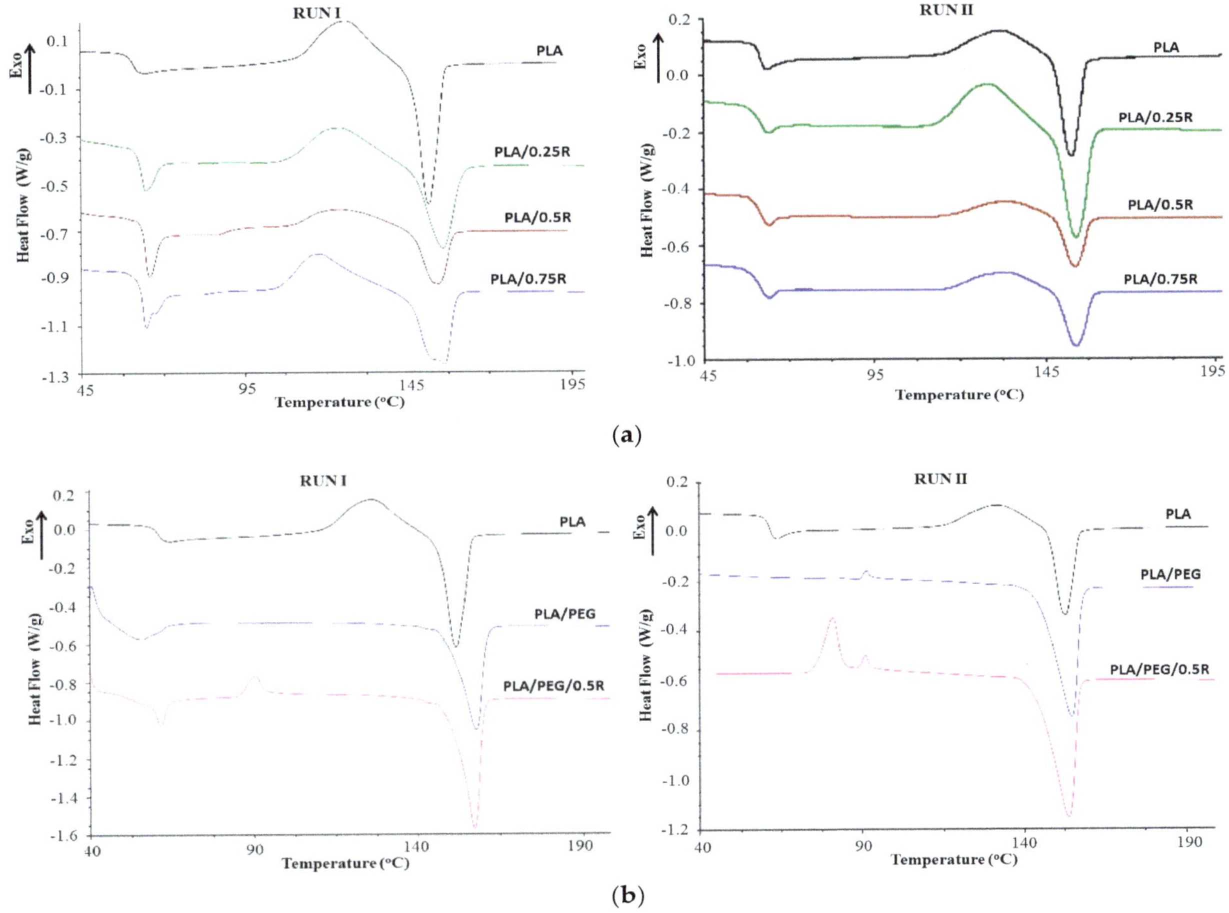

Figure 8. Differential scanning calorimetry (DSC) curves of the PLA and PLA/R (**a**) and plasticized PLA/R systems (**b**).

There is a strong dependence of the thermal properties of the powdered rosemary ethanolic extract containing samples on their particle distribution into the PLA matrix. In the binary systems, because of the agglomeration of particles, the majority of the properties decrease or remain unchanged while in the binary and ternary blend containing plasticizer the thermal characteristics are changed, eventually resulting in an increased degree of crystallization. The thermal characteristics recorded in the first run are a little different from those recorded in the second run, but the variation with sample composition is almost similar. T_g increases with R content and decreases as is expected when plasticizer is incorporated, and increases again for the ternary system PLA/PEG/0.5R. The T_{cc} recorded in both runs decreases for all studied systems, especially for PLA/PEG/0.5R. The decrease is less significant for values corresponding to the second run. The melting temperature T_m, increases in all cases; the most important increase was observed for plasticized PLA systems. As the plasticized systems are highly ordered the melting peak is very pronounced, especially in the second run because the plasticizer favors arrangement of the chains during cooling and pre-melting processes, increasing the degree of crystallinity. The results are in accordance with those found in other papers [16,17]. The enthalpy of the cold crystallization (ΔH_{cc}) is approximately constant for binary PLA/R systems and is very small for plasticized PLA systems, while the enthalpy of melting process (ΔH_m) decreases by R incorporation and increases in the cases of systems containing plasticizers because of the good distribution of particles and better mobility of the chains which favors the three-dimensional arrangements. As a result, the crystallization degree (X_{cr}) decreases by R incorporation and increases for plasticized PLA systems.

3.7. TG-FTIR/MS Data

As can be seen from the TG/DTG curves in Figure 9, the PLA and PLA/R samples decompose in a single step observed from 310–380 °C, evolving more than 98% as volatile products, see Table 6. Under an inert atmosphere, a lower thermal stability of the samples containing R was found. The plasticized PLA systems decomposed in two steps, the first one as the main step occurring in the temperature range of 280–370 °C with a mass loss of 82–83%, indicating a significant influence of the plasticizer on the thermal decomposition. The second step was assigned to the decomposition of the residue formed in the first step. It takes place in the temperature region of 365–430 °C with a mass loss of 15–16%. Comparing the onset and T_{10} decomposition temperature of the PLA/PEG and PLA/PEG/0.5R, results indicated that the incorporation of R led to an increased thermal stability of the system.

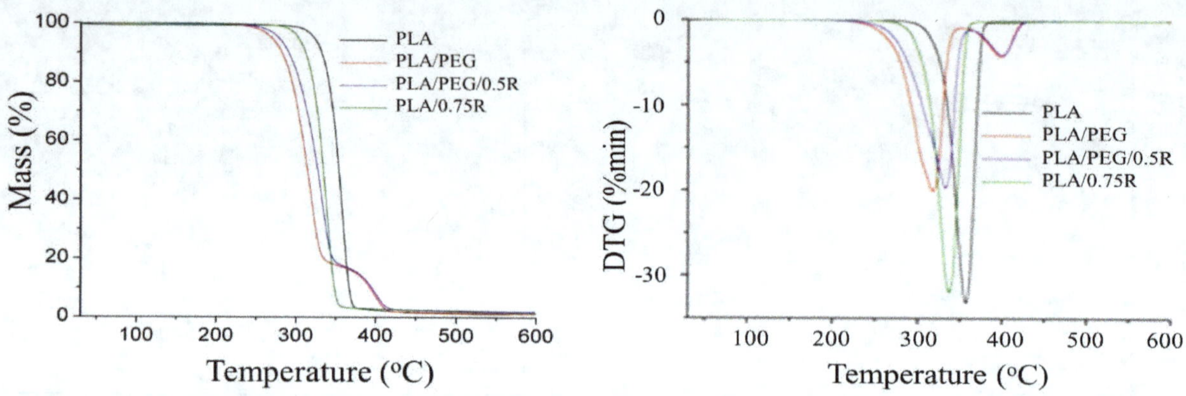

Figure 9. TG/DTG curves of the PLA, PLA/R, and PLA/PEG/R systems.

Table 6. TG data for PLA containing powdered rosemary ethanolic extract.

Sample	Degradation Stage	T_{onset} (°C)	T_{peak} (°C)	ΔW (%)	T_{10} (°C)	T_{20} (°C)
PLA	I residue	339	358	98.72 1.28	332.5	341.5
PLA/0.75R	I residue	310	338	98.50 1.50	312.5	321.5
PLA/PEG	I II residue	280 373	322 402	82.51 16.31 1.18	285.5	298
PLA/PEG/0.5R	I II residue	285 365	336 403	83.03 15.04 1.93	294	207.5

T_{onset}—the temperature at which the thermal degradation start; T_{peak}—the temperature at which the degradation rate is maximum; T_{10}, T_{20}—the temperatures corresponding to 10 wt % and 20 wt % mass losses; ΔW—residual mass at 600 °C.

As it concerns the type of decomposition products, some information has been obtained by coupling TG with FTIR/MS spectroscopies. Looking at the results from the data in Figure 10 showing the 3D FTIR spectra, the change in the thermal decomposition products after incorporation of R is evident.

The FTIR spectra show differences both in the intensity of the bands and also in the PLA/0.75R system. The number of bands is increased or important shifts in the PLA bands are observed due to the possible interaction between components of the system.

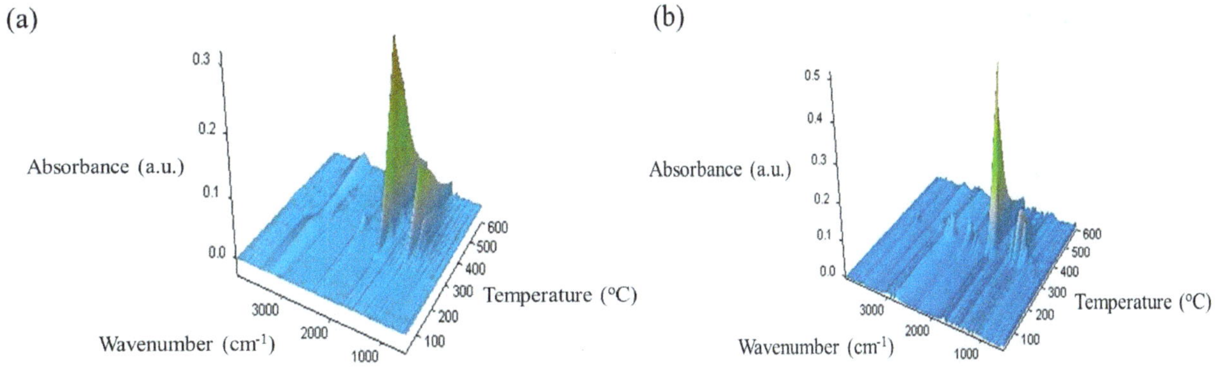

Figure 10. 3D-FTIR spectra of the volatile decomposition products resulting from the thermal degradation of PLA (**a**) and PLA/0.75R (**b**) samples.

The new bands in the 2D spectra of the thermal degradation products resulting from PLA/0.75R residue, see Figure 11, have been identified as: 2949.8 cm^{-1} (C–H stretching of CH_2 and CH_3); 2313.7, 1631.9 cm^{-1} (C=O, C–N and COO^- stretching); 1490.4 cm^{-1} (aromatic domain and N–H bending, C–O stretching vibration (amide), and C–C stretching from phenyl groups, COO– stretching, CH_2 bending; 861; 747.9 and 602.5 cm^{-1} (C–H out-of-plane bending vibration from isoprenoids, etc.) [65].

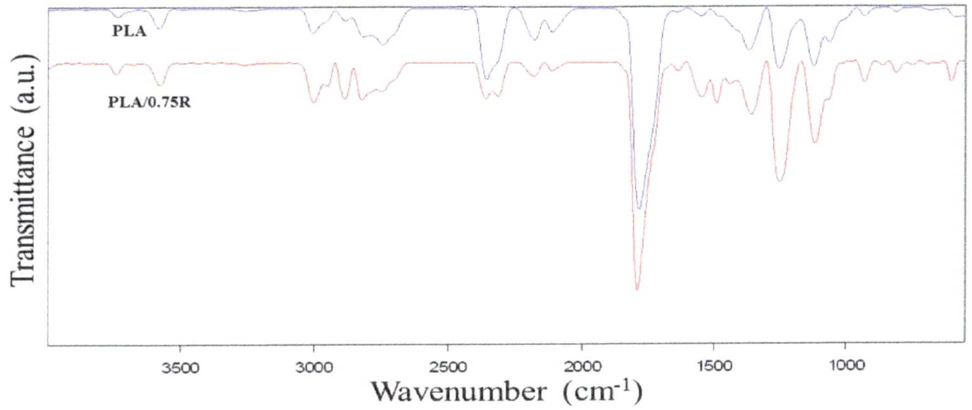

Figure 11. 2D FTIR spectra of the thermal degradation products resulting from PLA and PLA/0.75R.

The important shifts of the following bands are found from 2812 cm^{-1} in PLA to 2820 cm^{-1} in PLA/0.75R; 2841 cm^{-1} to 2845 cm^{-1}; 1785 cm^{-1} to 1791 cm^{-1}; 1367.7 cm^{-1} to 1359 cm^{-1}; 1122 to 1117.2 cm^{-1} and 680.6 to 691.6 cm^{-1}. In the FTIR spectrum of degradation products, the following new bands and shifts of the bands appeared at approximately: 1736, 1673, 1366, 1243, 862, and 750 cm^{-1} are found. These bands probably correspond to the fragments resulting from α—pinene and 1,8-cineole products which may evolve from rosemary ethanolic extract. These products appear together with those resulting from the PLA matrix.

At $T > 420$ °C all samples are almost totally degraded according to results from the TG/DTG curves, with a residual amount of ~2%. Therefore, by FTIR/MS some of the thermal degradation products from the resulting mixtures at the final temperature are identified. Comparing these results with those obtained from the MS in Figure 12 indicated different products resulted by thermal degradation of PLA and its blends containing alcoholic extract of rosemary.

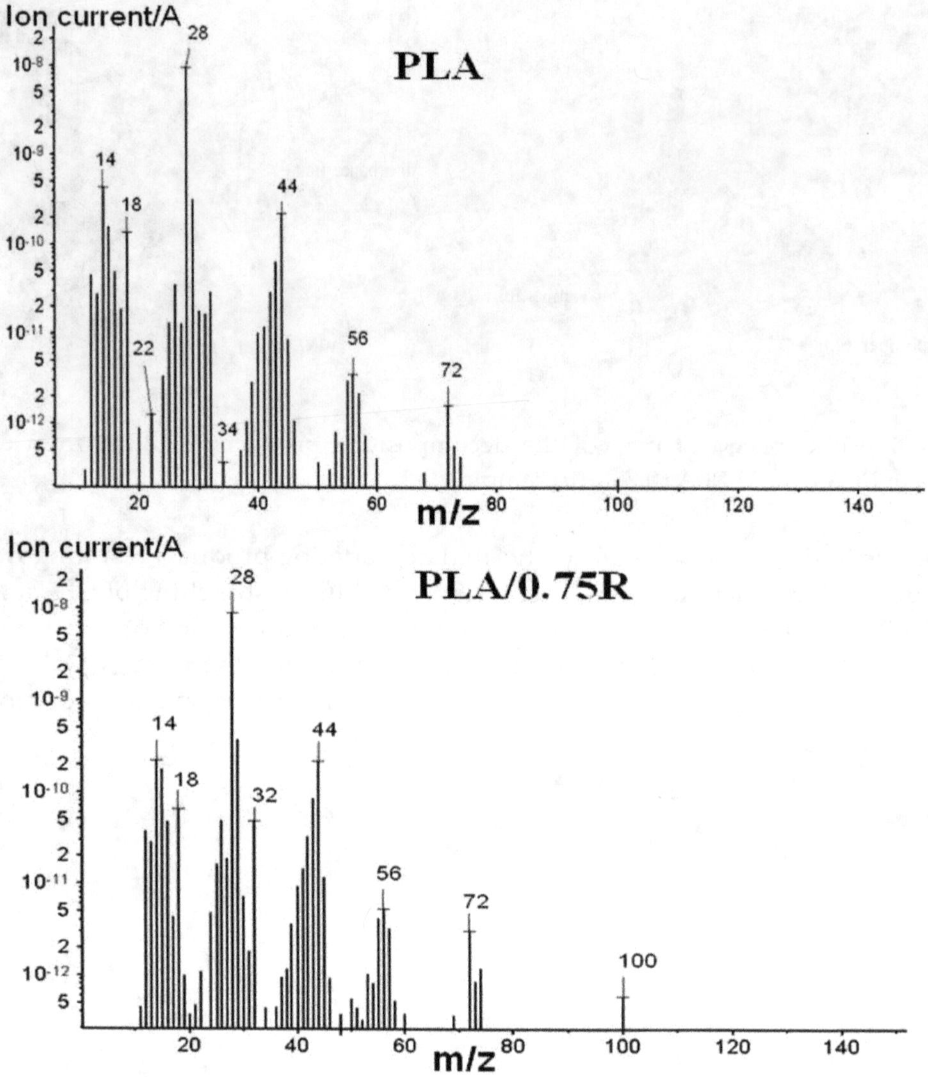

Figure 12. MS spectra of the thermal degradation products resulted from PLA and PLA/0.75R.

The pyrolysis of neat PLA results in the production of a large number of cyclic oligomers through the random degradation process. In accordance with the literature data, the main fragments of lactide meso-form or DL-form [66,67] are identified for m/z = 32, 43, 45 and 56. PLA, in general, shows a dominant series of signals with m/z = 56 + (n × 72) in which n assumed values of 1, 2, 3, and 4. Acetaldehyde (m/z 15, 26 and 43), 2,3-pentadione, and acrylic acid were also identified. Further decomposition products were H_2O, CO_2, and hydrocarbons (m/z of 18, 44, 12–17, etc., respectively) [68]. The presence of the plasticizer led to very complex spectra.

3.8. Chemiluminescence

The stability of PLA is related to the formation of lactide structures [69]. Under gamma irradiation poly(L-lactic acid), the random scission of molecular chains takes place, which results in the sharp decrease in the numerical average molecular weight [70]. Moreover, due to the exposure to gamma radiation, several free radicals centered on carbon atoms are formed which release volatile products or restore polymeric configurations. The radiochemical behavior is the mirror of radiation processing during sterilization and grafting, when the weaker sites, α-methyl positions and branching places, become the main sources of degradation initiators [71].

The addition of natural phenolic antioxidants like powdered rosemary ethanolic extract increases the thermoxidative and radiation stabilities by scavenging degradation precursors. The contribution revealed by the components of this natural stabilizing mixture is observed by means of the differences between the CL intensities recorded for stabilized systems. The sharp progress in the radiation-stimulated degradation of PLA, see Figure 13a, is the result of the fast energy transfer from incidental radiation onto macromolecules followed by a high radiochemical yield of scission. In Figure 13b, the blocking action of oxidation is demonstrated by the diminution in the evolution of emission intensities, which drops as a parabolic decrease in the non-irradiated sample and the sudden hindrance of oxidation in the irradiated PLA/rosemary specimens.

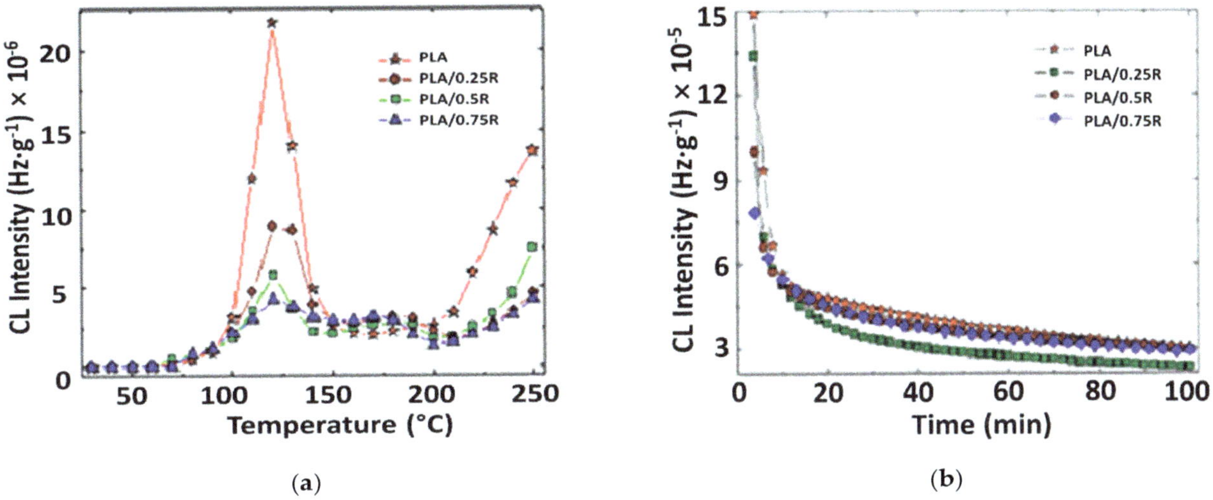

Figure 13. Chemiluminescence (CL) spectra recorded on PLA-based samples irradiated at 20 kGy: (**a**) nonisothermal measurements, heating rate: 3.7 °C min^{-1}; (**b**) isothermal measurements: testing temperature 180 °C

The comparison between progress in the oxidation of irradiated neat and stabilized PLA emphasizes that the scavenging action of free radicals is selective because the deceleration of oxidation does not happen similarly. The chemiluminescence intensity is gradually diminished over time in protected polymer samples, while the radicals are suddenly oxidized in pristine PLA.

The improvement in the thermal resistance of PLA in the presence of R is revealed in Figure 14. The linear modification of this parameter sustains the proportionality between the increase in the protection activity and material loading with rosemary ethanolic extract.

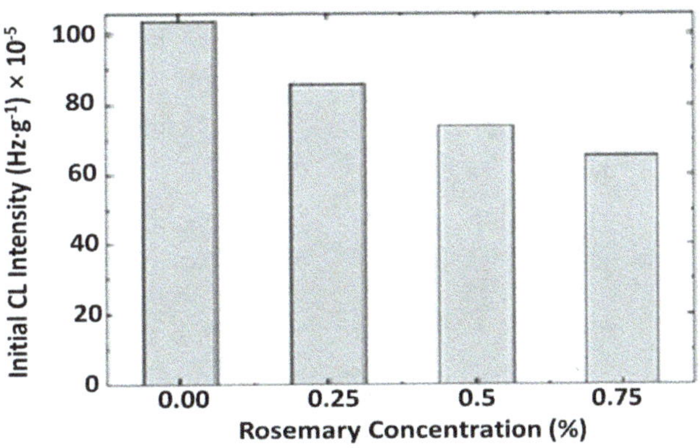

Figure 14. The dependence of the CL intensity on the concentration of powdered rosemary ethanolic extract.

The modification in the rosemary loading can provide a remarkable effect on stabilization. If the concentrations of 0.25% and 0.50% bring small contributions at medium oxidation temperatures, the degradation is effectively delayed at temperatures exceeding 170 °C. The concentration of 0.75% R, delivers a significant effect toward the preservation of the oxidation state in PLA samples. The low CL intensities recorded for these last-mentioned samples in comparison with the low-loaded material is evidence for the involvement of the powder component in the breaking action on the oxidation chain process.

The stabilization profiles for PLA mixed with PEG in the presence of rosemary are different if the concentrations of radicals are changed. As the literature data mentioned, "the components of rosemary extracts are very efficient stabilizers against oxidation by blocking free radicals towards reaction with oxygen according to the mechanism proposed by Pospišil" [72]. Thus, a cascade mechanism is responsible for the oxidation of carnosic acid, forming other intermediates with active antioxidant structures showing a high level in stabilization activity. The stabilization effectiveness depends on the active phenolic structures and the flavonoid components contained in rosemary extract [27,73].

The initial intensity of CL emission is related to the concentration of peroxy species [74]. A value of 1.63×10^{-3} Hz/g was found for the neat PLA/PEG system, while that of PLA/PEG/0.5R was 0.18×10^{-3} Hz/g, about nine times lower than that of the system without R, indicating a good oxidative stabilization of the system is achieved through this additive.

It can be concluded, that the non-isothermal oxidation starts at high temperatures, decreasing with R content. The isothermal CL determinations point out an evident contribution of powdered rosemary ethanolic extract to the polymer stability and the retardation of oxidation to long exposure times.

3.9. Antioxidant Activity Evaluation

The radical scavenging activity of the powdered rosemary ethanolic extract was determined by ABTS•$^+$ method, and an IC50 (the concentration of the sample required to inhibit 50% of radicals) of 26 μg/mL was obtained, which proved that the powdered rosemary ethanolic extract has very good antioxidant activity.

In Figure 15, the ABTS•$^+$ radical scavenging activity of PLA blends with different contents of powdered rosemary ethanolic extract is presented. The free radical scavenging activity in polymeric systems is directly proportional with the content of powdered rosemary ethanolic extract of the blend and is again higher for the plasticized PLA. The antioxidant activity of plasticized PLA/R samples is comparable to other systems based on plasticized PLA. For example, Lopusiewicz et al. [75] found an antioxidant activity value of 23% for PLA/melanin composites, and Byun et al. [76] determined an antioxidant activity of 90% and 14% for PLA-based blends with α-tocopherol or buthylated hydroxytoluene (BHT), respectively.

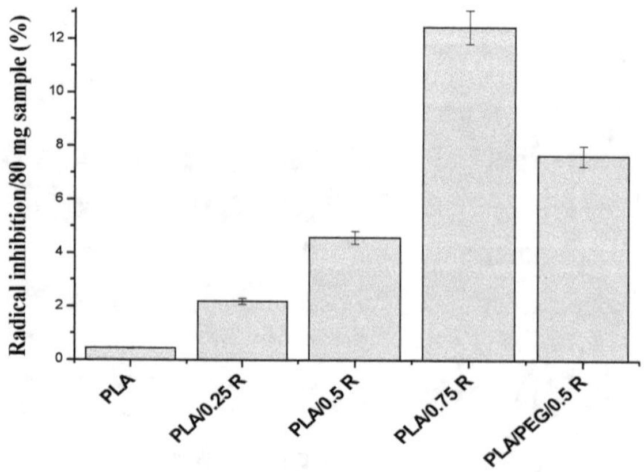

Figure 15. ABTS•$^+$ radical inhibition activity of the polymeric blends containing powdered rosemary ethanolic extract.

The obtained results are in accordance with those obtained by the chemiluminescence method.

3.10. Antibacterial Activity

Escherichia coli and *Salmonella typhimurium* are Gram-negative bacteria while *Bacillus cereus* or *B. cereus* is a Gram-positive, rod-shaped, aerobic, facultative anaerobic, motile, beta hemolytic bacterium commonly found in soil and food that produces toxins. These toxins can cause two types of illness: One type is characterized by diarrhea and the other, called emetic toxin, by nausea and vomiting. These bacteria can multiply quickly at room temperature [77]. Both Gram-negative bacteria and Gram-positive bacteria are susceptible to the antibacterial activity of powdered rosemary ethanolic extract, and their sensitivity is highly variable, see Table 7. The antimicrobial activity at 24 h increases with an increasing concentration of the powdered rosemary ethanolic extract, reaching values of about 100% for a concentration of 0.75%, and is higher in the case of the plasticized PLA systems because of better distribution of the powder into the matrix, as the SEM results demonstrated. It is worthwhile to mention the promising results obtained against *B. cereus*, as it is known that very serious problems are caused by this bacterium in association with food poisoning. Also, this bacterium has been associated with a multitude of other clinical conditions such as anthrax-like progressive pneumonia, severe eye infections, and devastating central nervous system infections, etc. Its role in nosocomial acquired bacteremia and wound infections in postsurgical patients has also been demonstrated, especially when intravascular devices such as catheters are inserted. *B. cereus* produces a potent β-lactamase conferring a marked resistance to β-lactam antibiotics [78].

Table 7. Antibacterial activity of the alcoholic extract from rosemary incorporated into PLA-based materials against *Bacilus cereus*, *Salmonella typhymurium*, and *Escherichia coli*.

Sample	ATCC *Bacillus cereus* 14579		ATCC *Salmonella typhymurium* 14028		ATCC *Escherichia coli* 25922	
	Inhibition %/24 h	**Inhibition %/48 h**	**Inhibition %/24 h**	**Inhibition %/48 h**	**Inhibition %/24 h**	**Inhibition %/48 h**
PLA	5	59	32	61	53	71
PLA/0.25 R	59	100	52	87	61	86
PLA/0.5 R	91	100	52	84	71	100
PLA/0.75 R	100	100	55	87	94	100
PLA/PEG	45	91	29	77	69	94
PLA/PEG/0.5R	86	100	48	100	76	100

The activity is lower in the case of *Salmonella typhymurium*. These results are in accordance with those found by other authors. Golshani and Sharifzadeh tested the rosemary alcoholic extract against *Staphylococcus aureus*, *Escherichia coli*, *Bacillus cereus*, and *Pseudomonas aeruginosa*, and proved its inhibitory effect for all these strains [79]. Some inhibitory effect for bacteria growth of PLA could be explained by the acidic pH imparted by it to the culture medium (pH 2), which is not favorable to bacteria growth and is due to the residual content of lactic acid.

3.11. Gas Permeability

It is well-known that the properties of PLA are compared to other commodity plastics, and the PLA permeation is shown to closely resemble that of polystyrene. Crystallinity was found to dominate permeation properties in a biaxially-oriented film [80]. The permeability depends on the film thickness and polymer morphology. Improvement of the gas permeability of the PLA-based films of approximately the same thickness in the presence of natural additive R is evident by comparing the data of Table 8.

Table 8. Permeability of tested films to CO_2 (~4 h to equilibrium) and O_2 (~7 h to equilibrium).

Sample	Thickness (mm)	CO_2 (mL/m^2/day)	O_2 (mL/m^2/day)
PLA	0.151	873	1308
PLA/0.25R	0.120	588	487
PLA/0.5R	0.122	535	273
PLA/0.75R	0.130	412	201
PLA/PEG	0.128	524	455
PLA/PEG/0.5R	0.126	489	278
Food freezing bag	0.020	64,601	50,266
Food packaging foil (LDPE/PP)	0.009	128,374	35,629

The gas barrier improvement depends on the amount of R in the film's composition both in PLA/R and plasticized PLA/R blends. The oxygen barrier properties are better than those of CO_2. The results are in accordance with those obtained by Yuniarto et al. [81]. They found that the plasticized PLA by PEG enhanced the permeability value by about 20%, while the largest fraction PEG400 reduced the ability to prevent oxygen from passing through the film. In this case, the barrier properties were significantly affected by the degree of crystallinity in the film with a correlation number of 0.85. The gas permeability of two commercial food packages (a foil and a freezing bag) was also measured for comparative purposes. An unexpectedly high permeability for CO_2 and for O_2 was found for both packages. However, taking into consideration the difference in film thickness between our samples and the commercial ones, we do consider that PLA-based composites could be suitable for use as packages for food applications.

3.12. Overall Migration of Active Components from PLA/R Formulations into 50% Ethanol Solution as a Food Simulant Medium

As mentioned in the experimental section, the overall migration of active components from PLA/R-based formulations into the simulant medium was determined according to EU regulations, and the results are summarized in Figure 16 and Table 9.

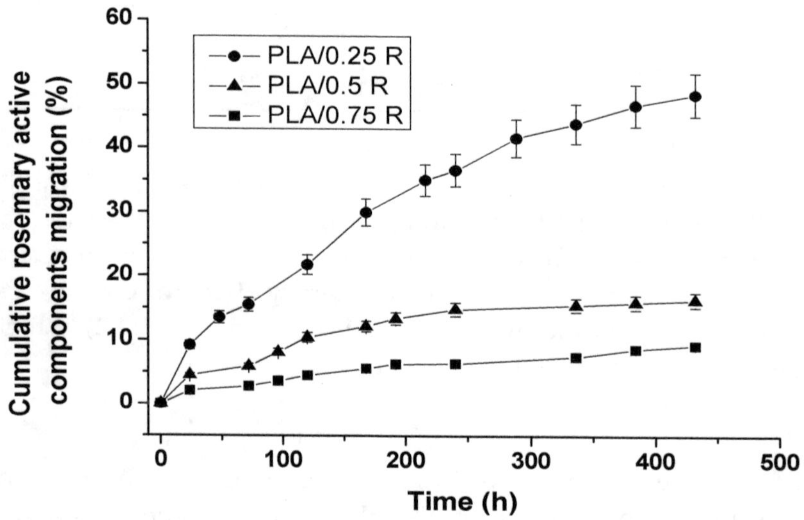

Figure 16. Migration profiles for the active ingredients of the powdered rosemary ethanolic extract (R) from PLA-based samples into 50% ethanol solution as a food simulant medium.

Table 9. Migration kinetic parameters for the active ingredients of the powdered rosemary ethanol extract from PLA-based materials into 50% ethanol solution as a food simulant medium.

Sample	Korsmeyer-Peppas Model				Higuchi Model	
	n	R^2	$K \times 10^{-3}$ (h)$^{-n}$	R^2	$k_H \times 10^{-3}$ (h)$^{-n}$	R^2
PLA/0.25R	0.61	0.99	14.1	0.989	27.57	0.98
PLA/0.5R	0.42	0.98	12.82	0.988	8.28	0.98
PLA/0.75R	0.54	0.985	3.52	0.993	4.48	0.99

The migration profiles from PLA/R films with three different R compositions show a quite similar release behavior of the rosemary ethanolic extract active ingredients, but it is quantitatively influenced by the amount of R incorporated in the film samples. Thus, in the 18-day interval investigated, the active ingredients of R were released faster and in the highest amount from the PLA/0.25R sample—reaching 48.4% release, followed by the PLA/0.5R sample with 16.3% released, and the PLA/0.75R sample with only 9.2% released. The increase of the R content leads to a slower release of the active components.

The calculated kinetic parameters and the corresponding correlation coefficient values (R^2) are summarized in Table 9.

The release exponent, n values, obtained by fitting the data to the Korsmeyer-Peppas equation, see Equation (2) [36], are 0.61 for PLA/0.25R, 0.42 for PLA/0.5R, and 0.55 for PLA/0.75R, all these values are situated around 0.5, indicating a behavior closer to Fickian diffusion, thus, in general, a diffusion-controlled mechanism characterizes the migration process of the bioactive components from the polymeric films. A decrease of the release rate constant values with an increasing amount of R incorporated (the lowest k = 3.52×10^{-3} for PLA/0.75R sample) was observed, which corresponds to a slower release, as shown by the migration profiles.

A good fit was obtained for the Higuchi model, see Equation (3) [37], as the R^2 values suggest, indicating diffusion as the preferred mechanism for migration, with the same trend of decreasing k_H values: $k_{PLA/0.25R} > k_{PLA/0.5R} > k_{PLA/0.75R}$ was found through the application of both models.

These results indicate that the migration rate of some components of the rosemary ethanolic extract from PLA/R films directly depends on their permeability and degree of crystallization. For films of approximately the same thickness, the gas permeability decreased by two or six times, indicating difficult diffusion of components through the materials. Also, the interaction between the PLA/R blends components can slow the migration into the food simulant.

3.13. Biocompatibility Evaluation

3.13.1. In Vitro Biocompatibility Evaluation Based on Surface Properties

The contact angles decrease for all systems with respect to that of PLA, see Table 10, indicating that an increased hydrophilicity was achieved through R incorporation. This increase is also proved by variations recorded in the surface tension and its components. The total surface tension γ_{sv}^{TOT} increases with R incorporation, and, as was the case for other properties, the highest values were registered for the PLA/PEG/0.5R system.

The most important change was found in the γ_{sv}^{-} electron donor (Lewis base) component which is 0.19 mN/m for PLA and increases to 38.05 mN/m for PLA/PEG/0.5R while γ_{sv}^{AB}, polar Lewis acid-base interaction varied from 0.86 mN/m to 13.99 mN/m. This phenomenon is related by the cohesion between molecules at the surface of the liquid. It is known that the stronger the intermolecular interactions, the greater the surface tension [82]; therefore, rosemary extract improved the interactions of the liquids with surfaces, obtaining composites with better water wettability. The presence of PEG as a plasticizer into the polymeric system enhances the surface hydrophilicity; the lowest CA was recorded for PLA/PEG (50°) compared with that of PLA (84.7°), see Table 10. The decrease of CA is directly correlated with the content of rosemary ethanolic extract, indicating that by increasing the content of rosemary extract, more hydrophilic groups are accessible at the top surface, obtaining a material with a moderate surface wettability (64.4° for 0.75% content rosemary extract).

Table 10. The contact angle values (θ) between liquids (water, formamide, or diiodomethane) and surfaces of the sample of different compositions.

Samples	Contact Angles Values (degrees)		
	Water	**Formamide**	**Diiodomethane**
PLA	84.7	70.9	65.0
PLA/0.25R	81.9	62.7	60.1
PLA/0.5R	78.8	59.0	56.7
PLA/0.75R	64.4	42.8	38.7
PLA/PEG	50.0	46.4	40.7
PLA/PEG/0.5R	63.7	71.0	51.0

In Table 11, the values of the surface tension components (determined by acid/base method) and work of spreading for red blood cells and platelets are presented, indicating that with the addition of rosemary ethanolic extract and PEG into the PLA, the total SFE increases. This change of the surface free energy may lead to the improvement of the bio-adhesive ability of the material. In the interaction of biomaterials with tissues, their polar character has an important role, which is reflected in the SFE's polar component (γ_{sv}^{AB}), see Table 11. At the surface of PLA, Lifshitz-van der Waals interactions prevail over acid-base interactions, by comparison with the surfaces of its composites. Blending PLA with rosemary ethanolic extract and PEG determines an increase in the contribution of the polar components to the total surface energy, having a more pronounced polar character than PLA, which is indicated by the higher value of γ_{sv}^{AB} component. The basic component (γ_{sv}^{-}) of the surface free energy is much higher than the acidic component (γ_{sv}^{+}), suggesting that the surface of the composites is mainly monopolar with a Lewis base character, due to the surface being enriched with electron donor functional groups (–OH, O–C=O) from rosemary ethanolic extract. If polymeric materials are used as implant biomaterials, strong cell adhesion and rapid cell growth on the polymer surface will generally be beneficial for the incorporation of the device into the body. Van der Valk et al. [83] have shown that cell (fibroblast) spreading appeared to be dependent on the polar surface free energy. Cell spreading is low when the SFE's polar part of the material is lower than 5 mN/m, and marked spreading occurs when γ_{sv}^{AB} is higher than 15 mN/m [43]. Based on the values of the SFE polar part determined for the PLA/R composites, it can be assumed that the sample incorporating all the components (PLA/PEG/0.5R) will give better spreading and division of the fibroblasts because their γ_{sv}^{AB} takes an intermediary value.

Table 11. Surface tension parameters (mN/m) and work of spreading for red blood cells and platelets of composites samples.

Samples	γ_{sv}^{LW}	γ_{sv}^{AB}	γ_{sv}^{+}	γ_{sv}^{-}	γ_{sv}^{TOT}	$W_{s/rbc}$	$W_{s/p}$
PLA	25.65	1.75	0.06	12.45	27.40	−8.95	−109.56
PLA/0.25R	28.46	3.31	0.62	7.22	32.69	1.43	−107.13
PLA/0.5R	30.41	4.23	0.11	24.01	33.72	−2.11	−90.51
PLA/0.75R	40.17	5.41	0.33	22.22	45.58	10.83	−74.12
PLA/PEG	39.17	1.92	0.03	34.91	41.09	4.54	−69.58
PLA/PEG/0.5R	33.64	13.95	1.27	38.25	47.59	12.27	−71.51

θ = contact angle; γ_{sv}^{TOT} = liquid's total surface tension; γ_{lv}^{LW} and γ_{sv}^{LW} = apolar Lifshitz—van der Waals components of the liquid and the solid $\gamma_{sv}^{+}\gamma_{lv}^{-}\gamma_{lv}^{-}$ and $\gamma_{sv}^{-}\gamma_{lv}^{+}$ = Lewis acid-base contributions of either the solid or the liquid phase; γ_{sv}^{-}= electron donor (Lewis base) and γ_{sv}^{+} = electron acceptor (Lewis acid) components; "lv" and "sv" = denote the interfacial liquid-vapor and surface-vapor tensions; "p" and "d" = denote the polar and disperse components of total surface tension, γ_{lv}^{TOT};γ_{sv}^{AB} = polar Lewis acid-base interaction.

When blood comes into contact with a biomaterial, the surface properties, especially its wettability, play a crucial role. The response of platelets towards the hydrophobic or hydrophilic materials is different [84]. In the presence of a biomaterial, platelets' adherence to the foreign material and activation, usually results in the initiation of platelet aggregation, further initiating thrombosis by secreting prothrombotic factors that lead to clotting [85]. Under normal conditions, thrombosis and complement activation are favorable responses which prevent blood loss. However, in the presence of a foreign material such as a blood-contacting device, thrombosis, and complement activation are unfavorable, leading to blood clots and hence a low biocompatibility [84]. Thus, a blood-contacting device that does not elicit thrombosis would be considered biocompatible [86]. As revealed in Table 11, the work of spreading of the red blood cells, $W_{s/rbc}$, has positive values (except for the PLA and PLA/0.5R samples) and the work of the spreading of platelets, $W_{s/p}$, has negative values, which suggests that the work of adhesion is higher than that of cohesion for the red blood cells, but, comparatively, a smaller work of adhesion than that of cohesion for platelets. These findings denote that the contact of blood with PLA/R-based composites causes an increase in the work of cohesion for platelets; hence the platelets will not adhere easily onto the biomaterial's surface, thus avoiding the appearance of thrombosis [87].

3.13.2. In Vivo Biocompatibility Evaluation

No significant modifications between the percentage values of blood leucocyte formula elements were discovered in rats with PLA/0.25R, PLA/0.5R, PLA/0.75R, PLA/PEG, and PLA/PEG/0.5R implants and those of control group, PLA, and R, respectively, at 24 h and 7 days in the experiment, see Table 12.

Table 12. The effects of the PLA-based materials containing rosemary ethanolic extract on the differential white cell count. Values were expressed as mean ± S.D. for six rats in a group.

Group		Leucocyte Formula				
		% Values				
		PMN	Ly	E	M	B
Control	24 h	29.5 ± 0.83	66.3 ± 2.11	0.6 ± 0.08	3.4 ± 0.10	0.2 ± 0.10
	7 days	29.7 ± 0.47	65.9 ± 1.93	0.7 ± 0.10	3.5 ± 0.10	0.2 ± 0.05
R	24 h	29.5 ± 0.69	66.1 ± 1.75	0.8 ± 0.06	3.4 ± 0.06	0.2 ± 0.10
	7 days	29.6 ± 0.73	66.2 ± 1.89	0.6 ± 0.08	3.4 ± 0.10	0.2 ± 0.05
PLA	24 h	29.6 ± 0.89	66.1 ± 2.13	0.6 ± 0.05	3.5 ± 0.05	0.2 ± 0.05
	7 days	29.7 ± 1.13	65.9 ± 1.55	0.7 ± 0.05	3.5 ± 0.05	0.2 ± 0.04
PLA/0.25R	24 h	29.6 ± 0.21	66.0 ± 1.73	0.7 ± 0.12	3.5 ± 0.08	0.2 ± 0.04
	7 days	29.8 ± 1.13	65.7 ± 1.29	0.7 ± 0.05	3.6 ± 0.10	0.2 ± 0.05
PLA/0.5R	24 h	29.7 ± 0.29	65.9 ± 2.14	0.6 ± 0.10	3.6 ± 0.08	0.2 ± 0.10
	7 days	29.7 ± 1.17	65.7 ± 1.33	0.8 ± 0.06	3.6 ± 0.05	0.2 ± 0.05
PLA/0.75R	24 h	29.6 ± 0.98	65.8 ± 1.67	0.8 ± 0.12	3.6 ± 0.10	0.2 ± 0.10
	7 days	29.8 ± 0.73	65.8 ± 1.75	0.6 ± 0.05	3.6 ± 0.08	0.2 ± 0.05
PLA/PEG	24 h	29.8 ± 0.89	65.7 ± 1.39	0.7 ± 0.10	3.6 ± 0.05	0.2 ± 0.04
	7 days	29.9 ± 0.55	65.5 ± 1.63	0.7 ± 0.05	3.7 ± 0.12	0.2 ± 0.05
PLA/PEG/0.5R	24 h	29.8 ± 0.27	65.7 ± 1.98	0.6 ± 0.10	3.7 ± 0.10	0.2 ± 0.05
	7 days	29.9 ± 1.63	65.4 ± 1.47	0.8 ± 0.13	3.7 ± 0.05	0.2 ± 0.05

PMN—polymorphonuclear leukocytes; Ly—lymphocytes; E—eosinophils; M—monocytes; B—basophils.

Laboratory investigations did not reveal substantial variations of AST, ALT, and LDH activity between groups with the PLA-based materials containing rosemary ethanolic extract pellets and the control, PLA, and R groups, respectively, 24 h and 7 days after the implantation, see Table 13.

Table 13. The effects of PLA-based materials containing rosemary ethanolic extract on the activity of AST, ALT, and LDH. Values were expressed as mean ± S.D. for six rats in a group.

Group		AST (U/mL)	ALT (U/mL)	LDH (U/mL)
Control	24 h	41.7 ± 2.72	95.3 ± 4.14	342.29 ± 44.55
	7 days	42.5 ± 3.07	96.5 ± 3.89	344.33 ± 41.37
PLA	24 h	42.3 ± 3.14	95.8 ± 4.46	342.17 ± 39.64
	7 days	42.9 ± 2.33	97.6 ± 3.55	345.25 ± 40.89
R	24 h	41.6 ± 1.89	96.2 ± 3.37	343.42 ± 42.14
	7 days	42.7 ± 2.64	98.7 ± 5.07	346.67 ± 38.33
PLA/0.25R	24 h	42.1 ± 3.14	97.5 ± 4.27	343.54 ± 43.46
	7 days	43.9 ± 3.46	97.9 ± 3.64	346.81 ± 41.37
PLA/0.5R	24 h	43.2 ± 3.33	97.6 ± 3.37	344.55 ± 39.89
	7 days	44.1 ± 3.27	98.8 ± 5.14	347.19 ± 44.14
PLA/0.75R	24 h	43.4 ± 2.37	98.3 ± 5.55	344.29 ± 43.72
	7 days	44.6 ± 2.55	98.7 ± 6.07	348.46 ± 45.07
PLA/PEG	24 h	43.7 ± 3.37	98.5 ± 5.46	345.72 ± 44.37
	7 days	44.8 ± 3.89	98.8 ± 4.33	349.15 ± 40.46
PLA/PEG/0.5R	24 h	43.5 ± 3.14	97.2 ± 3.89	345.45 ± 39.89
	7 days	44.6 ± 3.64	98.6 ± 5.72	347.83 ± 42.27

AST—aspartate transaminase, ALT—alanine aminotransferase and LDH—lactic dehydrogenase.

The implantation of pellets with PLA/0.25R, PLA/0.5R, PLA/0.75R, PLA/PEG and PLA/PEG/0.5R, did not induce significant differences in the serum levels of urea and creatinine compared with the groups receiving PLA, R, and the control, see Table 14.

Table 14. The effects of PLA-based materials containing rosemary ethanolic extract on the serum urea and creatinine concentration. Values were expressed as mean ± S.D. for six rats in a group.

Group		Urea (mg/dL)	Creatinine (mg/dL)
Control	24 h	37.2 ± 3.37	<0.1
	7 days	37.9 ± 4.55	<0.1
PLA	24 h	37.7 ± 3.89	<0.2
	7 days	38.1 ± 3.64	<0.2
R	24 h	37.9 ± 5.07	<0.1
	7 days	38.5 ± 3.46	<0.1
PLA/0.25R	24 h	38.6 ± 4.37	<0.1
	7 days	38.9 ± 4.33	<0.2
PLA/0.5R	24 h	38.8 ± 3.64	<0.2
	7 days	39.3 ± 3.72	<0.2
PLA/0.75R	24 h	39.1 ± 5.14	<0.2
	7 days	39.4 ± 4.46	<0.2
PLA/PEG	24 h	38.8 ± 3.55	<0.2
	7 days	39.6 ± 3.27	<0.2
PLA/PEG/0.5R	24 h	39.2 ± 4.64	<0.2
	7 days	39.5 ± 3.37	<0.2

No major dissimilarities were observed in the values of serum complement and the phagocytic capacity of peripheral neutrophils between the groups with PLA-based materials containing rosemary ethanolic extract implants, and PLA, R, and the control group, respectively, at 24 h and 7 days after the implantation see Table 15.

Table 15. The effects of PLA-based materials containing rosemary ethanolic extract on the serum complement activity and the NBT test. Values were expressed as mean ± S.D. for six rats in a group.

Group		Complement	NBT Test
Control	24 h	16.33 ± 1.55	53.73 ± 3.46
	7 days	16.48 ± 1.37	53.65 ± 3.55
PLA	24 h	16.41 ± 1.46	53.85 ± 4.14
	7 days	16.65 ± 1.14	53.49 ± 3.72
R	24 h	16.39 ± 1.55	52.63 ± 3.50
	7 days	16.47 ± 1.33	52.45 ± 3.14
PLA/0.25R	24 h	16.43 ± 1.64	52.77 ± 3.67
	7 days	16.85 ± 0.89	52.39 ± 3.25
PLA/0.5R	24 h	16.56 ± 1.72	53.68 ± 3.46
	7 days	16.89 ± 1.37	52.55 ± 3.83
PLA/0.75R	24 h	16.77 ± 1.64	53.73 ± 3.46
	7 days	17.11 ± 0.72	52.61 ± 3.37
PLA/PEG	24 h	17.07 ± 0.89	53.85 ± 4.05
	7 days	16.63 ± 1.55	53.37 ± 3.64
PLA/PEG/0.5R	24 h	17.03 ± 0.83	54.19 ± 4.17
	7 days	17.13 ± 1.46	53.49 ± 3.55

All these results indicate a good and increased biocompatibility of the PLA/R which recommends them as biomaterials for various applications because rosemary ethanolic extract offers both an improved biocompatibility of PLA-based materials and it also confers the different biological activities mentioned above.

4. Conclusions

New multifunctional materials based on PLA-containing additives derived from natural resources were obtained by melt mixing.

Incorporation of powdered rosemary ethanolic extract into PLA improved the elongation at break, rheological and thermal properties, and antibacterial and antioxidant activities. Additionally, the novel materials showed a good compatibility and in vitro and in vivo biocompatibility. The contribution of additive to the protection of PLA against oxidative degradation is discussed with respect to the phenolic compounds contained in rosemary powder.

A good agreement between results was found, such as between those of chemiluminescence and radicals' scavenging activity determination via a chemical method that evidenced the increased thermoxidative stability of the PLA biocomposites containing powdered rosemary ethanolic extract which acts as an antioxidant. These biocomposites show low migration rates of the bioactive compounds from matrices and permeability to gases, and, therefore, can be considered as high-performance materials for food packaging.

In vitro biocompatibility based on the determination of surface properties demonstrated a good hydrophilicity, better spreading and division of fibroblasts, and an increase of platelets' cohesion. Therefore, they will not adhere easily on the biomaterial's surface thus avoiding the appearance of thrombosis.

It was demonstrated that implantation of PLA-based materials containing rosemary ethanolic extract resulted in similar blood parameter changes and biochemical responses with the control group, the group treated with pellets impregnated with R solution, and with PLA pellets, respectively.

The subcutaneous application of PLA-based materials containing rosemary ethanolic extract pellets did not significantly influence the immune reactivity of rats, compared with PLA, R, and the control group.

It can be concluded that, in our laboratory conditions, the implantation of PLA/0.25R, PLA/0.5R, PLA/0.75R, PLA/PEG, and PLA/PEG/0.5R pellets, proved a good in vivo biocompatibility, suggesting that these PLA-based materials containing rosemary ethanolic extract show very good properties as potential biomaterials, which could be further used for evaluating their possible pharmacodynamics effects in different experimental models in animals.

The PLA/R-based materials show promising properties for application both in biodegradable food packaging and as biomaterials with bioactive activities.

Author Contributions: In this study, the concepts and designs for the experiment were supervised by C.V. and E.S. Manuscript text and results analysis were performed by C.V., R.N.D.-N. and E.S. Mechanical and rheological experiments and data processing were conducted by R.N.D.-N. The experimental results were examined by C.V., E.S., R.N.D.-N., D.P., R.P.D., L.T., T.Z., M.A.B., N.T., G.M.P., K.L. All authors read and approved the final manuscript.

References

1. Scaffaro, R.; Lopresti, F.; Marino, A.; Nostro, A. Antimicrobial additives for poly(lactic acid) materials and their applications: Current state and perspectives. *Appl. Microbiol. Biotechnol.* **2018**, *102*, 7739–7756. [CrossRef] [PubMed]
2. Bitchagno, G.T.; Sama Fonkeng, L.; Kopa, T.K.; Tala, M.F.; Kamdem Wabo, H.; Tume, C.B.; Tane, P.; Kuiate, J.R. Antibacterial activity of ethanolic extract and compounds from fruits of *Tectona grandis* (Verbenaceae). *BMC Complement. Altern. Med.* **2015**, *15*, 1–6. [CrossRef] [PubMed]
3. Alara, O.R.; Abdurahman, N.H.; Olalere, O.A. Ethanolic extraction of flavonoids, phenolics and antioxidants from vernonia amygdalina leaf using two-level factorial design. *J. King Saud Univ. Sci.* **2017**, in press. [CrossRef]
4. Madsen, H.L.; Andersen, L.; Christiansen, L.; Brockhoff, P.; Bertelsen, G. Antioxidative activity of summer savory (*Satureja hortensis* L.) and rosemary (*Rosemarinus officinalis* L.) in minced, cooked pork meat. *Z. Lebensm. Unters. Forsch.* **1996**, *203*, 333–338. [CrossRef]
5. Erkan, N.; Ayranci, G.; Ayranci, E. Antioxidant activities of rosemary (*Rosmarinus officinalis* L.) extract, blackseed (*Nigella sativa* L.) essential oil, carnosic acid, rosmarinic acid and sesamol. *Food Chem.* **2008**, *110*, 76–82. [CrossRef] [PubMed]
6. Borrás-Linares, I.; Stojanović, Z.; Quirantes-Piné, R.; Arráez-Román, D.; Švarc-Gajić, J.; Fernández-Gutiérrez, A.; Segura-Carretero, A. Rosmarinus Officinalis Leaves as a Natural Source of Bioactive Compounds. *Int. J. Mol. Sci.* **2014**, *15*, 20585–20606. [CrossRef] [PubMed]
7. Bolumar, T.; LaPeña, D.; Skibsted, L.H.; Orlien, V. Rosemary and oxygen scavenger in active packaging for prevention of high-pressure induced lipid oxidation in pork patties. *Food Packag. Shelf Life* **2016**, *7*, 26–33. [CrossRef]
8. Cuvelier, M.-E.; Richard, H.; Berset, C. Antioxidative activity and phenolic composition of pilot-plant and commercial extracts of sage and rosemary. *J. Am. Oil Chem. Soc.* **1996**, *73*, 645–652. [CrossRef]
9. Aguilar, F. Scientific Opinion of the Panel on Food Additives, Flavourings, Processing aids and materials in contact with food on a request from the commission on the use of rosemary extracts as a food additive. *EFSA J.* **2008**, *721*, 1–29. Available online: http://www.efsa.europa.eu/sites/default/files/scientific_output/files/main_documents/721.pdf (accessed on 1 August 2018).
10. Matejczyk, M.; Swisłocka, R.; Golonko, A.; Lewandowski, W.; Hawrylik, E. Cytotoxic, genotoxic and antimicrobial activity of caffeic and rosmarinic acids and their lithium, sodium and potassium salts as potential anticancer compounds. *Adv. Med. Sci.* **2018**, *63*, 14–21. [CrossRef] [PubMed]
11. Lo, A.H.; Liang, Y.C.; Lin-Shiau, S.Y.; Ho, C.T.; Lin, J.K. Carnosol, an antioxidant in rosemary, suppresses inducible nitric oxide synthase through down-regulating nuclear factor-κB in mouse macrophages. *Carcinogenesis* **2002**, *23*, 983–991. [CrossRef] [PubMed]
12. Rašković, A.; Milanović, I.; Pavlović, N.; Ćebović, T.; Vukmirović, S.; Mikov, M. Antioxidant activity of rosemary (*Rosmarinus officinalis* L.) essential oil and its hepatoprotective potential. *BMC Complement. Altern. Med.* **2014**, *14*, 225–237. [CrossRef] [PubMed]

13. Mohapatra, A.K.; Mohanty, S.; Nayak, S.K. Effect of PEG on PLA/PEG blend and its nanocomposites: A study of thermo-mechanical and morphological characterization. *Polym. Compos.* **2013**, *35*, 283–293. [CrossRef]
14. Arrieta, M.P.; Samper, M.D.; Aldas, M.; López, J. On the Use of PLA-PHB Blends for Sustainable Food Packaging Applications. *Materials* **2017**, *10*, 1008. [CrossRef] [PubMed]
15. Hansen, M.C. Polymer additives and solubility parameters. *Prog. Org. Coat.* **2004**, *51*, 109–112. [CrossRef]
16. Darie-Nita, R.N.; Vasile, C.; Irimia, A.; Lipsa, R.; Rapa, M. Evaluation of some eco-friendly plasticizers for PLA films processing. *J. Appl. Polym. Sci.* **2016**, *133*, 43223. [CrossRef]
17. Pillin, I.; Montrelay, N.; Bourmaud, A.; Grohens, Y. Thermo-mechanical characterization of plasticized PLA: Is the miscibility the only significant factor. *Polymer* **2006**, *47*, 4676–4682. [CrossRef]
18. Zahed, E.; Ansari, S.; Wu, B.M.; Bencharit, S.; Moshaverinia, A. Hydrogels in craniofacial tissue engineering. In *Biomaterials for Oral and Dental Tissue Engineering*; Tayebi, L., Moharamzadeh, K., Eds.; Woodhead Publishing Elsevier Ltd.: Amsterdam, The Netherlands, 2018; pp. 47–64, ISBN 978-0-08-100961-1.
19. Hamad, K.; Kaseem, M.; Yang, H.W.; Deri, F.; Ko, Y.G. Properties and medical applications of polylactic acid: A review. *Express Polym. Lett.* **2015**, *9*, 435–455. [CrossRef]
20. Pawar, R.P.; Tekale, S.U.; Shisodia, S.U.; Totre, J.T.; Domb, A.J. Biomedical Applications of Poly(Lactic Acid). *Recent Pat. Regen. Med.* **2014**, *4*, 40–51. [CrossRef]
21. Davachi, S.M.; Kaffash, B. Polylactic Acid in Medicine. *Polym. Plast. Technol. Eng.* **2015**, *54*, 944–967. [CrossRef]
22. Abd Alsaheb, R.A.; Aladdin, A.; Othman, N.Z.; Malek, R.A.; Leng, O.M.; Aziz, R.; El Enshasy, H.A. Recent applications of polylactic acid in pharmaceutical and medical industries. *J. Chem. Pharm. Res.* **2015**, *7*, 51–63.
23. Commission, E. Commission regulation (EU) no. 10/2011 of 14 January 2011 on plastic materials and articles intended to come into contact with food. *J. Eur. Commun.* **2011**, *45*, 2–89.
24. Kuorwel, K.K.; Cran, M.J.; Sonneveld, K.; Miltz, J.; Bigger, S.W. Migration of antimicrobial agents from starch-based films into a food simulant. *LWT Food Sci. Technol.* **2013**, *50*, 432–438. [CrossRef]
25. Tawakkal, I.S.M.A.; Cran, M.J.; Bigger, S.W. Release of thymol from poly(lactic acid)-based antimicrobial films containing kenaf fibres as natural filler. *LWT Food Sci. Technol.* **2016**, *66*, 629–637. [CrossRef]
26. Mascheroni, E.; Guillard, V.; Nalin, F.; Mora, L.; Piergiovanni, L. Diffusivity of propolis compounds in polylactic acid polymer for the development of antimicrobial packaging films. *J. Food Eng.* **2010**, *98*, 294–301. [CrossRef]
27. Zaharescu, T.; Jipa, S.; Mariş, A.D.; Mariş, M.; Kappel, W.; Mantsch, A. Effect of rosemary extract on the radiation stability of UHMWPE. *ePolymers* **2009**, *149*, 1–9. [CrossRef]
28. Scalbert, A.; Monties, B.; Janin, G. Tannins in wood-comparison of different estimation methods. *Agric. Food Chem.* **1989**, *37*, 1324–1329. [CrossRef]
29. Zhishen, J.; Mengcheng, T.; Jianming, W. The determination of flavonoid contents in mulberry and their scavenging effect on superoxide radicals. *Food Chem.* **1999**, *64*, 555–559. [CrossRef]
30. Kim, I.S.; Yang, M.R.; Lee, O.H.; Kang, S.H. Antioxidant activities of hot water extracts from various spices. *Int. J. Mol. Sci.* **2011**, *12*, 4120–4131. [CrossRef] [PubMed]
31. Liu, H.; Hsieh, C.T.; Hu, D.S.G. Solute diffusion through degradable semicrystalline polyethylene glycol/poly(L-lactide) copolymers. *Polym. Bull.* **1994**, *32*, 463–470. [CrossRef]
32. Commission, E. Commission directive 97/48/EC of 29 July 1997 amending for the second time council directive 82/711/EEC laying down the basic rules necessary for testing migration of the constituents of plastic materials and articles intended to come into contact with foodstuffs (97/48/EC). *J. Eur. Commun.* **1997**, *222*, 210–215.
33. McClements, D.J.; Decker, E.A. Lipid Oxidation in Oil-in-Water Emulsions: Impact of Molecular Environment on Chemical Reactions in Heterogeneous Food Systems. *J. Food Sci.* **2000**, *65*, 1270–1282. [CrossRef]
34. Mulinacci, N.; Innocenti, M.; Bellumori, M.; Giaccherini, C.; Martini, V.; Michelozzi, M. Storage method, drying processes and extraction procedures strongly affect the phenolic fraction of rosemary leaves: An HPLC/DAD/MS study. *Talanta* **2011**, *85*, 167–176. [CrossRef] [PubMed]
35. Zhang, Y.; Smuts, J.P.; Dodbiba, E.; Rangarajan, R.; Lang, J.C.; Armstrong, D.W. Degradation study of carnosic acid, carnosol, rosmarinic acid, and rosemary extract (*Rosmarinus officinalis* L.) assessed using HPLC. *J. Agric. Food Chem.* **2012**, *60*, 9305–9314. [CrossRef] [PubMed]

36. Peppas, N.A.; Sahlin, J.J. A simple equation for the description of solute release. III. Coupling of diffusion and relaxation. *Int. J. Pharm.* **1989**, *57*, 169–172. [CrossRef]
37. Peppas, N.A. Analysis of Fickian and non-Fickian drug release from polymers. *Pharm. Acta Helv.* **1985**, *60*, 110–111. [PubMed]
38. Pâslaru, E.; Fras Zemljic, L.; Bračič, M.; Vesel, A.; Petrinić, I.; Vasile, C. Stability of a chitosan layer deposited onto a polyethylene surface. *J. Appl. Polym. Sci.* **2013**, *130*, 2444–2457. [CrossRef]
39. Stoleru, E.; Baican, M.C.; Coroaba, A.; Hitruc, G.E.; Lungu, M.; Vasile, C. Plasma-activated fibrinogen coatings onto poly(vinylidene fluoride) surface for improving biocompatibility with tissues. *J. Bioact. Compat. Polym.* **2016**, *31*, 91–108. [CrossRef]
40. Pascu, M. Contact angle method. In *Surface Properties of Polymers*, 1st ed.; Vasile, C., Pascu, M.C., Eds.; Research Signpost: Kerala, India, 2007; pp. 179–201.
41. Vasile, C.; Pascu, M.; Bumbu, G.G.; Cojocariu, A. Surface properties of polymers. In *Surface Properties of Polymers*, 1st ed.; Vasile, C., Pascu, M.C., Eds.; Research Signpost: Kerala, India, 2007; pp. 1–64.
42. Yuan, Y.; Lee, T.R. Contact angle and wetting properties. In *Surface Science Techniques*; Bracco, G., Holst, B., Eds.; Springer Series in Surface Sciences; Springer: Berlin/Heidelberg, Germany, 2013; pp. 3–34.
43. Stoleru, E.; Munteanu, B.S.; Darie-Niţă, R.N.; Pricope, G.M.; Lungu, M.; Irimia, A.; Râpă, M.; Lipşa, R.D.; Vasile, C. Complex poly(lactic acid)-based biomaterial for urinary catheters: II. Biocompatibility. *Bioinspir. Biomim. Nanobiomater.* **2016**, *5*, 152–166. [CrossRef]
44. Van Oss, C.J.; Chaudhury, M.K.; Good, R.J. Interfacial Lifshitz-Van der Waals and polar interactions in macroscopic systems. *Chem. Rev.* **1988**, *88*, 927–941. [CrossRef]
45. Pascu, M.C.; Popescu, M.C.; Vasile, C. Surface modifications of some nanocomposites containing starch. *J. Phys. D Appl. Phys.* **2008**, *41*, 175407. [CrossRef]
46. Popescu, M.C.; Vasile, C.; Macocinschi, D.; Lungu, M.; Craciunescu, O. Biomaterials based on new polyurethane and hydrolyzed collagen, k-elastin, hyaluronic acid and chondroitin sulphate. *Int. J. Biol. Macromol.* **2010**, *47*, 646–653. [CrossRef] [PubMed]
47. Onofrei, M.D.; Dobos, A.M.; Dunca, S.; Ioanid, E.G.; Ioan, S. Biocidal activity of cellulose materials for medical implants. *J. Appl. Polym. Sci.* **2015**, *132*, 41932. [CrossRef]
48. Macocinschi, D.; Filip, D.; Butnaru, M.; Dimitriu, C.D. Surface characterization of biopolyurethanes based on cellulose derivatives. *J. Mater. Sci. Mater. Med.* **2009**, *20*, 775–783. [CrossRef] [PubMed]
49. Vijayanand, K.; Deepak, K.; Pattanayak, D.K.; Rama Mohan, T.R.; Banerjee, R. Interpenetring blood-biomaterial interactions from surface free energy and work of adhesion. *Trends Biomater. Artif. Organs.* **2005**, *182*, 73–83.
50. Albu, R.M.; Avram, E.; Stoica, I.; Ioanid, E.G.; Popovici, D.; Ioan, S. Surface properties and compatibility with blood of new quaternized polysulfones. *J. Biomater. Nanobiotechnol.* **2011**, *2*, 114–124. [CrossRef]
51. Hodgson, E. *A Textbook of Modern Toxicology*, 3rd ed.; John Wiley & Sons: Hoboken, NJ, USA, 2004; ISBN 0-471-26508-X.
52. Botham, P.A. Acute systemic toxicity-prospects for tiered testing strategies. *Toxicol. In Vitro* **2004**, *18*, 227–230. [CrossRef]
53. Wolf, M.F.; Andwraon, J.M. Practical approach to blood compatibility assessments: General considerations and standards. In *Biocompatibility and Performance of Medical Devices*, 1st ed.; Boutrand, J.P., Ed.; Woodhead Publishing Ltd.: Cambridge, UK, 2012; pp. 159–200, ISBN 9780857090706.
54. Lupuşoru, C.E. *Imunofarmacologie*, 1st ed.; ALFA: Iaşi, Romania, 2001; pp. 288–294.
55. Peacman, M. *Clinical & Experimental Immunology*; Wiley Library, British Society of Immunology: Hoboken, NJ, USA, 2011.
56. Zimmerman, M. Ethical guidelines for investigations of experimental pain in conscious animals. *Pain* **1983**, *16*, 109–110. [CrossRef]
57. Protocole D'amendement à la Convention Européenne sur la Protection des Animaux Vertébrés Utilisés à des Fins Expérimentales ou à D'autres Fins Scientifiques. Available online: https://rm.coe.int/CoERMPublicCommonSearchServices/DisplayDCTMContent?documentId=090000168007f2e3 (accessed on 1 August 2018). (In French)
58. Arrieta, M.P.; Samper, M.D.; López, J.; Jiménez, A. Combined effect of poly(hydroxybutyrate) and plasticizers on polylactic acid properties for film intended for food packaging. *J. Polym. Environ.* **2014**, *22*, 460–470. [CrossRef]

59. Ionita, P.; Dinoiu, V.; Munteanu, C.; Turcu, I.M.; Tecuceanu, V.; Zaharescu, T.; Oprea, E.; Ilie, C.; Anghel, D.; Ionita, G. Antioxidant activity of rosemary extracts in solution and embedded in polymeric systems. *Chem. Pap.* **2015**, *69*, 872–880. [CrossRef]
60. Beauchamp Spectroscopy Tables. Available online: http://www.cpp.edu/~psbeauchamp/pdf/spec_ir_nmr_spectra_tables.pdf (accessed on 1 August 2018).
61. Ozkoc, G.; Kemalogu, S. Morphology, biodegradability, mechanical, and thermal properties of nanocomposite films based on PLA and plasticized PLA. *J. Appl. Polym. Sci.* **2009**, *114*, 2481–2487. [CrossRef]
62. Li, D.; Jiang, Y.; Lv, S.; Liu, X.; Gu, J.; Chen, Q.; Zhang, Y. Preparation of plasticized poly (lactic acid) and its influence on the properties of composite materials. *PLoS ONE* **2018**, *13*, e0193520. [CrossRef] [PubMed]
63. Zeid, A.M. Preparation and Evaluation of Polylactic Acid Antioxidant Packaging Films Containing Thyme, Rosemary and Oregano Essential Oils. Master's Thesis, The American University in Cairo (AUC), Cairo, Egypt, May 2015.
64. Petkova-Parlapanska, K.; Nancheva, V.; Diankov, S.; Hinkov, I.; Karsheva, M. Rheological properties of cosmetic compositions containing rosemary and grapefruit pulp and seeds extracts. *J. Chem. Technol. Metall.* **2014**, *49*, 487–493.
65. Topala, C.M.; Tataru, L.D. ATR-FTIR Study of thyme and rosemary oils extracted by supercritical carbon dioxide. *Rev. Chim. (Bucharest)* **2016**, *67*, 842–846.
66. Fan, Y.; Nishida, H.; Hoshihara, S.; Shirai, Y.; Endo, T. Effects of tin on poly(l-lactic acid) pyrolysis. *Polym. Degrad. Stab.* **2003**, *81*, 515. [CrossRef]
67. Tsuge, S.; Ohtani, H.; Watanabe, C. *Pyrolysis—GC/MS Data Book of Synthetic Polymers: Pyrograms, Thermograms and MS of Pyrolyzates*; Elsevier: Amsterdam, The Netherlands, 2011; p. 252.
68. Lin, H.; Han, L.; Dong, L. Thermal degradation behavior and gas phase flame retardant mechanism of polylactide/PCPP blend. *J. Appl. Polym. Sci.* **2014**, *131*, 40480. [CrossRef]
69. Rychý, J.; Rychlá, L.; Stloukal, P.; Koutný, M.; Pekařová, S.; Verney, V.; Fiedlorová, A. UV initiated oxidation and chemiluminescence from aromatic-aliphatic copolyesters and polylactic acid. *Polym. Degrad. Stab.* **2013**, *98*, 2556–2563. [CrossRef]
70. Nugroho, P.; Mitoshi, H.; Yoshii, F.; Kume, T. Degradation of poly(L-lactic acid). *Polym. Degrad. Stab.* **2001**, *72*, 337–343. [CrossRef]
71. Babanalbandi, A.; Hill, D.J.T.; Whittaker, A.K. Volatile products and new polymer structures formed on ^{60}Co γ-radiolysis of poly(lactic acid) and poly(glycolic acid). *Polym. Degrad. Stab.* **1997**, *58*, 203–214. [CrossRef]
72. Pospisil, J. Chemical and photochemical behaviour of phenolic antioxidants in polymer stabilization: A state of the art report, part II. *Polym. Degrad. Stab.* **1993**, *39*, 103–115. [CrossRef]
73. Vasile, C.; Sivertsvik, M.; Miteluţ, A.C.; Brebu, M.A.; Stoleru, E.; Rosnes, J.T.; Tănase, E.E.; Khan, W.; Pamfil, D.; Cornea, C.P.; et al. Comparative analysis of the composition and active property evaluation of certain essential oils to assess their potential applications in active food packaging. *Materials* **2017**, *10*, 45. [CrossRef] [PubMed]
74. Setnescu, R.; Kaci, M.; Jipa, S.; Setnescu, T.; Zaharescu, T.; Hebal, G.; Benhamida, A.; Djedjelli, H. Chemiluminescence study on irradiated low-density polyethylene containing various photo-stabilisers. *Polym. Degrad. Stab.* **2004**, *84*, 475–481. [CrossRef]
75. Łopusiewicz, Ł.; Jędra, F.; Mizielińska, M. New poly(lactic acid) active packaging composite films incorporated with fungal melanin. *Polymers* **2018**, *10*, 386. [CrossRef]
76. Byun, Y.; Kim, Y.T.; Whiteside, S. Characterization of an antioxidant polylactic acid (PLA) film prepared with α-tocopherol, BHT and polyethylene glycol using film cast extruder. *J. Food Eng.* **2010**, *100*, 239–244. [CrossRef]
77. Bacillus Cereus. Available online: https://www.foodsafety.gov/poisoning/causes/bacteriaviruses/bcereus/index.html (accessed on 1 August 2018).
78. Bottone, E.J. Bacillus cereus, a volatile human pathogen. *Clin. Microbiol. Rev.* **2010**, *23*, 382–398. [CrossRef] [PubMed]
79. Golshani, Z.; Sharifzadeh, A. Evaluation of Antibacterial Activity of Alcoholic Extract of Rosemary Leaves against Pathogenic Strains. *Zahedan J. Res. Med. Sci. (ZJRMS)* **2014**, *16*, 12–15. Available online: http://zjrms.ir/browse.php?a_code=A-10-2066-2&slc_lang=en&sid=1&sw=antibacterial+activity (accessed on 1 August 2018).

80. Lehermeier, H.J.; Dorgan, J.R.; Way, J.D. Gas Permeation Properties of Poly(Lactic Acid). (R826733). *J. Membr. Sci.* **2001**, *190*, 243–251. [CrossRef]
81. Yuniarto, K.; Welt, B.A.; Irawan, C. Morphological, thermal and oxygen barrier properties plasticized film polylactic acid. *J. Appl. Packag. Res.* **2017**, *9*, 6.
82. Intermolecular Forces in Action: Surface Tension, Viscosity, and Capillary Action. Available online: https://chem.libretexts.org/Textbook_Maps/General_Chemistry_Textbook_Maps/Map%3A_A_Molecular_Approach_(Tro)/11%3A_Liquids%2C_Solids%2C_and_Intermolecular_Forces/11.04%3A_Intermolecular_Forces_in_Action%3A_Surface_Tension%2C_Viscosity%2C_and_Capillary_Action#title (accessed on 8 May 2018).
83. Van der Valk, P.; van Pelt, A.W.J.; Busscher, H.; de Jong, H.P.; Wildevuur, C.R.; Arends, J. Interaction of fibroblasts and polymer surfaces: Relationship between surface free energy and fibroblast spreading. *J. Biomed. Mater. Res.* **1983**, *17*, 807–817. [CrossRef] [PubMed]
84. Spijker, H.T.; Graaff, R.; Boonstra, P.W.; Busscher, H.J.; van Oeveren, W. On the influence of flow conditions and wettability on blood material interactions. *Biomaterials* **2003**, *24*, 4717–4727. [CrossRef]
85. Courtney, J.M.; Lamba, N.M.; Sundaram, S.; Forbes, C.D. Biomaterials for blood-contacting applications. *Biomaterials* **1994**, *15*, 737–744. [CrossRef]
86. Menzies, K.L.; Jones, L. The Impact of Contact Angle on the Biocompatibility of Biomaterials. *Optom. Vis. Sci.* **2010**, *87*, 387–399. [CrossRef] [PubMed]
87. Nita, L.E.; Chiriac, A.P.; Stoleru, E.; Diaconu, A.; Tudorachi, N. Tailorable polyelectrolyte protein complex based on poly(aspartic acid) and bovine serum albumin. *Des. Monomers Polym.* **2016**, *19*, 596–606. [CrossRef]

PERMISSIONS

All chapters in this book were first published by MDPI; hereby published with permission under the Creative Commons Attribution License or equivalent. Every chapter published in this book has been scrutinized by our experts. Their significance has been extensively debated. The topics covered herein carry significant findings which will fuel the growth of the discipline. They may even be implemented as practical applications or may be referred to as a beginning point for another development.

The contributors of this book come from diverse backgrounds, making this book a truly international effort. This book will bring forth new frontiers with its revolutionizing research information and detailed analysis of the nascent developments around the world.

We would like to thank all the contributing authors for lending their expertise to make the book truly unique. They have played a crucial role in the development of this book. Without their invaluable contributions this book wouldn't have been possible. They have made vital efforts to compile up to date information on the varied aspects of this subject to make this book a valuable addition to the collection of many professionals and students.

This book was conceptualized with the vision of imparting up-to-date information and advanced data in this field. To ensure the same, a matchless editorial board was set up. Every individual on the board went through rigorous rounds of assessment to prove their worth. After which they invested a large part of their time researching and compiling the most relevant data for our readers.

The editorial board has been involved in producing this book since its inception. They have spent rigorous hours researching and exploring the diverse topics which have resulted in the successful publishing of this book. They have passed on their knowledge of decades through this book. To expedite this challenging task, the publisher supported the team at every step. A small team of assistant editors was also appointed to further simplify the editing procedure and attain best results for the readers.

Apart from the editorial board, the designing team has also invested a significant amount of their time in understanding the subject and creating the most relevant covers. They scrutinized every image to scout for the most suitable representation of the subject and create an appropriate cover for the book.

The publishing team has been an ardent support to the editorial, designing and production team. Their endless efforts to recruit the best for this project, has resulted in the accomplishment of this book. They are a veteran in the field of academics and their pool of knowledge is as vast as their experience in printing. Their expertise and guidance has proved useful at every step. Their uncompromising quality standards have made this book an exceptional effort. Their encouragement from time to time has been an inspiration for everyone.

The publisher and the editorial board hope that this book will prove to be a valuable piece of knowledge for researchers, students, practitioners and scholars across the globe.

LIST OF CONTRIBUTORS

Hsin-Yu Wu and Yeu-Long Jiang
Graduate Institute of Optoelectronic Engineering and Department of Electrical Engineering, National Chung Hsing University, Taichung 40227, Taiwan

Ting-Xuan Liu
Department of Electrical Engineering, Da-Yeh University, Chunghua 51591, Taiwan

Chia-Hsun Hsu, Yun-Shao Cho, Zhi-Jia Xu and Shui-Yang Lien
Department of Materials Science and Engineering, Da-Yeh University, Chunghua 51591, Taiwan

Bo-Han Zeng
Bachelor Program for Design and Materials for Medical Equipment and Devices, Da-Yeh University, Changhua 51591, Taiwan

Shu-Chuan Liao
Bachelor Program for Design and Materials for Medical Equipment and Devices, Da-Yeh University, Changhua 51591, Taiwan
Institute of Biomedical Engineering, National Taiwan University, Taipei 106, Taiwan

Vanessa Dias
Centro de Ciência e Tecnologia de Plasmas e Materiais (PlasMat), Instituto Tecnológico de Aeronáutica (ITA), São José dos Campos, SP 12228-900, Brazil
Laboratório de Nanotecnologia Biomédica, Universidade do Vale do Paraíba (Univap), São José dos Campos, SP 12244-000, Brazil

Homero Maciel and Rodrigo Pessoa
Centro de Ciência e Tecnologia de Plasmas e Materiais (PlasMat), Instituto Tecnológico de Aeronáutica (ITA), São José dos Campos, SP 12228-900, Brazil

Fernanda R. Marciano
Laboratório de Nanotecnologia Biomédica, Universidade do Vale do Paraíba (Univap), São José dos Campos, SP 12244-000, Brazil
Instituto Científico e Tecnológico, Universidade Brasil, São Paulo, SP 08230-030, Brazil

Mariana Fraga
Instituto de Ciência e Tecnologia, Universidade Federal de São Paulo (Unifesp), São José dos Campos, SP 12231-280, Brazil

Anderson O. Lobo
LIMAV-Laboratório Interdisciplinar de Materiais Avançados, Universidade Federal do Piauí (UFPI), Teresina, PI 64049-550, Brazil

Cristina Mellinas, Marina Ramos, Alfonso Jiménez and María Carmen Garrigós
Department of Analytical Chemistry, Nutrition & Food Sciences, University of Alicante, ES-03690 Alicante, San Vicente del Raspeig, Spain

Alejandro Aragón-Gutierrez and Mar López-González
Instituto de Ciencia y Tecnología de Polímeros, ICTP-CSIC, Juan de la Cierva 3, 28006 Madrid, Spain

Marina P. Arrieta
Departamento de Ingeniería Química y del Medio Ambiente, Escuela Politécnica Superior de Ingenieros Industriales, Universidad Politécnica de Madrid (ETSII-UPM), Calle José Gutiérrez Abascal 2, 28006 Madrid, Spain
Grupo de Investigación: Polímeros, Caracterización y Aplicaciones (POLCA), 28006 Madrid, Spain

Marta Fernández-García and Daniel López
Instituto de Ciencia y Tecnología de Polímeros, ICTP-CSIC, Juan de la Cierva 3, 28006 Madrid, Spain
Interdisciplinary Platform for Sustainable Plastics towards a Circular Economy-Spanish National Council (SusPlast-CSIC), 28006 Madrid, Spain

Bogdanel Silvestru Munteanu
Faculty of Physics, Alexandru Ioan Cuza University, 11 Carol I bvd, 700506 Iasi, Romania

Liviu Sacarescu, Ana-Lavinia Vasiliu and Gabriela Elena Hitruc
"P. Poni" Institute of Macromolecular Chemistry, Romanian Academy, 41A Grigore GhicaVoda Alley, 700487 Iasi, Romania

Gina M Pricope
Veterinary and the Food Safety Laboratory, Food Safety Department, 700489 Iasi, Romania

Morten Sivertsvik and Jan Thomas Rosnes
Nofima AS, Deptartment of Processing Technology, Muninbakken 9-13, Tromsø 9291, Norway

Stanislav Kotsilkov and Nikolay Kolev Vitanov
Institute of Mechanics, Bulgarian Academy of Sciences, Acad. G. Bonchev, Block 4, 1113 Sofia, Bulgaria

Evgeni Ivanov
Institute of Mechanics, Bulgarian Academy of Sciences, Acad. G. Bonchev, Block 4, 1113 Sofia, Bulgaria
Research and Development of Nanomaterials and Nanotechnologies (NanoTechLab Ltd.), Acad. G. Bonchev, Block 4, 1113 Sofia, Bulgaria

Sandra Madalina Constantin, Lenuța Profire, Ioana Mirela Vasincu, Maria Apotrosoaei and Florentina Lupașcu
Department of Pharmaceutical Chemistry, Faculty of Pharmacy, University of Medicine and Pharmacy "Grigore T. Popa", 16 University Street, 700115 Iasi, Romania

Frederic Buron and Sylvain Routier
Institut de Chimie Organique et Analytique, Univ Orleans, CNRS, ICOA, UMR 7311, F-45067 Orléans, France

Luminița Confederat and Cristina Tuchilus
Department of Microbiology, Faculty of Pharmacy, University of Medicine and Pharmacy "Grigore T. Popa", 16 University Street, 700115 Iasi, Romania

Marta Teodora Constantin
University of Kent, School of Physical Sciences, Canterbury CT2 7NH, UK

Alexandru Sava
Department of Analytical Chemistry; Faculty of Pharmacy, University of Medicine and Pharmacy "Grigore T. Popa", 16 University Street, 700115 Iasi, Romania

Valentina Siracusa
Department of Chemical Science, University of Catania, Viale A. Doria 6, 95125 Catania (CT), Italy

Cinzia Mannozzi, Urszula Tylewicz and Santina Romani
Department of Agricultural and Food Sciences-DISTAL, Campus of Food Science, University of Bologna, P.zza Goidanich 60, 47521 Cesena, Italy

Matteo Gigli
Department of Chemical Science and Technologies, University of Rome Tor Vergata, Via della Ricerca Scientifica 1, 00133 Rome, Italy
Department of Civil, Chemical, Environmental and Materials Engineering, University of Bologna, via Terracini 28, 40131 Bologna, Italy

Juan Pablo Cecchini
Universidad Nacional de Rafaela, Bv. Roca 989, Rafaela, 3000 Santa Fe, Argentina

Nadia Lotti
Department of Civil, Chemical, Environmental and Materials Engineering, University of Bologna, via Terracini 28, 40131 Bologna, Italy

Lucía Rodríguez-Parada, Pedro F. Mayuet and Antonio J. Gámez
Department of Mechanical Engineering & Industrial Design, Faculty of Engineering, University of Cadiz, Av. Universidad de Cádiz 10, E-11519 Puerto Real-Cadiz, Spain

Raluca Nicoleta Darie-Niță, Cornelia Vasile, Elena Stoleru, Daniela Pamfil, Niță Tudorachi, Raluca Petronela Dumitriu and Mihai Adrian Brebu
Department of Physical Chemistry of Polymers, "Petru Poni" Institute of Macromolecular Chemistry, 41A Gr. Ghica Voda Alley, 700487 Iasi, Romania

Traian Zaharescu
National Institute for Electrical Engineering (INCDIE ICPE CA), 313 Splaiul Unirii, 030138 Bucharest, Romania

Liliana Tarțău
Grigore T. Popa University of Medicine and Pharmacy Iasi, 16 University Street, 700115 Iasi, Romania

Gina Mihaela Pricope
Veterinary and Food Safety Laboratory, Department of Food Safety, 700115 Iasi, Romania

Karol Leluk
Institute of Environmental Protection Engineering, Wroclaw University of Technology, Plac Grunwaldzki 9, 50-377 Wroclaw, Poland

Index

mcontent.com/pod-product-compliance
Source LLC
TN
948210926
V00004B/803